Plant Cryopreservation

Plant Cryopreservation

Editor
Carla Benelli

MDPI • Basel • Beijing • Wuhan • Barcelona • Belgrade • Manchester • Tokyo • Cluj • Tianjin

Editor
Carla Benelli
National Research Council of
Italy (CNR/IBE)
Italy

Editorial Office
MDPI
St. Alban-Anlage 66
4052 Basel, Switzerland

This is a reprint of articles from the Special Issue published online in the open access journal *Plants* (ISSN 2223-7747) (available at: https://www.mdpi.com/journal/plants/special_issues/plant_cryopreservation).

For citation purposes, cite each article independently as indicated on the article page online and as indicated below:

LastName, A.A.; LastName, B.B.; LastName, C.C. Article Title. *Journal Name* **Year**, *Volume Number*, Page Range.

ISBN 978-3-0365-2788-8 (Hbk)
ISBN 978-3-0365-2789-5 (PDF)

Cover image courtesy of Carla Benelli

© 2021 by the authors. Articles in this book are Open Access and distributed under the Creative Commons Attribution (CC BY) license, which allows users to download, copy and build upon published articles, as long as the author and publisher are properly credited, which ensures maximum dissemination and a wider impact of our publications.

The book as a whole is distributed by MDPI under the terms and conditions of the Creative Commons license CC BY-NC-ND.

Contents

About the Editor vii

Carla Benelli
Plant Cryopreservation: A Look at the Present and the Future
Reprinted from: *Plants* 2021, 10, 2744, doi:10.3390/plants10122744 1

Alois Bilavcik, Milos Faltus and Jiri Zamecnik
The Survival of Pear Dormant Buds at Ultra-Low Temperatures
Reprinted from: *Plants* 2021, 10, 2502, doi:10.3390/plants10112502 7

Milos Faltus, Alois Bilavcik and Jiri Zamecnik
Vitrification Ability of Combined and Single Cryoprotective Agents
Reprinted from: *Plants* 2021, 10, 2392, doi:10.3390/plants10112392 15

Hsing-Hui Li, Jia-Lin Lu, Hui-Esther Lo, Sujune Tsai and Chiahsin Lin
Effect of Cryopreservation on Proteins from the Ubiquitous Marine Dinoflagellate *Breviolum* sp.
(Family Symbiodiniaceae)
Reprinted from: *Plants* 2021, 10, 1731, doi:10.3390/plants10081731 35

**Michelle Issac, Princy Kuriakose, Stacie Leung, Alex B. Costa, Shannon Johnson,
Kylie Bucalo, Jonathan M. Stober, Ron O. Determann, Will L. Rogers,
Jenifer M. Cruse-Sanders and Gerald S. Pullman**
Seed Cryopreservation, Germination, and Micropropagation of Eastern Turkeybeard,
Xerophyllum asphodeloides (L.) Nutt.: A Threatened Species from the Southeastern United States
Reprinted from: *Plants* 2021, 10, 1462, doi:10.3390/plants10071462 51

**Saija Rantala, Janne Kaseva, Anna Nukari, Jaana Laamanen, Merja Veteläinen,
Hely Häggman and Saila Karhu**
Successful Cryopreservation of Dormant Buds of Blackcurrant (*Ribes nigrum* L.) by Using
Greenhouse-Grown Plants and In Vitro Recovery
Reprinted from: *Plants* 2021, 10, 1414, doi:10.3390/plants10071414 73

**Fionna M. D. Samuels, Dominik G. Stich, Remi Bonnart, Gayle M. Volk
and Nancy E. Levinger**
Non-Uniform Distribution of Cryoprotecting Agents in Rice Culture Cells Measured by CARS
Microscopy
Reprinted from: *Plants* 2021, 10, 589, doi:10.3390/plants10030589 91

Carla Benelli, Lara S. O. Carvalho, Soumaya EL merzougui and Raffaella Petruccelli
Two Advanced Cryogenic Procedures for Improving *Stevia rebaudiana* (Bertoni)
Cryopreservation
Reprinted from: *Plants* 2021, 10, 277, doi:10.3390/plants10020277 101

**Mariam Gaidamashvili, Eka Khurtsidze, Tamari Kutchava, Maurizio Lambardi
and Carla Benelli**
Efficient Protocol for Improving the Development of Cryopreserved Embryonic Axes of
Chestnut (*Castanea sativa* Mill.) by Encapsulation–Vitrification
Reprinted from: *Plants* 2021, 10, 231, doi:10.3390/plants10020231 113

Jiri Zamecnik, Milos Faltus and Alois Bilavcik
Vitrification Solutions for Plant Cryopreservation: Modification and Properties
Reprinted from: *Plants* 2021, 10, 2623, doi:10.3390/plants10122623 123

Chris O'Brien, Jayeni Hiti-Bandaralage, Raquel Folgado, Alice Hayward, Sean Lahmeyer, Jim Folsom and Neena Mitter
Cryopreservation of Woody Crops: The Avocado Case
Reprinted from: *Plants* **2021**, *10*, 934, doi:10.3390/plants10050934 . . . . **141**

About the Editor

Carla Benelli is a Researcher of the National Research Council of Italy (CNR), at the Institute of BioEconomy. She has more than 20 years of research experience in plant cryopreservation and is Head of the Cryopreservation and Cryo-banking laboratory. Other research activities and expertise are tissue culture (micropropagation, synthetic seed technology, TIS bioreactor), in vitro conservation and slow-growth storage of plant biodiversity. She has produced about 100 scientific contributions published in reviewed journals with Impact Factors and in national and international journals, as well as chapters in scientific international books.

Editorial

Plant Cryopreservation: A Look at the Present and the Future

Carla Benelli

Institute of BioEconomy, National Research Council (CNR/IBE), Sesto Fiorentino, 50019 Florence, Italy; carla.benelli@ibe.cnr.it

Citation: Benelli, C. Plant Cryopreservation: A Look at the Present and the Future. *Plants* **2021**, *10*, 2744. https://doi.org/10.3390/plants10122744

Received: 2 December 2021
Accepted: 10 December 2021
Published: 13 December 2021

Publisher's Note: MDPI stays neutral with regard to jurisdictional claims in published maps and institutional affiliations.

Copyright: © 2021 by the author. Licensee MDPI, Basel, Switzerland. This article is an open access article distributed under the terms and conditions of the Creative Commons Attribution (CC BY) license (https://creativecommons.org/licenses/by/4.0/).

Cryopreservation is known as an applied aspect of cryobiology or the study of life at low temperatures. Plant cryopreservation, specifically, is a process of cooling and storing vegetal structure as plant cells, tissues, or organs in liquid nitrogen (LN; −196 °C) or LN vapor (−160 °C). This methodology ensures the maintenance of samples' viability after thawing, and indefinite storage is possible. The cryopreservation technique is based on the removal of all freezable water from tissues by physical or osmotic dehydration, followed by ultrarapid freezing. The ultralow temperature stops metabolic and biochemical reactions in the cell, after adequate dehydration of plant tissues, to prevent the formation of intracellular ice crystals, which can cause cell death and destruction of cell organelles during the freezing process. Cryopreservation is currently the most innovative and affordable biotechnological approach that allows safe long-term conservation of plant biodiversity without risk of genetic modifications. Few events and reports are available in the literature about genetic modifications in cryopreservation and, if they occurred, the exact mechanism and elucidation of the nature of genetic instability was not clarified, considering the multiple stages involved in the process (in vitro culture-cryoprotection-regeneration).

A pioneer in the plant cryopreservation was Prof. Akira Sakai, who reported the survival of mulberry twigs after exposure to liquid nitrogen [1]. However, plant cryopreservation studies took another 20 years to become established as an area of investigation. Since then, the cryopreservation has been disseminated by the development and application of cryogenic procedures, based on the slow cooling system first and then on the two-step cooling system. In particular, the two-step cooling system took over for its easy application, low cost, and less time-consuming procedure aiming at the direct immersion in liquid nitrogen of plant specimens from tissue culture, without resorting to expensive apparatus and with a considerable simplification of procedures. All this has allowed a large-scale application on plant species, with suitable cryopreservation protocols that can provide high plant regrowth after thawing, thus facilitating the establishment of organized and strategic cryobanks of plant genetic resources.

The two-step cooling process is based on the induction of explant "vitrification" during a very fast decrease in temperature [2]."Vitrification" of cells and tissues is the physical process, which avoids intracellular ice crystallization, during ultra-freezing, by the transition of the aqueous solution of the cytosol into an amorphous, glassy state. As a consequence of this process, plant tissues are protected from damage and remain viable during their long-term storage at −196 °C.

Various vitrification-based techniques have been developed and are available for different plant species such as vitrification, encapsulation-dehydration, encapsulation-vitrification, desiccation, [3] and, more recently, droplet vitrification and D or V cryo-plate [4,5], but the techniques are an ever-changing skill to improve the plant recovery rates, to expand the number of the cryopreserved species and, above all, working on the species, which are still hard to process with the cryopreservation. For example, in this Special Issue, present is an optimized cryopreservation protocol for embryonic axes of chestnut (*Castanea sativa* Mill.) developed based on the encapsulation–vitrification procedure; furthermore, the addition of activated charcoal (AC) as a component of the artificial matrix of synthetic seeds promoted growth by shortening the development times and limiting the loss of cryopreserved explants [6].

Cryogenic protocols are multi-stage, every step requires care to assure the successful of cryopreservation along with all the investigations connected to them, such as histo-anatomical, molecular, and physiological studies and in vitro culture procedures necessary to support recovery of cryopreserved explants.

The continuous research and technological evolution can markedly improve the cryogenic methodologies, allowing to enhance the recovery percentage of the species, as has occurred over the years. The new knowledge offers the prospect of bringing the cryopreservation technique to a superior level for preserving the vitality and integrity of the samples before and after storage. In *Stevia rebaudiana*, for example, the effectiveness of shoot tips cryopreservation increased with the application of V cryo-plate procedure, resulting in superior regrowth of 93%, [7] compared to the vitrification procedure applied by Shatnawi et al. [8], which obtained 68% of shoot tips regrowth. Moreover, new information has been acquired based on investigations in order to improve the explant physiological state, pre-treatment conditions, time and conditions of the cryoprotectant treatments, enhance the cooling and warming rates, and the recovery medium to achieve successful viability and regrowth of cryopreserved species. A contribution to the arduous process of optimizing cryoprotectant formulations in this Special Issue was given. Faltus at al. [9] has described in detail the thermal characteristics of two important plant vitrification solutions (PVS2 and PVS3) and their components depending on their concentration and temperature, while the opportunity of PVS2 modification to provide better application to new species has been dealt with by Zamecnik et al. [10]; support can also come from the Coherent Anti-Stokes Raman Scattering (CARS) microscopy by facilitating the visualization of deuterated cryoprotectants within living cells [11].

Distinct and adapted cryogenic techniques have been applied to a wide range of explants, including pollen, seeds, somatic and zygotic embryos, suspension or callus culture, apical buds, shoot tips, and dormant buds. The explant choice is also connected to the best in vitro recovery method for a specific genotype after cryopreservation. Hence, it is necessary to have an efficient in vitro regeneration system for a wider application of plant cryopreservation. For vegetatively propagated species, the most widely used organs are shoot tips excised from the in vitro plant [12,13]. Using this type of explant, the somaclonal variation is less probable to occur with respect to direct or indirect organogenesis [14]. In some woody species (e.g., apple, pear), using dormant buds as explants has also been developed and applied to cryopreservation protocols, assessing the recovery by grafting [15,16]. The dormant bud cryopreservation technique is an efficient alternative to the labor-intensive in vitro shoot tip cryopreservation process, allowing the preservation of large quantities of germplasm in a season [17]. Recently, in the 23 cryopreserved blackcurrant cultivars, using non-desiccated dormant buds collected from a greenhouse, the estimated recovery ranged between 42 and 90% [18].

Over time, cryopreservation protocols have been established for several hundreds of plant species [3] and further research is being conducted to enable adoption of this approach even more broadly. Currently, over 10,000 accessions starting from in vitro cultures are preserved through cryopreservation methods, and more than 80% of these belong to five crops: potato, cassava, bananas, mulberry, and garlic. Other important plant cryopreservation collections representing thousands of accessions are those of dormant apple buds [19–21]. Cryopreservation techniques are now used for plant germplasm storage in many institutes around the world [22,23]. The preservation of plant genetic resources (PGRs) is highly important for food security and agrobiodiversity, in breeding programs to obtain new or more productive plants, but also to have plants resistant to abiotic and biotic stresses. The application of advanced biotechnology, such as cryopreservation, represents an efficient alternative method for ex situ conservation of germplasm, and helps overcome several limitations of storage by conventional methods (seed banks and clonal orchards) [24–26]. Cryopreservation can be considered a safe strategy for long-term conservation and a backup to field collections to reduce the loss of plant germplasm. In this Special Issue, a protocol for seed cryopreservation and following in vitro germination has been reported for the first time in Eastern Turkeybeard (*Xerophyllum asphodeloides* (L.) Nutt.), a threatened species that has responded positively to the cryogenic technique [27].

On the other hand, a clear indication that the cryopreservation is a useful and necessary tool for conservation of plant species has also been underlined in the Plant Conservation Report 2020 [28]; this report mentioned the cryopreservation among alternative conservation methods. Several cryopreservation germplasm repositories (cryobanks) have been established for various plant species in different countries (e.g., cassava, potato, banana, apple, pear, coffee, mulberry, garlic), applying different cryopreservation techniques, and this strategy represents a guide for conservation in the future.

However, a few remarks should be made to face this new challenge; specifically, some critical issues need to be overcome and they will be part of the strategy to be pursued over the coming years. Detailed and exhaustive reviews [3,12,29–31] and articles described the various cryogenic methods applied to plants, but although much progress has been accomplished in the last years, some drawbacks still limit the wide use of cryopreservation, and the difficulties and challenges with the aim to further expand its frontiers should be considered.

The cryopreservation practice requires an initial technological investment but the maintenance costs for the application of the different techniques will later be lower, considering that the use of cryopreservation facilitates the storage and rapid multiplication of plant germplasm in a pathogen-free aseptic environment as well as optimization of physical space and labor. In a perspective of conservation strategy, the introduction of an accession into cryopreserved storage is more expensive than establishing an accession in in vitro culture or in the field, but the cryopreservation costs for the long-term vision (over 20 years) are considerably lower than those of maintenance in the field or in vitro, particularly when many accessions are preserved.

The difficulty to transfer technology and validating protocols between laboratories is a key issue [32]. There are some critical factors that involve all the cryopreservation steps such as type of plant materials, conditions of preculture, cryopreservation technique, cooling, warming, and regrowth conditions [13]. The lack of reproducibility available protocols, or the difficulty in adapting them, can be due to numerous causes, ranging from different sources of laboratory supplies to the different equipment and the different levels of technical skills found in cryopreservation laboratories.

The development and dissemination of increasingly simple and well-described protocols [33], with adequate facilities and trained personnel, will allow new challenges in each cryopreservation laboratory or institution as well as the implementation of the cryopreservation procedures in cryobanks and for biological materials and organisms important for research and in applications, including algae [34].

For this reason, it will therefore be essential to carry out joint strategies and programs among different countries to overcome some critical issues and, above all, find large-scale, simple and effective protocols and easily replicable in all laboratories. Accumulating experience in routine procedures, enhancing the skills of staff can be helpful, in addition to proceed with basic research, which should be encouraged (or rather financed) to achieve a better understanding of some topics less investigated.

In conclusion, since there is an increasing importance of plant preservation technology in the modern world, a clear need exists to have reliable methodologies as well as strong research and development activity in cryopreservation to ensure the applications are fit for purpose. The development of simple, reliable and cost-effective methods is essential and advances will be made faster if the know-how will be wide and shared, all this will support the cryopreservation, which is potentially the safest method to maintain vegetative germplasm and recalcitrant seed for a long-time life.

Funding: This research received no external funding.

Institutional Review Board Statement: Not applicable.

Informed Consent Statement: Not applicable.

Conflicts of Interest: The author declares no conflict of interest.

References

1. Sakai, A. Survival of plant tissue of super-low temperatures. *Low. Temp. Sci. B* **1956**, *14*, 17–23.
2. Benson, E.E. Cryopreservation theory. In *Plant Cryopreservation: A Practical Guide*; Reed, B.M., Ed.; Springer: New York, NY, USA, 2008; pp. 15–32.
3. Reed, B.M. (Ed.) Cryopreservation—Pratical considerations. In *Plant Cryopreservation: A Practical Guide*; Springer: New York, NY, USA, 2008; ISBN 978-0-387-72275-7.
4. Yamamoto, S.I.; Rafique, T.; Priyantha, W.S.; Fukui, K.; Matsumoto, T.; Niino, T. Development of a cryopreservation procedure using aluminium cryo-plates. *Cryo-Letters* **2011**, *32*, 256–265.
5. Niino, T.; Yamamoto, S.I.; Fukui, K.; Martińez, C.R.C.; Valle Arizaga, M.; Matsumoto, T.; Engelmann, F. Dehydration improves cryopreservation of mat rush (*Juncus decipiens* Nakai) basal stem buds on cryo-plates. *Cryo-Letters* **2013**, *34*, 549–560. [PubMed]
6. Gaidamashvili, M.; Khurtsidze, E.; Kutchava, T.; Lambardi, M.; Benelli, C. Efficient Protocol for Improving the Development of Cryopreserved Embryonic Axes of Chestnut (*Castanea sativa* Mill.) by Encapsulation–Vitrification. *Plants* **2021**, *10*, 231. [CrossRef]
7. Benelli, C.; Carvalho, L.S.O.; EL merzougui, S.; Petruccelli, R. Two Advanced Cryogenic Procedures for Improving Stevia rebaudiana (Bertoni) Cryopreservation. *Plants* **2021**, *10*, 277. [CrossRef]
8. Shatnawi, M.A.; Shibli, R.A.; Abu-Romman, S.M.; Al-Mazra'awi, M.S.; Al Ajlouni, Z.I.; Shatanawi, W.A.; Odeh, W.H. Clonal propagation and cryogenic storage of the medicinal plant Stevia rebaudiana. *Spanish J. Agric. Res.* **2011**, *9*, 213–220. [CrossRef]
9. Faltus, M.; Bilavcik, A.; Zamecnik, J. Vitrification Ability of Combined and Single Cryoprotective Agents. *Plants* **2021**, *10*, 2392. [CrossRef]
10. Zamecnik, J.; Faltus, M.; Bilavcik, A. Vitrification Solutions for Plant Cryopreservation: Modification and Properties. *Plants* **2021**, *10*, 2623. [CrossRef]
11. Samuels, F.M.D.; Stich, D.G.; Bonnart, R.; Volk, G.M.; Levinger, N.E. Non-uniform distribution of cryoprotecting agents in rice culture cells measured by cars microscopy. *Plants* **2021**, *10*, 589. [CrossRef] [PubMed]
12. Benelli, C.; De Carlo, A.; Engelmann, F. Recent advances in the cryopreservation of shoot-derived germplasm of economically important fruit trees of Actinidia, Diospyros, Malus, Olea, Prunus, Pyrus and Vitis. *Biotechnol. Adv.* **2013**, *31*, 175–185. [CrossRef]
13. Bettoni, J.C.; Bonnart, R.; Volk, G.M. Challenges in implementing plant shoot tip cryopreservation technologies. *Plant Cell. Tissue Organ Cult.* **2021**, *144*, 21–24. [CrossRef]
14. Scowcroft, W.R. Genetic variability in tissue culture: Impact on germplasm conservation and utilization. In *Proceedings of the International Board for Plant Genetic Resources (IBPGR) Technical Report AGPGIBPGR/84/152*; IBPGR: Rome, Italy, 1984.
15. Vogiatzi, C.; Grout, B.W.W.; Toldam-Andersen, T.B.; Green, J. Cryopreservation of dormant buds from temperate fruit crops to optimise working collection resources. *Acta Hortic.* **2011**, *908*, 477–482. [CrossRef]
16. Bilavcik, A.; Faltus, M.; Zamecnik, J. The Survival of Pear Dormant Buds at Ultra-Low Temperatures. *Plants* **2021**, *10*, 2502. [CrossRef]
17. Tanner, J.D.; Chen, K.Y.; Bonnart, R.M.; Minas, I.S.; Volk, G.M. Considerations for large-scale implementation of dormant budwood cryopreservation. *Plant Cell. Tissue Organ Cult.* **2021**, *144*, 35–48. [CrossRef]
18. Rantala, S.; Kaseva, J.; Nukari, A.; Laamanen, J.; Veteläinen, M.; Häggman, H.; Karhu, S. Successful Cryopreservation of Dormant Buds of Blackcurrant (*Ribes nigrum* L.) by Using Greenhouse-Grown Plants and In Vitro Recovery. *Plants* **2021**, *10*, 1414. [CrossRef]
19. Volk, G.M.; Waddell, J.; Bonnart, R.; Towill, L.; Ellis, D.; Luffman, M. High viability of dormant Malus buds after 10 years of storage in liquid nitrogen vapour. *Cryo-Letters* **2008**, *29*, 89–94.
20. Höfer, M. Cryopreservation of winter-dormant apple buds: Establishment of a duplicate collection of Malus germplasm. *Plant Cell. Tissue Organ Cult.* **2015**, *121*, 647–656. [CrossRef]
21. Jenderek, M.M.; Reed, B.M. Cryopreserved storage of clonal germplasm in the USDA National Plant Germplasm System. *Vitr. Cell. Dev. Biol.-Plant* **2017**, *53*, 299–308. [CrossRef]
22. Niino, T. Developments in plant genetic resources cryopreservation technologies. *Proc. APEC Work. Eff. Gene Bank Manag. APEC Memb. Econ. Suwon Korea* **2006**, 197–217.
23. Malik, S.K.; Chaudhury, R.; Pritchard, H.W. Long-term, large scale banking of citrus species embryos: Comparisons between cryopreservation and other seed banking temperatures. *Cryo-Letters* **2012**, *33*, 453–464. [PubMed]
24. González-Benito, M.E.; Ramírez, I.C.; Aranda, J.M.L. The use of cryopreservation for germplasm conservation of vegetatively propagated crops. *Spanish J. Agric. Res.* **2004**, *3*, 341–352. [CrossRef]
25. Paunescu, A. Biotechnology for endangered plant conservation: A critical overview. *Rom. Biotechnol. Lett.* **2009**, *14*, 181–202.
26. Ruta, C.; Lambardi, M.; Ozudogru, E.A. Biobanking of vegetable genetic resources by in vitro conservation and cryopreservation. *Biodivers. Conserv.* **2020**, *29*, 3495–3532. [CrossRef]
27. Issac, M.; Kuriakose, P.; Leung, S.; Costa, A.B.; Johnson, S.; Bucalo, K.; Stober, J.M.; Determann, R.O.; Rogers, W.L.; Cruse-Sanders, J.M.; et al. Seed Cryopreservation, Germination, and Micropropagation of Eastern Turkeybeard, *Xerophyllum asphodeloides* (L.) Nutt.: A Threatened Species from the Southeastern United States. *Plants* **2021**, *10*, 1462. [CrossRef] [PubMed]

28. Sharrock, S. *Plant Conservation Report 2020: A review of progress in implementation of the Global Strategy for Plant Conservation 2011-2020*. Secretariat of the Convention on Biological Diversity, Montréal, Canada and Botanic Gardens Conservation International; Technica: Richmond, UK, 2020; ISBN 9789292257057.
29. Kulus, D.; Zalewska, M. Cryopreservation as a tool used in long-term storage of ornamental species-A review. *Sci. Hortic.* **2014**, *168*, 88–107. [CrossRef]
30. O'Brien, C.; Hiti-Bandaralage, J.; Folgado, R.; Hayward, A.; Lahmeyer, S.; Folsom, J.; Mitter, N. Cryopreservation of Woody Crops: The Avocado Case. *Plants* **2021**, *10*, 934. [CrossRef]
31. Engelmann, F. Use of biotechnologies for the conservation of plant biodiversity. *Vitr. Cell. Dev. Biol.-Plant* **2011**, *47*, 5–16. [CrossRef]
32. Reed, B.M.; Kovalchuk, I.; Kushnarenko, S.; Meier-Dinkel, A.; Schoenweiss, K.; Pluta, S.; Straczynska, K.; Benson, E.E. Evaluation of critical points in technology transfer of cryopreservation protocols to international plant conservation laboratories. *Cryo-Letters* **2004**, *25*, 341–352.
33. Engelmann, F. Plant cryopreservation: Progress and prospects. *Vitr. Cell. Dev. Biol.-Plant* **2004**, *40*, 427–433. [CrossRef]
34. Li, H.-H.; Lu, J.-L.; Lo, H.-E.; Tsai, S.; Lin, C. Effect of Cryopreservation on Proteins from the Ubiquitous Marine Dinoflagellate *Breviolum* sp. (Family Symbiodiniaceae). *Plants* **2021**, *10*, 1731. [CrossRef]

Communication

The Survival of Pear Dormant Buds at Ultra-Low Temperatures

Alois Bilavcik *, Milos Faltus and Jiri Zamecnik

Crop Research Institute, Drnovska 507, 16106 Prague, Czech Republic; faltus@vurv.cz (M.F.); zamecnik@vurv.cz (J.Z.)
* Correspondence: bilavcik@vurv.cz

Abstract: Currently, there is a varietal diversity decline in pear orchards of the Czech Republic. Thus, the safe storage of their gene pool collections is becoming increasingly important. Therefore, the ultra-low temperature survival after two-step cryopreservation treatment of dormant buds was tested for a safe and rapid way to conserve pear germplasm in a broader range of varieties. The following varieties crucial for cultivation in the Czech Republic were tested; 'Amfora', 'Beurré Hardy', 'Bosc', 'Clapp's Favourite', 'Conference', 'Dicolor', 'Erika', 'Lucas', 'Williams' and 'Williams Red'. In 2011 and 2012, dormant pear buds were dehydrated to 40.1% and 36.0% water content, respectively, before cryopreservation. The average regeneration of the dormant pear buds after cryopreservation by the two-step cryoprotocol in 2011 and 2012 was 54.3% and 16.1%, respectively. The mentioned cryopreservation procedure is suitable for the safe storage of dormant buds in most tested pear varieties.

Keywords: cryopreservation; dormant buds; low-temperature resistance; pear tree

1. Introduction

The number of cultivated varieties of individual fruit trees decreased significantly with the development of intensive orchard management. Only 11 pear varieties are significantly used in the current production of pear trees in the Czech Republic [1]. From a pomological or agrotechnical point of view, pear orchard collections of individual curators are considered old and unattractive. Currently, unsuitable pear varieties are kept in situ in the Czech Republic. These in situ collections are exposed to various adverse effects that negatively affect the number of accessions stored, such as pests, diseases, climate change effects, etc. Therefore, a safe backup of in situ collections is one of the primary tasks of current programs for preserving the genetic diversity of vegetatively propagated crops. Cryopreservation is currently one of the only practical, safe and long-term conservation techniques used as an alternative method to safeguard these species. Pear is an important fruit species not only in the Czech Republic but also in the world. Therefore, there is a need to test its safe storage by cryopreservation and ensure a reliable and secure alternative for the long-term backup of the species.

Cryopreservation is a method of storing plants at ultra-low temperatures. It is a technology enabling long-term storage of biological material at a very low temperature (usually in liquid nitrogen at −196 °C or in liquid nitrogen vapours below −130 °C), while maintaining its viability after transfer to normal conditions [2]. During the cryopreservation procedure, it is necessary to induce processes and conditions in parts of plants that increase their natural resistance to the formation of ice crystals in their tissues and related ice induced dehydration [3,4]. For the cryopreservation of fruit trees, methods based on biological glass formation of in vitro cultures are used [5–7]. In addition, the recently reintroduced two-step cryopreservation—mainly in dormant buds of fruit trees are also used [2,8].

The first report of the survival of pear dormant bud in liquid nitrogen was by Sakai and Nishiyama [9]. Oka et al. [10] were able to regenerate plants from cryopreserved dormant buds of *Pyrus serotina* Rehder, 'Senryo' by in vitro transfer of meristematic tissues, but achieved a rate of regeneration of less than 8%. Using a similar procedure of in vitro regeneration, Suzuki et al. [11] obtained 88.6% survival of dormant buds of *Pyrus communis* L. 'Beurre d'Amanlis' after dehydration to 41% water content. The other 12 cryopreserved pear varieties by Suzuki et al. [11] had survival from 55.5% to 92.5%. However, after they directly micrografted the cryopreserved buds onto rootstocks, the survival was only 30% because of injury of vascular tissues. Guyader et al. [12] adapted the protocol developed at NCGRP, Fort Collins, USA [2,13] with uninodal dormant pear segment pre-freezing at $-5\ °C$ until they reached 23% water content. They used slow cooling at a rate of $1\ °C\ h^{-1}$ to $-30\ °C$ and 24 h annealing before placing in vapour phase over liquid nitrogen in a liquid nitrogen freezer. After slow rewarming and rehydration, the chip budding on Kirchensaller pear rootstocks was used. The average regeneration percentage of the 15 cryopreserved cultivars was 26.9%. Additionally, the pre-treatment of dormant buds prior to the two-step cryopreservation with different cryoprotectant solutions was applied. Based on 2,3,5-Triphenyltetrazolium chloride (TTC) staining viability tests, Zhumagulova et al. [14] evaluated six cryoprotective solutions and found that PVS3 solution gave the best survival rates. Recently, forced bud development [15] may also be used as a recovery system if followed by tissue culture regeneration after two-step cryopreservation. This post-cryo recovery approach led to a successful in vitro shoot tip establishment by using an antimicrobial forcing solution (8-hydroxyquinoline citrate and sucrose).

In summary, various methods of cryoconservation of pear according to a type of source material, in vitro or dormant buds, have been published. Cryopreservation of in vitro cultures has its advantages in the possibility of a well-defined treatment on relatively homogeneous plant material, dissected shoot tips. However, these procedures are laborious and time-consuming. When cryopreserving dormant buds by two-step freezing, it is difficult to prepare such homogeneous plant material in terms of size, maturation, or hardening, and this material is only available during a limited winter period. However, a larger number of varieties can be frozen in a relatively short time. Therefore, this work aimed to evaluate a two-step cryopreservation protocol for dormant pear buds with direct regeneration via chip budding onto pear rootstock to contribute to the safe storage of the pear gene pool.

2. Results

The average water content of dormant buds of the tested pear varieties in 2011 was $40.1 \pm 1.46\%$. In 2012, pear varieties were freeze-dehydrated to an average water content of $36.0 \pm 1.73\%$. The highest dehydration occurred in 2011 for the variety 'Lucas' (37.1%) and the lowest for the variety 'Erika' (46.2%). In 2012, the variety 'Erika' was the most dehydrated (32.3%) and the variety 'Conference' the least dehydrated (37.9%). Thus, the dehydration of pear varieties in 2012 was $4.1 \pm 2.59\%$ higher than in 2011. The highest difference in dehydration was in the variety 'Erika' (9.2%) and the lowest in 'Amfora' (1.0%), see Table 1.

The average regeneration of pear dormant buds after cryopreservation by a two-step cryoprotocol in 2011 was $54.3 \pm 23.71\%$. The highest regeneration showed 'Clapp's Favourite' (83.3%) and the lowest 'Williams Red' (8.3%). In 2012, the average regeneration of pear varieties was $16.1 \pm 12.33\%$. The highest regeneration showed 'Clapp's Favourite' (37.5%) and the lowest 'Erika' (0%). There was a high degree of variability in regeneration percentage between the two years: the varieties 'Lucas' and 'Conference' showed the most significant difference between the two years. Although 'Williams Red' and 'Dicolor' showed the least difference between the two years, they also showed the worst regeneration in 2011. 'Bosc' and 'Erika' showed a smaller decrease in regeneration percentage, but they both fell to zero in 2012. On average, regeneration of $37.8 \pm 17.87\%$ was achieved in dormant pear buds after cryopreservation in both years, see Table 2.

Table 1. The water content of dormant pear buds after freeze-drying before a two-step cryopreservation cryoprotocol. The water content of varieties in 2011, 2012, the average water content and the difference between 2011 and 2012 is shown.

Variety	Water Content (%)			
	2011	2012	Average	Δ 2011–2012
'Amfora'	38.7	37.7	38.2	1.0
'Beurré Hardy'	40.5	34.5	37.5	6.0
'Bosc'	38.3	37.2	37.7	1.1
'Clapp's Favourite'	40.8	37.9	39.4	2.9
'Conference'	40.8	37.9	39.3	3.0
'Dicolor'	41.5	35.0	38.2	6.6
'Erika'	41.5	32.3	36.9	9.2
'Lucas'	37.1	35.6	36.3	1.5
'Williams'	40.8	37.1	38.9	3.8
'Williams Red'	41.2	35.3	38.3	5.8
Average	40.1	36.0	38.1	4.1
SD	1.46	1.73	0.95	2.59

Table 2. Regeneration of dormant pear buds after a two-step dehydration cryoprotocol.

Variety	Regeneration (%)				
	2011	SD	2012	SD	Δ 2011–2012
'Amfora'	65.8 [cde]	15.9	29.2 [cd]	11.79	36.2
'Beurré Hardy'	79.2 [de]	5.9	29.2 [cd]	5.89	48.7
'Bosc'	33.3 [ab]	21.2	0.0 [a]	0.00	33.3
'Clapp's Favourite'	83.3 [de]	11.8	37.5 [d]	10.21	45.8
'Conference'	71.7 [cde]	24.6	12.5 [abc]	10.21	56.7
'Dicolor'	29.2 [ab]	15.6	16.2 [abc]	15.73	13.2
'Erika'	46.7 [bc]	11.2	0.0 [a]	0.00	46.2
'Lucas'	75.8 [cde]	18.3	11.1 [abc]	15.71	64.9
'Williams'	50.0 [bcd]	10.2	20.8 [bcd]	5.89	29.2
'Williams Red'	8.3 [a]	5.9	4.2 [ab]	5.89	4.0
Average	54.3		16.1		37.8
SD	23.71		12.33		17.87

[a–e] averages with the same index do not differ significantly ($\alpha = 0.05$, analysis of variance—LSD test).

Survival and regeneration of dormant pear buds after two-step dehydration cryopreservation is presented in Figure 1.

(**a**)   (**b**)

Figure 1. Regeneration of dormant buds of pear variety 'Clapp's Favourite' after two-step dehydration cryopreservation. The grafting was done by chip budding in the spring period: (**a**) Arrows show the dormant buds sprouted in new shoots from meristematic parts of the chip; (**b**) the upper bud (**a**) survived but did not sprout into a new shoot. The bud sprouted into a new shoot at the second sap during the summer period. The lower bud (**b**) did not survive and was released. The buds were chip budded on pear seedlings in an orchard during the spring sap.

3. Discussion

A two-step cryopreservation protocol that includes an initial dehydration step is important to reduce ice crystal formation in tissues and increase survivability and regeneration. Within both years of the study, the difference in the dehydration degree of the varieties used was kept to a minimum (SD up to 1.7%). In the first experiments carried out in 2011, the dehydration level of dormant buds was set to the 40% level [11]. The aim of higher dehydration in 2012 compared to 2011 was to approach the values used for apple trees [16] and comparable levels of pears [12], to reduce the lethal effect of ice crystal formation in tissues and thus to increase regeneration [17]. Both of the above-mentioned authors [12,16] used the dehydration level of 30%. According to our preliminary studies, the level of 30% dehydration was too low, so in 2012 we decided to choose dehydration at approximately half the level, close to 36%. We maximally standardized the sampling and preparation of the dormant buds and their cold hardening conditions so that they were as similar as possible in both years. The regeneration of dormant buds differed significantly from year to year. In 2011, the regeneration of all varieties except 'Williams Red' was over the acceptable rate (29%). In the following year (2012), when the dormant buds were intentionally more dehydrated, the regeneration decreased significantly in all varieties. In 2012, regeneration decreased the most for the varieties that showed the highest regeneration in 2011, 'Lucas', 'Conference', and the least for the low-regeneration varieties, 'Williams Red' and 'Dicolor'.

For 'Erika', which was the most dehydrated of all varieties (32.3%), regeneration fell to zero in 2012.

A similarly significant decrease in the regeneration of dehydrated dormant pear buds below 34% was found by Suzuki et al. [11]. Regeneration of dormant pear buds frozen in 2011 reached (except for two varieties ('Williams Red' and 'Dicolor')) similar values to Suzuki et al. [11]. These authors achieved regeneration in 12 pear varieties in the range of 55–92%. The dormant bud regeneration procedure may have caused their slightly higher values of regeneration. They dissected shoot tips from cryopreserved dormant buds and regenerated them in vitro. In vitro conditions can be better standardized for regeneration, and it is possible to eliminate adverse conditions compared to regeneration by chip budding in an orchard. Differences in the regeneration of the same varieties in different years were also found in fruit trees by other authors, such as Höfer [18]. These differences can be justified by the differences in seasons and the response of varieties to them [2], if there is not precise pre-treatment of plant material by cold hardening at below zero temperatures [13]. Jenderek et al. [19] tested if the physical geographic location of the apple dormant buds, and by interference the preharvest temperature, compromised cryotolerance. Their data showed that for three locations tested, the geographic location of the apple dormant bud harvest did not adversely affect the bud cryopreservability. They also did not find a significant difference in cryopreservability of tested apple varieties in two from three seasons and the only different season was probably caused by an equipment malfunction. When using the same protocol in different series of tests at different dates in one season, Guyader et al. [12] found regeneration after cryopreservation fluctuating from 11.1% to 91.7% for pear variety 'Williams' and they identified several factors, which seemed to significantly influence the results: bud morphotypes, rehydration phase (technique used and duration), rootstock calibre, grafting technique, etc. On the other hand, an internal physiological state, such as endodormancy, does not affect the cryopreservability of dormant buds themselves [20]. The above publications led us to believe that by maintaining the same conditions for sampling, pre-treatment, hardening, and post-cryopreservation regeneration we could separate the extent of dehydration in different years. A comparison of published cryopreservation results of selected pear varieties (*Pyrus communis* L.) with the results obtained in this work is in Table 3. The first report on cryopreservation of pear varieties 'Amfora', 'Dicolor', 'Erika', and 'Williams Red' is presented. With the exception of the varieties 'Bosc' and 'Williams', we achieved higher regeneration in all comparable varieties. On average, the varieties frozen by encapsulation-dehydration of in vitro cultures achieved a 40% reduction in regeneration compared to our results [21–23]. The only higher regeneration of cryopreserved in vitro variety, 'Bosc', was obtained by controlled two-step freezing of shoot tips pre-treated with a cryoprotectant mixture (polyethylene glycol, glucose, and DMSO) [24]. The DMSO as a cryoprotectant was used by Dereuddre et al. [25] for 'Beurré Hardy' with 60% regeneration compared to our 79%. Due to the potential mutagenic effects of the cryoprotectant DMSO [26], there have been attempts to omit the DMSO during the cryopreservation process. A slightly lower average regeneration, by 10% compared to our results was achieved by Guyaeder et al. [12] with a similar method, see Table 3. On the other hand, the two-step freezing cryopreservation of dormant buds modified by introducing sprouting shoots from dormant buds into in vitro had 20% less regrowth in one comparable variety [27], see Table 3. Although the above method was less reliable, it could have potential in a regeneration system eliminating the environmental risk of grafting cryopreserved buds under orchard conditions.

Table 3. Reports on cryopreservation of selected pear varieties (*Pyrus communis* L.) and comparison with the obtained results presented in this work (dormant bud, two step-freezing, slow-cooling prior to storage in LN).

Variety	Max. Regrowth [a] (%)	Published Results			
		Max. Regrowth [%]	Source [b]	Method [c]	Ref.
'Amfora'	66				
'Beurré Hardy'	79	40	iv	En-Dehy	[21]
'Beurré Hardy'	79	60	iv	DMSO/TSF	[25]
'Bosc'	33	90	iv	PGD/TSF	[24]
'Clapp's Favourite'	83	14	iv	En-Dehy	[22]
'Clapp's Favourite'	83	~33	db	Dehy-TSF-Graft	[12]
'Conference'	72	50	db	Dehy-TSF-Graft	[12]
'Dicolor'	29				
'Erika'	47				
'Lucas'	76	44	iv	En-Dehy	[22]
'Williams'	50	26	iv	En-Dehy	[23]
'Williams' [d]	50	~30	db	Dehy-TSF-iv	[27]
'Williams'	50	92	db	Dehy-TSF-Graft	[12]
'Williams Red'	8				

[a] The obtained results presented by this work. Freeze dehydration followed by two-step freezing, slow cooling prior to storage in LN, regeneration by grafting (Dehy-TSF-Graft). [b] Type of the cryopreserved plant tissue, iv = shoot tips from in vitro culture, db = nodal segments with dormant buds. [c] En-Dehy = encapsulation–dehydration; DMSO/TSF = DMSO pre-treatment followed by two-step freezing, slow cooling prior to storage in LN; PGD/TSF = PGD cryoprotectant mixture pre-treatment followed by two-step freezing, slow cooling prior to storage in LN, Dehy-TSF-Graft = freeze dehydration followed by two-step freezing, slow cooling prior to storage in LN, regeneration by grafting; Dehy-TSF-iv = freeze dehydration followed by two-step freezing, slow cooling prior to storage in LN, regeneration by in vitro. [d] The 'Bartlett' pear in the United States and Canada.

According to our results, the optimal dehydration of dormant bud cryopreservation of selected pear varieties by the two-step freezing was 40%. It can be concluded that the tested procedure of cryopreservation of dormant pear buds can already be used and, after optimizing the dehydration conditions to extend it to hitherto less cryopreservable varieties.

4. Materials and Methods

One-year-old dormant shoots of the pear varieties 'Amfora', 'Beurré Hardy', 'Bosc', 'Clapp's Favourite', 'Conference', 'Dicolor', 'Erika', 'Lucas', 'Williams' and 'Williams Red' were used in the experiments. Shoots were taken from outdoor conditions during January 2010 and 2011 from the orchards of SEMPRA Litomerice Ltd., Litomerice, Czech Republic (USDA Plant Hardiness Zone 7a). The shoots were cut into uninodal segments with one bud in the middle and placed in a freezer at -4 °C. At this temperature, they were freeze-dehydrated for 4–7 weeks. The water content was determined gravimetrically on a fresh weight basis after drying a random sample of 5 segments at 85 °C for constant weight. In the first year, the dehydration level was set at 40% and values below 38% of the water content in the second year. After drying, a sufficient number of nodal segments (from 20 to 25 segments) were frozen in 50 mL tubes covered with aluminium foil, Kartell Conical Grad Test Tube, Kartell S.p.A., Italy, with a two-step cryoprotocol. In the first step of the cryoprotocol, the temperature was lowered from -4 °C to -25 °C (cooling rate 2 °C h^{-1}) in a computer-controlled freezer, Arctiko LTF 325, Denmark, and after equilibration for 12 h; the second step of the cryoprotocol was done; the tubes were immersed in liquid nitrogen. At least 120 buds from each variety were frozen for the cryobank storage,

LS4800 Taylor Wharton, USA, and 24 buds for evaluation of control sample regeneration. Buds for survival evaluation of the cryoprotocol were placed at +4 °C and allowed to slowly thaw spontaneously. After 2 weeks of rehydration of the buds in moist white peat, Baltic white peat, Hawita, Germany, at +4 °C, the buds were chip budded on pear seedlings, *Pyrus communis* L., in an orchard during the spring sap. After approximately two months, the survival of the buds was assessed. The sprouted buds were evaluated as regenerated and successfully cryopreserved, and the non-sprouted buds were evaluated as unregenerated and damaged by cryopreservation.

5. Conclusions

The paper presents the cryopreservation procedure of dormant pear buds tested in a broader range of varieties. It is evident that the successful cryopreservation of dormant pear bud depends both on the variety and, especially, on the acclimation of the buds. The acclimation not only depends on the specific course of the winter season, and thus, on the naturally induced frost resistance of dormant buds, but also on the cryopreservation procedure with artificial frost dehydration. According to the results, the frost dehydration to 40% of water content enabled successful cryopreservation for most pear varieties. The obtained results show the potential of introducing the tested cryopreservation procedure in the cryobanking of pear species, thus ensuring the safe backup of endangered in situ pear collections.

Author Contributions: Conceptualization, A.B., M.F. and J.Z.; methodology, A.B. and M.F.; investigation, A.B.; resources, J.Z.; data curation, A.B. and M.F.; writing—original draft preparation, A.B.; writing—review and editing, A.B. and M.F.; project administration, J.Z.; funding acquisition, J.Z. All authors have read and agreed to the published version of the manuscript.

Funding: This research was funded by the Ministry of Agriculture of the Czech Republic, grant number MZE RO0418.

Institutional Review Board Statement: Not applicable.

Informed Consent Statement: Not applicable.

Data Availability Statement: The data presented in this study are available on request from the corresponding author.

Acknowledgments: Many thanks to Stacy Denise Hammond Hammond for her English correction of the manuscript.

Conflicts of Interest: The authors declare no conflict of interest.

References

1. Anonym. Fruit Harvest 2020, Including 3 and 5 Year Comparisons. Sklizeň Ovoce 2020, Včetně 3 a 5letých Srovnání. Available online: http://eagri.cz/public/web/file/668403/Skliznen_ovoce_2020_a_3lete_a_5lete_srovnani.pdf (accessed on 30 August 2021). (In Czech)
2. Towill, L.E.; Forsline, P.L.; Walters, C.; Waddell, J.W.; Laufmann, J. Cryopreservation of *Malus* germplasm using a winter vegetative bud method: Results from 1915 accessions. *CryoLetters* **2004**, *25*, 323–334. [PubMed]
3. Bilavčík, A.; Zámečník, J.; Grospietsch, M.; Faltus, M.; Jadrná, P. Dormancy development during cold hardening of in vitro cultured *Malus domestica* Borkh. plants in relation to their frost resistance and cryotolerance. *Trees* **2012**, *26*, 1181–1192. [CrossRef]
4. Faltus, M.; Bilavčík, A.; Zámečník, J. Thermal analysis of grapevine shoot tips during dehydration and vitrification. *VITIS—J. Grapevine Res.* **2015**, *54*, 243–245.
5. Reed, B.; Denoma, J.; Luo, J.; Chang, Y.; Towill, L. Cryopreservation and long-term storage of pear germplasm. *Vitr. Cell. Dev. Biol.-Plant* **1998**, *34*, 256–260. [CrossRef]
6. Sedlak, J.; Paprstein, F.; Bilavcik, A.; Zamecnik, J. Adaptation of apple and pear plants to in vitro conditions and to low temperature. *Acta Hortic* **2001**, *560*, 457–460. [CrossRef]
7. Benelli, C.; De Carlo, A.; Engelmann, F. Recent advances in the cryopreservation of shoot-derived germplasm of economically important fruit trees of *Actinidia, Diospyros, Malus, Olea, Prunus, Pyrus* and *Vitis*. *Biotechnol. Adv.* **2013**, *31*, 175–185. [CrossRef] [PubMed]
8. Tyler, N.J.; Stushnoff, C. Dehydration of dormant apple buds at different stages of cold acclimation to induce cryopreservability in different cultivars. *Can. J. Plant Sci.* **1988**, *68*, 1169–1176. [CrossRef]

9. Sakai, A.; Nishiyama, Y. Cryopreservation of winter vegetative buds of hardy fruit trees in liquid nitrogen. *HortScience* **1978**, *13*, 225–227.
10. Oka, S.; Yakuwa, H.; Sate, K.; Niino, T. Survival and shoot formation in vitro of pear winter buds cryopreserved in liquid nitrogen. *HortScience* **1991**, *26*, 65–66. [CrossRef]
11. Suzuki, M.; Niino, T.; Akihama, T.; Oka, S. Shoot Formation and Plant Regeneration of Vegetative Pear Buds Cryopreserved at −150 °C. *J. Jpn. Soc. Hort. Sci.* **1997**, *66*, 29–34. [CrossRef]
12. Guyader, A.; Guisnel, R.; Simonneau, F.; Rocand, B.; Le Bras, C.; Grapin, A.; Chatelet, P.; Dussert, S.; Engelmann, F.; Feugey, L. First results on cryopreservation by dormant bud technique of a set of *Malus* and *Pyrus* cultivars from the INRA Biological Resources Centre. In Proceedings of the COST Action 871 Cryopreservation of Crop Species in Europe Final Meeting, Angers, France, 8–11 February 2011; Grapin, A., Keller, J., Lynch, P., Panis, B., Revilla, A., Engelmann, F., Eds.; OPOCE: Luxembourg, 2012; pp. 141–144. [CrossRef]
13. Towill, L.E.; Ellis, D.D. Cryopreservation of dormant buds. In *Plant Cryopreservation. A Practical Guide*; Reed, B.M., Ed.; Springer: New York, NY, USA, 2008; pp. 421–442. [CrossRef]
14. Zhumagulova, Z.B.; Kovalchuk, I.Y.; Reed, B.M.; Kampitova, G.A.; Turdiev, T.T. Effect of pretreatment methods of dormant pear buds on viability after cryopreservation. *World Appl. Sci. J.* **2014**, *30*, 330–334. [CrossRef]
15. Tanner, J.D.; Minas, I.S.; Chen, K.Y.; Jenderek, M.M.; Wallner, S.J. Antimicrobial forcing solution improves recovery of cryopreserved temperate fruit tree dormant buds. *Cryobiology* **2020**, *92*, 241–247. [CrossRef] [PubMed]
16. Vogiatzi, C.; Grout, B.W.W.; Wetten, A. Cryopreservation of winter-dormant apple: III—bud water status and survival after cooling −30 °C and during recovery from cryopreservation. *CryoLetters* **2012**, *33*, 160–168.
17. Stushnoff, C. Cryopreservation of apple genetic resources. *Can. J. Plant Sci.* **1987**, *67*, 1151–1154. [CrossRef]
18. Höfer, M. Cryopreservation of winter-dormant apple buds: Establishment of a duplicate collection of *Malus* germplasm. *Plant Cell Tissue Organ Cult. (PCTOC)* **2015**, *121*, 647–656. [CrossRef]
19. Jenderek, M.M.; Forsline, P.; Postman, J.; Stover, E.; Ellis, D. Effect of geographical location, year, and cultivar on survival of *Malus* sp. dormant buds stored in vapors of liquid nitrogen. *HortScience* **2011**, *46*, 1230–1234. [CrossRef]
20. Bilavčík, A.; Zámečník, J.; Faltus, M. Cryotolerance of apple tree bud is independent of endodormancy. *Front. Plant Sci.* **2015**, *6*, 695. [CrossRef]
21. Dereuddre, J.; Scottez, C.; Arnaud, Y.; Duron, M. Effects of cold hardening on cryopreservation of axillary pear (*Pyrus communis* L. cv Beurré Hardy) shoot tips of in vitro plantlets. *Comptes Rendus Acad. Sci. Ser. III-Sci. Vie-Life Sci.* **1990**, *310*, 265–272.
22. Sedlak, J.; Paprstein, F.; Bilavcik, A.; Zamecnik, J. In vitro cultures and cryopreservation as a tool for conserving of fruit species. *Bull. Bot. Gard.* **2004**, *13*, 65–67.
23. Damiano, C.; Caboni, E.; Frattarelli, A.; Condello, E.; Arias, M.; Engelmann, F. Cryopreservation of Fruit Tree Species through Encapsulation-Dehydration at the CRA—Fruit Research Centre of Rome. In Proceedings of the 1st International Symposium on Cryopreservation in Horticultural Species, Leuven, Belgium, 5–8 April 2009; pp. 187–190.
24. Chang, Y.J.; Reed, B.M. Extended alternating-temperature cold acclimation and culture duration improve pear shoot cryopreservation. *Cryobiology* **2000**, *40*, 311–322. [CrossRef]
25. Dereuddre, J.; Scottez, C.; Arnaud, Y.; Duron, M. Resistance of alginate-coated axillary shoot tips of pear tree (*Pyrus communis* L. cv Beurré Hardy) in vitro plantlets to dehydration and subsequent freezing in liquid nitrogen: Effects of previous cold hardening. *Comptes Rendus l'Académie Sci. Série III Sci. Vie* **1990**, *310*, 317–323.
26. Valencia-Quintana, R.; Gómez-Arroyo, S.; Waliszewski, S.M.; Sánchez-Alarcón, J.; Gómez-Olivares, J.L.; Flores-Márquez, A.R.; Villalobos-Pietrini, R. Evaluation of the genotoxic potential of dimethyl sulfoxide (DMSO) in meristematic cells of the root of Vicia faba. *Toxicol. Environ. Health Sci.* **2012**, *4*, 154–160. [CrossRef]
27. Tanner, J.D.; Chen, K.Y.; Jenderek, M.M.; Wallner, S.J.; Minas, I.S. Determining the effect of pretreatments on freeze resistance and survival of cryopreserved temperate fruit tree dormant buds. *Cryobiology* **2021**, *101*, 87–94. [CrossRef] [PubMed]

Article

Vitrification Ability of Combined and Single Cryoprotective Agents

Milos Faltus *, Alois Bilavcik and Jiri Zamecnik

Crop Research Institute, Drnovska 507, 16106 Prague, Czech Republic; bilavcik@vurv.cz (A.B.); zamecnik@vurv.cz (J.Z.)
* Correspondence: faltus@vurv.cz

Citation: Faltus, M.; Bilavcik, A.; Zamecnik, J. Vitrification Ability of Combined and Single Cryoprotective Agents. *Plants* **2021**, *10*, 2392. https://doi.org/10.3390/plants10112392

Academic Editor: Carla Benelli

Received: 7 October 2021
Accepted: 27 October 2021
Published: 6 November 2021

Publisher's Note: MDPI stays neutral with regard to jurisdictional claims in published maps and institutional affiliations.

Copyright: © 2021 by the authors. Licensee MDPI, Basel, Switzerland. This article is an open access article distributed under the terms and conditions of the Creative Commons Attribution (CC BY) license (https://creativecommons.org/licenses/by/4.0/).

Abstract: Cryoprotective agents (CPA) are an important part of many current vitrification methods. The vitrification ability of CPAs influences the probability of the glass transition and water crystallization occurrence. Thermal characteristics and the vitrification ability of two combined CPAs (PVS2 and PVS3), common plant vitrification solutions, and four single CPAs (ethylene glycol, DMSO, glycerol, and sucrose), the components of the mentioned PVSs, were evaluated utilizing a differential scanning calorimetry (DSC) during standard cooling/warming rates of 10 °C min^{-1}. The effect of solute concentration on their vitrification ability was shown in the CPAs tested. Four typical concentration regions at which the glassy state and/or crystallization occurred were defined. We suggest the solute concentration of 0.7 g g^{-1} as the universal vitrification concentration, characterized by an actual Tg of CPA solution and limited water crystallization. Knowledge of the thermal properties of CPAs allows the design of new combined CPAs with the required vitrification ability respecting the cryopreservation method used and the characteristics of the cryopreserved sample.

Keywords: cryopreservation; cryoprotectant; differential scanning calorimetry; glass transition; vitrification; water crystallization

1. Introduction

Cryopreserved organisms or their parts can be kept unchanged for many years. The reason for their stability is not only the low temperature itself, which slows down all processes or biochemical reactions but the specific state of matter [1]. During the cryopreservation process, the liquid state of the matter changes to a solid, so-called glassy state, and these conditions are maintained throughout the cryopreservation [2]. The solid-state of the matter is a source of stability in the cryopreserved material, due to its extreme viscosity [3]. There are no significant changes in the solid-state of the matter. The glassy state preserves the original structure of matter, in contrast to the growth of ice crystals [4]. Changes associated with the crystal arrangement of matter result in damage to the original structure and function and caused serious damage to living organisms [5]. The phase transition from the liquid to the glassy state of matter is called the glass transition and this process is called vitrification [6].

Dehydration and/or vitrification of a cryopreserved living organism is a prerequisite for successful cryopreservation and stability of cryopreserved organisms [7]. All cryoprotocols (Table 1), which lead to the dehydration or vitrification of living organisms, reduce the water content to prevent the occurrence of harmful ice crystals during cooling to ultra-low temperatures [8]. These procedures may utilize natural acclimatization processes leading to a moderate reduction in water content and/or increase in tolerance to dehydration [9], but a procedure leading to a dramatic reduction in water content (Table 1) is always included [10]. This dehydration takes place by treatment with dry air over the silica gel or in a laminar flow-bench, the freeze dehydration during controlled slow cooling in a two-step cryopreservation method or by the action of osmotically active solutions, CryoProtective

Agents (CPAs), which dehydrate living cells and help vitrify them during cooling [11]. The latter methods are usually called vitrification methods (Table 1) [12].

The CPAs can contain a mixture of non-penetrating and penetrating components [13]. The first acts outside the cells and osmotically dehydrates the protoplasts. The effect of penetrating cryoprotectants is combined. They act osmotically outside the cell and prevent the water crystallization after penetrating the cells, but usually show higher cytotoxicity compared to non-penetrating cryoprotectants [14].

CPA-based cryoprotocols are currently very popular [15–17], but most of them are developed more or less empirically by testing the effect of various CPA mixtures on cryopreserved organisms/organs/cells. In addition, cryoprotocols are unfortunately still not easily transferable to other laboratories [18] and this fact significantly limits their wider use.

Table 1. Characteristics of cryopreservation methods with respect to acclimation and dehydration.

Acclimation [1]	Pre-Treatment [2]	Extensive Dehydration [3]	Cooling Rate [4]	Cryopreservation Method [5]
Cold	None	Freezing	Slow	Two-step/slow-cooling/controlled-freezing
	Diluted CPAs	Freezing	Rapid	Droplet-freezing
	Loading solution	CPAs	Rapid	Vitrification
Osmotic	Osmotic solution	Air-dehydration	Slow	Encapsulation–dehydration
	Osmotic solution	Air-dehydration	Rapid	Encapsulation–dehydration
	Osmotic solution	CPAs	Rapid	Encapsulation–vitrification
	Loading solution	CPAs	Rapid	Vitrification
None	Loading solution	CPAs	Rapid	Vitrification
	Diluted CPAs	Freezing	Slow	Two-step/slow-cooling/controlled-freezing
	Diluted CPAs	Freezing	Rapid	Droplet-freezing

[1] Acclimation means the long-term action of low temperature and/or moderate osmotic stress. [2] Pre-treatment is the exposition of isolated shoot tips/meristems/cells to a diluted solution of osmotically active components: the osmotic solution is a common designation of diluted osmotic solutions, usually saccharides, the loading solution is a specific type of the osmotic solution, which usually consists of glycerol and sucrose. [3] Extensive dehydration is performed by either: extracellular freezing, air-dehydration above silicagel or in a laminar flow-bench, or CPAs (Cryoprotective Agents) representing a mixture of highly concentrated penetrating and nonpenetrating cryoprotectants. [4] Cooling rate influences the final state of matter: a slow cooling rate (degrees per hour) provides controlled (equilibrium) freezing conditions resulting in ice crystals production, a rapid cooling rate (usually thousands or hundreds of degrees per minutes) represents (nonequilibrium) vitrification conditions avoiding water crystallization. [5] Cryopreservation methods list general designations of cryopreservation protocols [3,7,10].

The routine use of cryopreservation methods can be supported by the development of new cryoprotocols or modification of currently used ones based on exact knowledge of the characteristics of CPAs used. When developing a new CPA, several key points need to be addressed: (1) selection of a suitable method of dehydration (air, freeze, osmotic), (2) the type of liquid phase transition (freezing and/or vitrification), and (3) the CPA treatment (none/single/combined) [19,20]. The choice of these options (Table 1) depends on the properties of the cryopreserved material, in particular its ability to withstand at least mild dehydration naturally by air dehydration or freezing without CPA treatment, typical of some plant species [21]. If a cryopreserved object does not tolerate dehydration, usually animal or sensitive plant cells or tissues, the use of CPAs is necessary for the successful application of the cryopreservation method. CPAs can be used either to modify or to prevent water freezing [12]. In the first case, the CPA concentration is low and the (equilibrium) freezing of water is controlled at a slow cooling rate; in the latter case, the CPA concentration is high to skip water freezing at rapid cooling/warming (C/W) rates (Table 1). The Critical Cooling Rates (CCRs) and Critical Warming Rates (CWRs) have already been defined for some CPAs [22–24].

Based on the CPA concentration, the thermodynamically non-equilibrium vitrification approach can occur at three different conditions: (1) Unstable vitrification happen when solute concentration is very low and must be overcome by extremely high C/W rates, (2) metastable vitrification occurred at medium CPA concentration and moderate C/W rates, and (3) stable vitrification depends on very high CPA concentration and it is independent on the C/W rates [6,13]. The use of high concentrated CPAs is limited by their toxicity to living cells and therefore the appropriate concentration of CPAs must be validated [25]. With regard to the term stable vitrification, it should be noted that the glassy state itself is not a thermodynamically equilibrium or stable [6,26]. The glassy state is metastable because it is not in its lowest energy form, in contrast to the crystalline form of matter [26], which is thermodynamically favored [6]. In this context, the stable vitrification is not thermodynamically correct term but it rather indicates the tendency of the solution to vitrify [27].

The effectiveness of CPA concentration on successful cryopreservation depends on is ability to induce the glassy state without water crystallization, which is a function of the CPA concentration and the proportion and characteristics of the CPA components [13,16]. At least two types of glass transitions can be identified: (1) the actual glass transition of solution (Tg) and (2) the glass transition of maximally freeze-concentrated solution (Tg'), which is defined by a concentration ($C_{g'}$), which occurred as a result of water freezing and progressive concentration of solution [28,29]. The Tg characterizes glass transition development without any water freezing but the Tg' detection always shows ice crystals occurred during the cooling period [28–30]. Whereas the Tg indicates a reliable vitrification process, the Tg' reveals the suboptimal vitrification conditions. Therefore, the development of new cryoprotocols can be improved through knowledge of the thermal characteristics of the CPAs, such as the melting/freezing point and the change in a specific heat capacity associated with matter phase transitions [31]. These parameters can be measured by differential scanning calorimetry [32–39]. The thermal characteristics of the cryoprotective solutions used and their components may help to select appropriate CPAs for specific cryopreservation methods concerning the cryopreserved material and the presence or avoidance of crystallization during C/W cycles [32,40].

This work solved the problem of selecting suitable CPAs for the development of reliable vitrification cryoprotocols, which are based on the ability of CPAs to induce a glassy state without harmful water crystallization, which we defined as the vitrification ability. This paper aimed to analyze the effect of solute concentration on thermal characteristics and the vitrification ability of two common multiple CPAs, the Plant Vitrification Solution 2 (PVS2) and the Plant Vitrification Solution 3 (PVS3), and four single CPAs: ethylene glycol (EG), dimethyl sulfoxide (DMSO), glycerol (Gly), and sucrose (Suc), which are the components of the combined CPAs tested.

2. Results

2.1. Vitrification Ability of PVS2-Based Solutions

The crystallization of water in the PVS2-based solutions during the cooling cycle occurred in a range of concentrations from 0.35 to 0.5 g g^{-1}, which corresponded to 40–70% of the PVS2 concentration (Figure 1). In this region, the percentage of water crystallization declined from 47% to 29% of the total mass of the sample with increasing solute concentration. No water crystallization occurred during the cooling cycle in solutions with a higher concentration of solutes. Water crystallization during the warming cycle was detected in solutions with concentrations of solutes of 0.55 and 0.6 g g^{-1} (80% PVS2 and 90% PVS2, respectively). The curve of crystallinity rapidly dropped to 23 and 11% of the total mass of the sample within these solute concentrations. No water crystallization was detected in the most concentrated solution containing 0.65 g g^{-1} (100% PVS2) during C/W cycles.

The occurrence and the value of the glass transition temperature of PVS2-based solutions tested were influenced by a solute concentration and two curves of Tg were detected (Figure 1). The first, the curve of Tg', was always accompanied by water crystallization during the cooling cycle and occurred at low solute concentrations (from 0.35 to 0.5 g g^{-1}). At the intersection with the Tg curve, the Tg' curve indicated a concentration ($C_{g'}$) of the solution originating from the water freezing in the diluted solutions, which occurred near 0.8 g g^{-1} (Figure 1). The Tg' was independent of original solute concentrations and it oscillated from −108 to −106 °C (Figure 1). The second, the curve of Tg, was detected at high solute concentrations (from 0.55 to 0.65 g g^{-1}) when no water crystallization occurred during the cooling cycle. It was dependent on solutes concentration when it increased from −120 to −115 °C with increasing solute concentration (Figure 1).

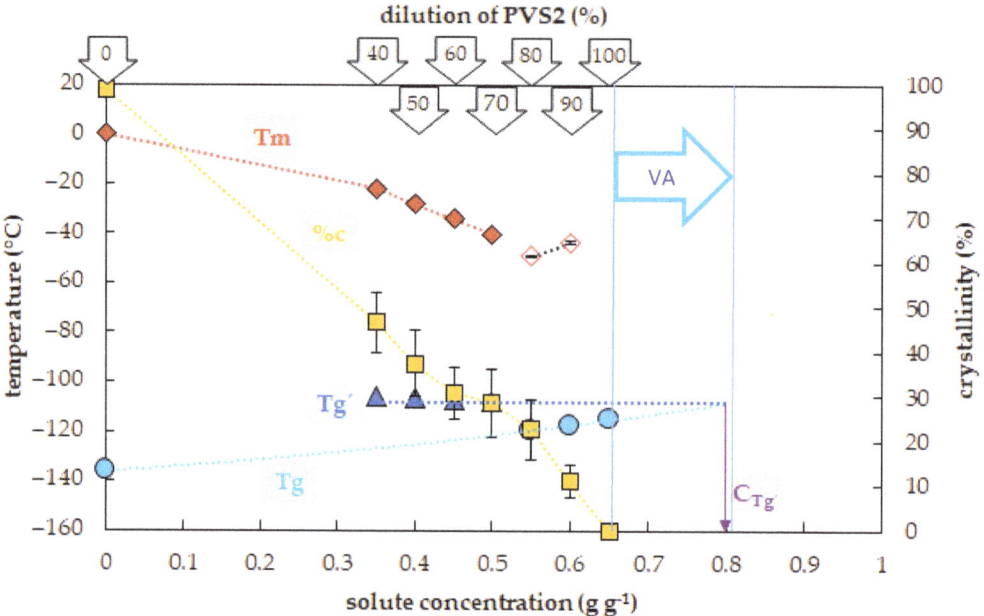

Figure 1. Thermal characteristics of osmotic components of the Plant Vitrification Solution 2 (PVS2). The x-axis expresses the total concentration of solutes (g) per mass of aqueous solution (g). The left y-axis indicates the temperature of thermal events, the right y-axis shows the percentage of water crystallinity on the total mass of solution. The temperature of the glass transition (Tg) of the solution is indicated by a cyan circle, the temperature of the glass transition Tg' of the freeze-concentrated solution is indicated by a blue triangle, the onset of the water melting peak (Tm) is indicated by a red diamond if the effect occurs during the cooling period, the onset of the water melting peak is indicated by a hollow diamond if the effect occurs during the warming period. The percentage of water crystallinity (%c) based on the total mass of the solution is indicated by an orange square. The Cg' value marks the concentration of the freeze-concentrated solution. The vitrification ability (VA) demonstrates the range of solute concentrations with the appropriate vitrification ability. All effects were measured during the warming period of C/W cycles at a rate of 10 °C min^{-1}. Results are presented as a mean of three repetitions and vertical bars represent standard error.

2.2. Vitrification Ability of PVS3-Based Solutions

Crystallization of water in the PVS3-based solutions during the cooling cycle occurred in the range of concentrations from 0.19 to 0.5 g g^{-1}, which corresponded to 20–60% of the PVS3 concentration (Figure 2). In this region, the percentage of water crystallization decreased from 69% to 21% of total sample mass with increasing solute concentration. No water crystallization occurred during the cooling cycle in solutions with higher concentrations of solutes. Negligible water crystallization during the warming cycle was detected in a solution with a solute concentration of 0.67 g g^{-1} (80% PVS3). At this solute concentration, the crystallinity curve dropped dramatically to 0.4% of the total sample mass. In the most concentrated solution (100% PVS3) containing 0.78 g g^{-1} of solutes, no water crystallization was detected during the C/W cycles.

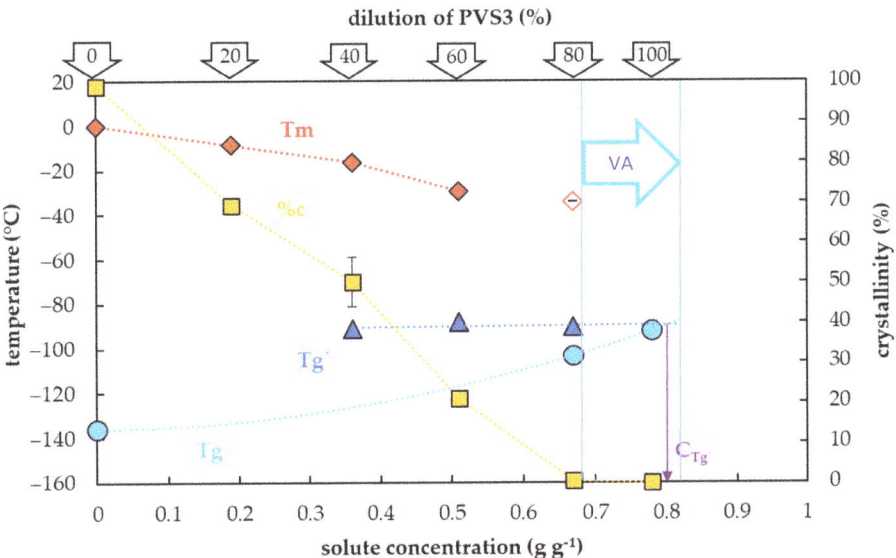

Figure 2. Thermal characteristics of osmotic components of the Plant Vitrification Solution 3 (PVS3) depending on their concentrations in aqueous solution. The total concentration of solutes is expressed as the mass of solutes (g) per mass of aqueous solution (g). The left y-axis indicates the temperature of thermal events, the right y-axis shows the percentage of water crystallinity based on the total mass of solution. The temperature of the glass transition (Tg) of the solution is indicated by a cyan circle, the Tg' originating from freeze-concentrated solution during the cooling period is indicated by a blue triangle, the onset of the water melting peaks (Tm) is indicated by a red diamond if the effect occurs during the cooling period, the onset of the water melting peak is indicated by a hollow diamond if the effect occurs during the warming period. The percentage of water crystallinity (%c) based on the total mass of the solution is indicated by an orange square. The vitrification ability (VA) demonstrates the range of solute concentrations with the appropriate vitrification ability. Cg' indicates a concentration of the freeze-concentrated solution. All effects were measured during the warming period of C/W cycles at a rate of 10 °C min^{-1}. Results are presented as a mean of three repetitions and vertical bars represent standard error.

The glass transition of PVS3-based solutions tested was also influenced by a solute concentration and the presence of water crystallization and two curves of glass transition temperatures were detected (Figure 2). The first, the curve of Tg', was always accompanied by water crystallization during the cooling cycle and occurred at low solute concentrations (from 0.19 to 0.5 g g^{-1}). It was independent of solute concentrations and oscillated from −90 to −88 °C (Figure 2). At the intersection with the Tg curve, the Tg' curve indicated a concentration ($C_{g'}$) of the solution originating from the water freezing in the diluted solutions, which occurred near 0.8 g g^{-1} (Figure 2). The second, the curve of Tg, was detected at high solute concentrations (from 0.67–0.78 g g^{-1}) when no water crystallization occurred during the cooling cycle. It was dependent on solutes concentration when increased from −103 to −92 °C with increasing solute concentration (Figure 2).

2.3. Vitrification Ability of the EG Solutions

Crystallization of water in the EG solutions during the cooling cycle occurred in a range of EG concentrations from 0.1 to 0.4 g g^{-1} (Figure 3). In this region, the percentage of water crystallization decreased from 67% to 34% of a total sample mass with the increasing EG concentration. No water crystallization occurred during the cooling cycle in solutions with a higher EG concentration. Crystallization of water during the warming cycle was detected in a solution with the EG concentration of 0.5 g g^{-1}.

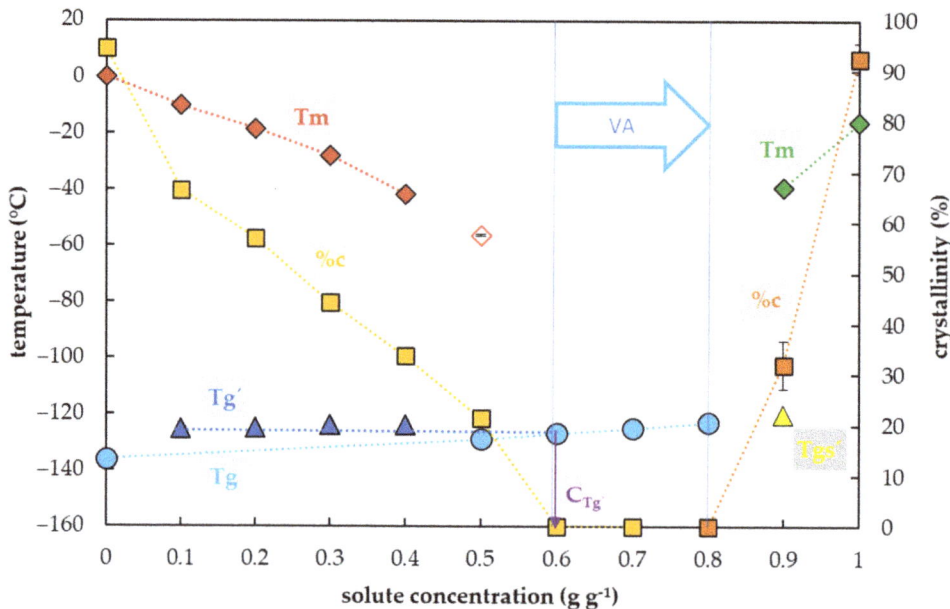

Figure 3. Thermal characteristics of ethylene glycol (EG) depending on its concentration in an aqueous solution. The total concentration of solutes is expressed as the mass of solutes (g) per mass of aqueous solution (g). The left y-axis indicates the temperature of the thermal events, the right y-axis shows the percentage of crystallinity of water or EG based on the total mass of the solution. The temperature of the glass transition (Tg) of the solution is indicated by a cyan circle, the Tg' originating from the freeze-concentrated solution during the cooling period is indicated by a blue triangle, the Tgs' originating from the EG crystallization is indicated by a yellow triangle, the onset of water melting peaks (Tm) is marked by a red diamond if the effect occurs during the cooling period and indicated by a hollow diamond if the effect occurs during the warming period. The onset of the EG melting peaks (Tm) is indicated by a green diamond. The percentage of crystallinity of the solvent or solute (%c) based on the total mass of the solution is indicated by an orange square for water and a brown square for EG. The vitrification ability (VA) demonstrates the range of solute concentrations with the appropriate vitrification ability. Cg' indicates a concentration of the freeze-concentrated solution. All effects were measured during the warming period of C/W cycles at a rate of 10 °C min^{-1}. Results are presented as a mean of three repetitions and vertical bars represent standard error.

The glass transition of the EG solutions tested was influenced by EG concentrations and two curves of glass transition temperatures were detected (Figure 3) as in the case of combined CPAs. The first, the curve of Tg', was always accompanied by water crystallization during the cooling cycle and occurred at low EG concentrations (from 0.1 to 0.4 g g^{-1}). It was independent of solute concentrations and oscillated from −125 to −124 °C (Figure 3). At the intersection with the Tg curve, the Tg' curve indicated a concentration (C$_{g'}$) of the solution originating from the water freezing in the diluted solutions, which occurred near 0.7 g g^{-1} (Figure 3). The second, the curve of Tg, was detected at high solute concentrations (from 0.5–0.9 g g^{-1}) when no water crystallization occurred during the cooling cycle. It was dependent on solutes concentration; it increased from −129 to −120 °C (Figure 3).

2.4. Vitrification Ability of the DMSO Solutions

Crystallization of water in the DMSO solutions during the cooling cycle occurred in a range of DMSO concentrations from 0.1 to 0.4 g g^{-1} (Figure 4). In this region, the percentage of water crystallization decreased from 80% to 25% of the total sample mass with an increasing DMSO concentration. No water crystallization occurred during the cooling cycle in solutions with higher concentrations of DMSO. No water crystallization was detected in solutions with 0.5 g g^{-1} or higher during the C/W cycles. However, a crystallization of DMSO was detected at concentrations of DMSO from 0.8 to 1.0 g g^{-1}, and a portion of the crystallinity ranged from 14 to 100% of DMSO.

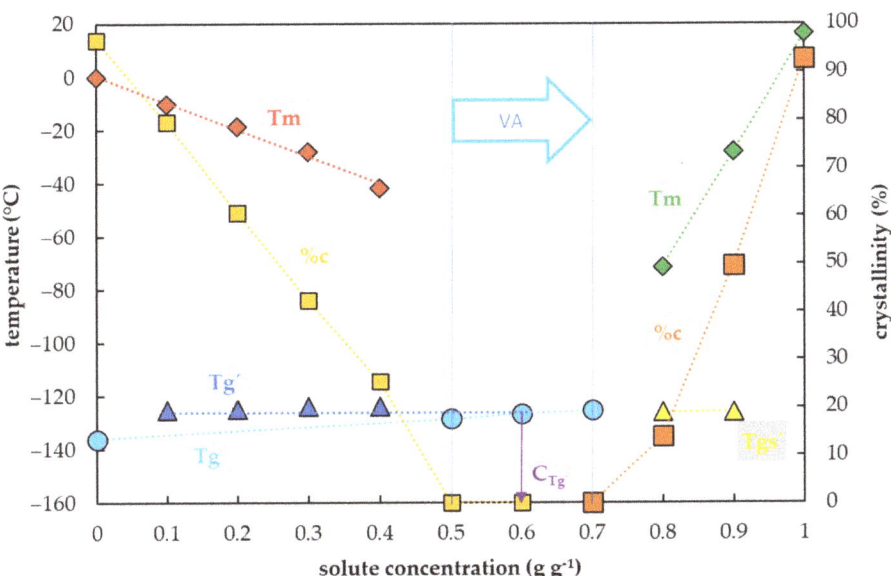

Figure 4. Thermal characteristics of dimethyl sulfoxide (DMSO) depending on its concentration in an aqueous solution. The total concentration of solutes is expressed as the mass of solutes (g) per mass of aqueous solution (g). The left y-axis indicates a temperature of thermal events, the right y-axis shows the percentage of crystallinity of water or DMSO based on the total mass of the solution. The temperature of the glass transition (Tg) of the solution is indicated by a cyan circle, the Tg′ originating from the freeze-concentrated solution during the cooling period is indicated by a blue triangle, the Tgs′ originating from the DMSO crystallization is indicated by a yellow triangle, the onset of water melting peaks (Tm) is indicated by a red diamond if the effect occurs during the cooling period, the onset of the DMSO melting peaks (Tm) is indicated by a green diamond. The percentage of crystallinity of the solvent or solute (%c) based on the total mass of the solution, is indicated by an orange square for water and a brown square for DMSO. The vitrification ability (VA) demonstrates the range of solute concentrations with the appropriate vitrification ability. Cg′ indicates a concentration of the freeze-concentrated solution. All effects were measured during warming period of the C/W cycles at a rate of 10 °C min^{-1}. Results are presented as a mean of three repetitions and vertical bars represent standard error.

The glass transition of DMSO solutions tested was influenced by a DMSO concentration, similarly to former CPAs. Three curves of glass transition were detected (Figure 4). The first, the curve of Tg′, was always accompanied by water crystallization during the cooling cycle and occurred at low DMSO concentrations (from 0.1 to 0.4 g g^{-1}). It was independent of solute concentrations and ranged from −126.2 to −126.5 °C (Figure 4). At the intersection with the Tg curve, the Tg′ curve indicated a concentration (C$_{g'}$) of the solution originating from the water freezing in the diluted solutions, which occurred below 0.7 g g^{-1} (Figure 4). The second, the curve of Tg, was detected at high solute concentrations (from 0.5–0.7 g g^{-1}) when no water crystallization occurred during the cooling cycle. It was dependent on solute concentrations; it increased from −128.4 to −125.3 °C. The third, the curve of Tgs′, was detected at the highest DMSO concentrations (from 0.8–0.9 g g^{-1}) when the DMSO crystallization occurred during the cooling cycle. It was independent of solutes concentrations; it oscillated from −125.9 to −125.8 °C (Figure 4).

2.5. Vitrification Ability of the Gly Solutions

Water crystallization in the Gly solutions during the cooling cycle occurred in a range of concentrations from 0.1 to 0.5 g g^{-1} (Figure 5). In this region, the percentage of water crystallization decreased from 82% to 38% of the total mass of the sample with increasing Gly concentration. No water crystallization occurred during the cooling cycle in solutions with a higher concentration of Gly. Water crystallization was detected during the warming cycle in a solution with a Gly concentration of 0.6 g g^{-1}. The curve of crystallinity dramatically dropped to 5% of the total mass of the sample in this Gly concentration. No water crystallization was detected in the Gly concentration of 0.7 g g^{-1} or higher during both the C/W cycles.

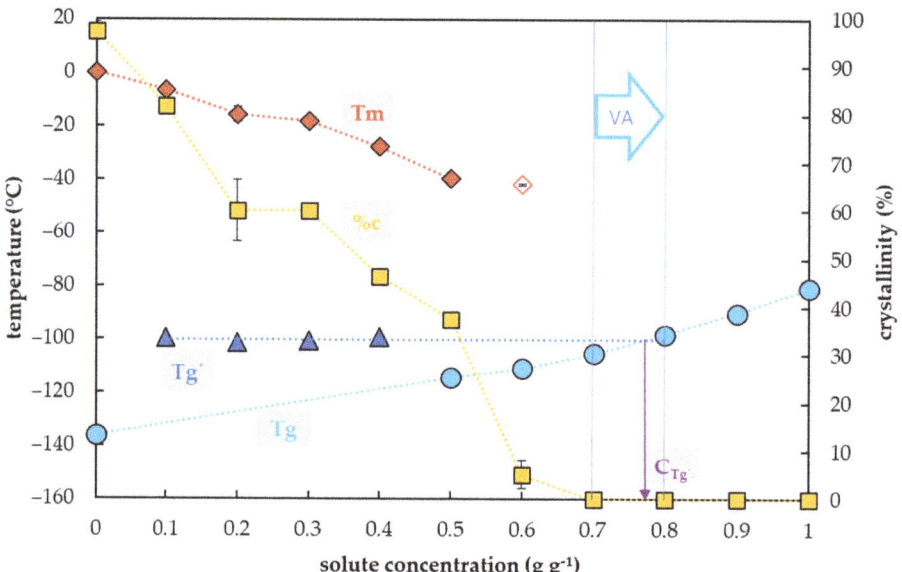

Figure 5. Thermal characteristics of glycerol (Gly) depending on its concentration in an aqueous solution. The total concentration of solutes is expressed as the mass of solutes (g) per mass of aqueous solution (g). The left y-axis indicates the temperature of thermal events and right y-axis shows the percentage of water crystallinity based on the total mass of solution. The temperature of the glass transition (Tg) of the solution is indicated by a cyan circle, the Tg' originating from the freeze-concentrated solution during the cooling period is indicated by a blue triangle, onset of the water melting peaks (Tm) is indicated by a red diamond if the effect occurs during the cooling period, the onset of the water melting peak is indicated by a hollow diamond if the effect occurs during the warming period. The percentage of crystallinity of the solute (%c) based on the total mass of the solution is indicated by an orange rectangle. The vitrification ability (VA) demonstrates the range of solute concentrations with the appropriate vitrification ability. Cg' indicates a concentration of the freeze-concentrated solution. All effects were measured during the warming period of the C/W cycles at a rate of 10 °C min^{-1}. Results are presented as a mean of three repetitions and vertical bars represent standard error.

The glass transition of Gly solutions was influenced by a Gly concentration and two curves of glass transition temperatures were detected (Figure 5). The first, the curve of Tg', was always accompanied by water crystallization during the cooling cycle and occurred at low Gly concentrations (from 0.1 to 0.4 g g^{-1}). It was independent of solute concentrations and oscillated from −101 to −99 °C (Figure 5). At the intersection with the Tg curve, the Tg' curve indicated a concentration (C$_{g'}$) of the solution originating from the water freezing in the diluted solutions, which occurred below 0.8 g g^{-1} (Figure 5). The second, the curve of Tg, was detected at high solute concentrations (from 0.5 to 1.0 g g^{-1}) and was dependent on solute concentration when it was increased from −114 to −81 °C (Figure 5).

2.6. Vitrification Ability of the Suc Solutions

Water crystallization in the solutions during the cooling cycle occurred in a range of Suc concentrations from 0.1 to 0.6 g g^{-1} (Figure 6). In this region, the percentage of water crystallization decreased from 91% to 38% of the total mass of the sample with increasing Suc concentration. No water crystallization occurred during the cooling cycle in solutions with higher Suc concentrations. Negligible water crystallization was detected in a solution with the Suc concentration of 0.7 g g^{-1} (Figure 6). The curve of crystallinity dramatically dropped to 0.6% of the total mass of the sample at this Suc concentration. No water crystallization was detected in the most concentrated solution (0.8 g g^{-1} of Suc) during the C/W cycles.

The Tgs of Suc solutions tested were influenced by a Suc concentration and two curves of glass transition temperature were detected (Figure 6). The first, the curve of Tg', was always accompanied by water crystallization during the cooling cycle and occurred at low Suc concentrations (from 0.1 to 0.5 g g^{-1}). It was independent of solute concentrations and ranged from −47 to −45 °C (Figure 6). At the intersection with the Tg curve, the Tg' curve indicated a concentration ($C_{g'}$) of the solution originating from the water freezing in the diluted solutions, which occurred near 0.8 g g^{-1} (Figure 6). The second, the curve of Tg, was detected at high solute concentrations (from 0.6–0.8 g g^{-1}) and was dependent on solutes concentration when increased from −87 to −43 °C (Figure 6).

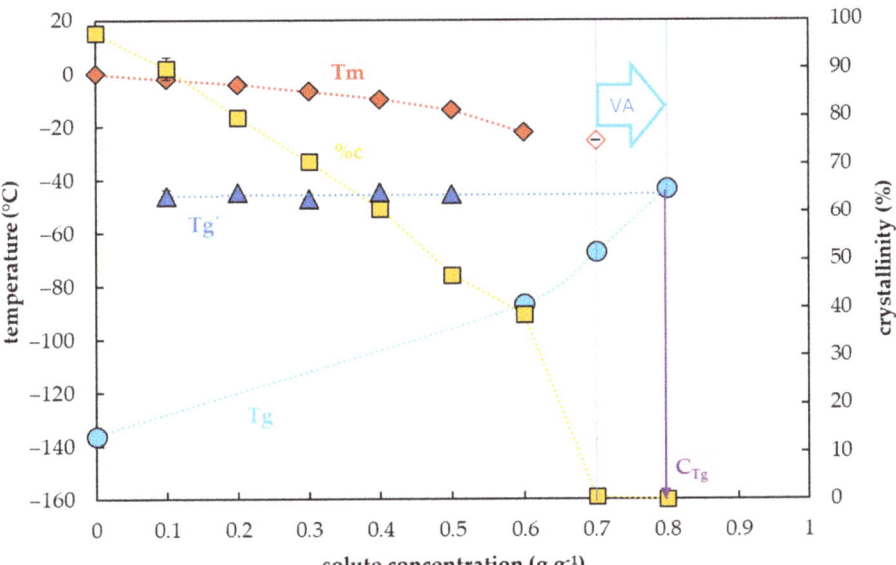

Figure 6. Thermal characteristics of sucrose (Suc) depending on its concentration in an aqueous solution. The total concentration of solutes is expressed as the mass of solutes (g) per mass of aqueous solution (g). The left y-axis indicates the temperature of thermal events, right y-axis shows the percentage of water crystallinity based on the total mass of the solution. The temperature of the glass transition (Tg) of the solution is indicated by a cyan circle, the Tg' originating from the freeze-concentrated solution during the cooling period is indicated by a blue triangle, onset of the water melting peaks (Tm) is indicated by a red diamond if the effect occurs during the cooling period, the onset of the water melting peak is indicated by a hollow diamond if the effect occurs during the warming period. The percentage of crystallinity of water (%c) based on the total mass of the solution is indicated by an orange square. The vitrification ability (VA) demonstrates the range of solute concentrations with the appropriate vitrification ability. Cg' indicates a concentration of the freeze-concentrated solution. All effects were measured during the warming period of the C/W cycles at a rate of 10 °C min^{-1}. Results are presented as a mean of three repetitions and vertical bars represent standard error.

3. Discussion

The thermal analysis of PVS2, PVS3, and their components demonstrated key features, the vitrification ability, and limits of these solutions for their use as CPAs in vitrification cryoprotocols concerning the CCR and CWR of 10 °C min^{-1}.

3.1. PVS2-Based Solutions

The results showed that all diluted PVS2 solutions (40–90%) can be used as a step of graduate dehydration [3,41–43] as they decreased the amount of frozen water but they did not have satisfactory vitrification ability (Figure 1) at the C/W rates of 10 °C min^{-1}. Only the most concentrated solution, the original (100%) PVS2 concentration, did not show any water crystallization in the C/W cycles (Figure 1) at the standard C/W rates. We proved that the PVS2 concentrations of 0.65 g g^{-1} or higher have an acceptable vitrification ability with no risk of water crystallization. We confirmed former findings that the CCR of PVS2 is less than 10 °C min^{-1} [44] but, moreover, we proved that the CWR of the PVS2 is less than 10 °C min^{-1} as well (Figure 1). A dilution of the PVS2 solution to 80–90% (0.55 and 0.6 g g^{-1}) resulted in an overcoming ice crystallization during the cooling cycle but the crystallization still occurred during the warming cycle (Figure 1). We conclude that the CCR of 80% PVS2 is less than 10 °C min^{-1} in contrast to the CWR, which is higher than 10 °C min^{-1}. The PVS2 solution diluted to 40–70% (0.35–0.5 g g^{-1}) still showed the osmotic effect of the solution due to decreasing water crystallization but the vitrification ability was weak (Figure 1). We conclude that the water freezing at PVS2 concentrations below 0.55 g g^{-1} risks sample damage during the C/W cycles because the successful sample vitrification strongly depends on very high values of both the CCRs and the CWRs.

Two types of Tg curves in the PVS2-based solutions were detected concerning the solute concentrations. The Tg' curve was a result of water crystallization [13] during the cooling cycle in the concentration range of 40–70% of PVS2 solution but it did not correspond to the Tg value of the original solutions. The Tg' value indicated the concentration of the freeze-concentrated solution (Cg') (Figure 1). The Cg' of the freeze-concentrated PVS2 solution was higher than the concentration of 100% PVS2 (Figure 1). The presence of Tg' in the 40–70% PVS2 (Figure 1) indicated higher CCR than the actual cooling rate used [13]. The second curve of Tg was detected in a concentration range of 80–100% PVS2 (Figure 1) and represented the Tgs of the original solutions. It was also developed during the cooling cycle and ranged in temperatures from −120 to −115 °C (Figure 1). The DSC measurement confirmed a satisfactory vitrification ability of the 100% PVS2 solution at the standard C/W rates. The Tg of this solution at −115 °C corresponds to the value of the PVS2 solutions already published [38,45,46].

We proved that the detection of the actual Tg was not crucial for the solution vitrification ability because water crystallization still occurred in 80% and 90% PVS2 during warming cycles simultaneously with Tg presence (Figure 1). Insufficient cryoprotectant concentration or insufficient warming rate may result in uncontrolled crystallization of water during thawing of samples [44,47,48], even though the samples were safely stored at liquid nitrogen temperature without any damage. This problem can also occur even with an appropriate concentration of CPA, but for an insufficient treatment period to properly dehydrate the sample [3].

Only the 100% PVS2 concentration showed sufficient vitrification ability, characterized by the presence of a glassy state without any water crystallization at the C/W rates of 10 °C min^{-1}. The CCR and CWR values of this solution are lower than 10 °C min^{-1}, which does not place too high demands on the temperature course during vitrification. Diluted PVS2 can be used as an osmotic agent to gradually dehydrate a sample before its vitrification or as CPA at higher C/W rates than 10 °C min^{-1}. This suggestion was successfully proved by the applicability of 80% PVS2 using the droplet-vitrification procedure [42]. The successful application of diluted PVS2 proved the higher C/W rates provided by the droplet-vitrification method, and thus is consistent with our results performed at the standard C/W rates of 10 °C min^{-1}. Insufficient dehydration can result in water crystallization mostly during the warming or even cooling cycle, in the case that the CWR or CCR are not adequate for the sample solute concentration. The presence of the Tg', which is associated with the freeze-concentrated solution, reveals an insufficient actual cooling rate, which is lower than the CCR of the solution, and simultaneously shows the solute concentration (Cg') providing conditions (~0.8 g g^{-1}) close to the stable vitrification conditions [6].

Two possible constraints of the CPA use should be taken into an account. A too extended period of dehydration can result in sample osmotic injury due to excessive water loss, and high concentration or presence of some penetrating components of CPAs can result in specific toxicity and injury of the cryopreserved material during vitrification [49,50]. To decrease the PVS2 toxicity the exposition at 0 °C is strongly recommended [46,51].

3.2. PVS3-Based Solutions

The results showed the vitrification ability of the PVS3-based solutions is satisfactory in a range from 80 to 10% of the original concentration (Figure 2). The most concentrated 100% PVS3 solution, did not show any water crystallization during the C/W cycles at the standard C/W rate. We can conclude that both the CCR and CWR are less than 10 °C min^{-1} in the 100% PVS3 solution. In the case of a solution diluted to 80% PVS3, no water crystallization occurred during cooling and only negligible water crystallization (0.4% w/w) was found during the warming cycle (Figure 2). This amount of frozen water was very close to the commonly agreed threshold value for the CCR at 0.2% of ice (w/w) [52] and the quantity of 0.5% w/w is not considered to be sufficient to cause any damage [52]. Moreover, real C/W rates are usually higher than 10 °C min^{-1} in the vitrification methods. Therefore, we consider 80–100% PVS3 solutions as appropriate CPAs with reliable vitrification ability for most vitrification methods. Our suggestions are in an agreement with the finding that the PVS3 solution in a concentration range from 80 to 100% resulted in acceptable cell survival (77.1–82.6%) [53]. More diluted PVS3 solutions (20%–60%) showed water crystallization during the cooling cycle (Figure 2) which indicated that the CCR of these solutions is higher than 10 °C min^{-1}. We conclude that diluted PVS3 solutions from 20% to 60% of PVS3 have very limited vitrification ability but they can be used as a step of graduate sample dehydration [43,54]. Insufficient vitrification ability confirmed previous results when the cell survival after 60%–70% PVS3 treatment ranged from 0 to 0.9% [53].

As in the case of the PVS2-based solutions, two types of Tg were detected in the PVS3-based solutions as well (Figure 2). As a result of water crystallization in diluted PVS3 solutions, the Tg's of freeze-concentrated solutions were found in a range from −91 to −88 °C (Figure 2). Secondly, the actual Tgs of solutions corresponding to 80 and 100% of PVS3 were detected at −103 and −92 °C, respectively (Figure 2). The Tg of the 100% PVS3 solution measured corresponds to the value of the PVS3 solution already published [38,53,55]. This Tg was close to the Tg' value of the freeze-concentrated solution detected at −91 °C (Figure 2). We conclude that the PVS3 solute concentration was almost identical to the freeze-concentrated solution (~0.8 g g^{-1}) and therefore the PVS3 contains almost no water, which can potentially be frozen at the stable vitrification conditions (6). In this respect, the PVS3 differs from the PVS2. A solute concentration in the original PVS2 (0.65 g g^{-1}) is similar to the solution of 80% PVS3 (0.67 g g^{-1}) but much less than in the original PVS3 solution (0.78 g g^{-1}). A higher concentration of solutes in the PVS3 solution indicates a higher vitrification capacity compared to PVS2, which is in line with previous results [56]. The advantage of the PVS3 solution is less cytotoxicity than in the case of PVS2 so it can be used at room temperature. The PVS3 constraint can be considered rather high viscosity, which can make PVS3 application difficult [57]. On the other hand, the diluted PVS3 solutions, with lower viscosity, in a range from 80 to 100% have acceptable vitrification ability (Figure 2) and have been successfully used as CPAs [42,43,53,54,58–60].

3.3. Single CPAs

The vitrification ability of the PVS components (a single CPA) had similar characteristics to the combined CPAs tested (PVS2- and PVS3-based solutions). For single

CPAs, the acceptable vitrification ability was found in the ranges of solute concentrations 0.5–0.7 g g^{-1} for DMSO, 0.7–0.8 g g^{-1} for EG or Suc, and 0.7–1.0 g g^{-1} for Gly. In these solute concentrations, typically, no water crystallization was found during the C/W cycles. The only 0.7 g g^{-1} Suc solution showed negligible water crystallization (0.6% w/w) during the warming cycle. It was close to the consent value (0.2%) for the CCR [52], corresponding to the 80% PVS3 solution (Figure 2). Therefore, we consider the 0.7 g g^{-1} sucrose aqueous solution a threshold value for the Suc vitrification ability. A lower solute concentration than 0.5 g g^{-1} resulted in an unreliable vitrification ability in all single CPAs tested. The vitrification ability was limited by solute crystallization at a concentration higher than 0.9 g g^{-1} in EG and 0.8 g g^{-1} in DMSO, and by the difficult solubility of Suc at a concentration higher than 0.8 g g^{-1}. The solute crystallization was described in EG aqueous solution under isothermal conditions [61]. It occurred at high solute concentrations and depended on the time of isothermal conditions. Therefore, we believe that the solute crystallization is not a serious problem of the vitrification method, as the individual components of CPAs cannot be used in a concentrated form due to their toxicity [62]. Glycerol was the only single CPA, which showed the vitrification ability up to 100% concentration.

The two previously identified types of Tg were found in all single CPAs tested. The Tg' presence was always detected together with the water crystallization during cooling and corresponded to the glass transition of the freeze-concentrated solutions. In diluted Suc solutions, the values for Tg' and C$_{g'}$ were −46 °C and ~0.8 g g^{-1}, respectively. The same values were defined for the Tg' and Cg' for the Suc maximal-freeze concentrated solution [29,63]. Therefore, we assume that all values of Tg' and Cg' identified in our work correspond, or at least were very close, to the maximally freeze-concentrated solution of the CPAs tested. Secondly, the actual Tg was detected when no or limited water content crystallized during the cooling cycle. Except for two mentioned Tgs, an additional Tg, marked as the Tgs' was detected as a result of solute crystallization in highly concentrated EG and DMSO solutions (Figures 3 and 4).

An appropriate vitrification ability of single CPAs was always determined by the presence of the Tg together with a limited crystallinity of water or solute (Figures 3–6). The lowest solute concentration (0.5 g g^{-1}) with the appropriate vitrification ability was detected in the DMSO solution. Our results of the vitrification ability (Figures 3–5) correspond to the achievements obtained earlier when the authors defined the glass-forming solute concentrations of 55, 46, and 65% (w/v) for EG, DMSO, and Gly, respectively [2]. We suggest the solute concentration 0.7 g g^{-1} as the universal value for obtaining optimal vitrification ability at the CCR and the CWR of 10 °C min^{-1} or higher. The CPAs at the concentration of 0.7 g g^{-1} content provided a small amount of water 0.42 g g^{-1} (water/solute CPAs). Similarly, the low water content of 0.4 g g^{-1} (water/dry mass) was defined as a border value for successful cryopreservation of encapsulated meristems [64]. This cryopreservation method used alginate beads moistened with 0.75 M sucrose, which were air dehydrated above silicagel for 6 h when they reached the final water content of 0.4 g g^{-1} and the corresponding Tg of −67.5 °C. We propose that this Tg value corresponds to the Tg of the concentrated Suc solution. In our study, a similar value of Tg at −67.1 °C was detected in 0.7 g g^{-1} Suc solution with a corresponding water content of 0.42 g g^{-1}. Therefore, we suggest the Tg value of −67.5 °C of the air-dehydrated alginate beads [63] was influenced by the Tg of the 0.71 g g^{-1} Suc solution with a corresponding water content of 0.4 g g^{-1}. This supports a former finding [65] that the minimal Suc concentration for a glass transition without water crystallization occurred at 70% (w/w). We assume that Suc solution acts as a single CPA in the encapsulation–dehydration method due to air dehydration of 0.75 M sucrose to 2.83 M (0.71 g g^{-1}). An appropriate osmotic acclimation increases the dehydration tolerance of encapsulated material and allows a decrease in alginate beads' water content up to 0.36 g g^{-1} [66] and decreases the risk of water crystallization. The high Tg value of Suc solution in comparison with other single cryoprotectants (Figures 3–6) should be taken into account when the combined CPAs are designed [2,28]. Including Suc or other saccharides [28] and its proportion in the CPA can influence the Tg of the designed

CPA mixture. Accordingly, the Tg of −115 °C in PVS2 with 0.4 M Suc significantly differed from Tg of −92 °C in PVS3 with 1.46 M Suc. The resulted Tg influences the course of sample C/W and the critical storage temperature [10,67].

The applicability of the suggested solute value of 0.7 g g^{-1} is very often limited by solute toxicity, especially in the penetrating CPA [25]. It was proved that the CPA become increasingly toxic as concentration increases [25]. On contrary, a cryopreservation method employing 10% DMSO combined with the rapid C/W rates was successfully used [68]. However, this method is not based on the sample vitrification because the low DMSO concentration did not avoid water freezing (Table 1), so the method was named the droplet-freezing method [69]. The DMSO induces some ultrastructural changes in treated cells that helped to overcome water freezing without fatal damage [70,71] even under nonequilibrium conditions [72]. This can be probably a significant factor that supports including DMSO in the combined CPAs and frequency of the DMSO-based solution used. Currently, more than four times number of papers are available for the PVS2 cryopreservation compared to the PVS3 in the WOS database, reflecting the generally higher efficiency of the PVS2 solution compared to the PVS3. This phenomenon can be explained by a different mechanism of actions of penetrating and non-penetrating CPA [73] related to the specific tolerance of cryopreserved material to CPA toxicity or osmotic stress in the treatments by PVS2 or PVS3, respectively [74]. The higher osmotic effect of PVS3 compared to PVS2 is probably due to the lower penetrability of the CPA component [73] together with the higher solutes concentration in the solution. Sensitivity to osmotic stress or insufficient osmotic acclimation of cryopreserved material can limit the PVS3 use [75]. Recently, some evidence of the harmful effect of DMSO on living organisms or genetic stability has put downward pressure on the proportion of DMSO in the CPA used [76].

Regardless of a possible DMSO controversy, we can generally conclude, the low CPA concentrations can be successfully used for the freezing cryopreservation methods, but not the vitrification ones because low concentrations of any single CPA did not show appropriate vitrification ability. A possible solution of the CPA toxicity is a design of the CPA mixture composition with the nontoxic effect of the components [13,25,77]. The crucial factor for the vitrification ability of CPAs is mostly a proportion of water in the solution, which can be up to 30% *w/w* according to our results (Figures 1 and 2), which is in an agreement with the previous findings [65]. On the other hand, based on our results and the results of other authors [2], the penetrating CPAs can be efficient in lower concentrations due to their higher vitrification ability than the non-penetrating CPAs. As we showed, the DMSO proved the best vitrification ability of the CPA tested at a concentration of 0.5 g g^{-1} (Figure 4). Although the 0.7 g g^{-1} solute concentration can be the universal vitrification concentration, the wider region from 0.5 to 0.8 g g^{-1} can conditionally be used concerning the proportion of single CPAs and the corresponding CCR or CWR.

Based on our results, four typical concentration regions with specific thermal characteristics can be identified (Figure 7) among the CPA concentrations. The first region was characterized by the water crystallization during the cooling cycle and the presence of Tg' of the freeze-concentrated solution originating from the water crystallization, and it usually ranged from 0 to 0.5 g g^{-1} (Figure 7). This region corresponds well to the unstable vitrification demonstrated earlier on glycerol solution [13]. The second region was typically characterized by water crystallization only during warming and the presence of the Tg corresponding to the actual solute concentration which ranged mostly from 0.5 to 0.6 g g^{-1} (Figure 7). This region corresponds to the metastable vitrification [13]. The third region was characterized by the presence of the actual solution Tg without significant water crystallization and ranged usually from 0.6 to 0.8 g g^{-1} (Figure 7). This region partly belongs to the metastable vitrification conditions [13], which moderately depends on C/W rates, and it ends at the stable vitrification conditions [13], which are independent of C/W rates. The $C_{g'}$ occurred in this range and represented the concentration of the freeze-concentrated phase. It indicated the proximity of the stable vitrification concentration. The fourth region represents sub-optimal conditions and was characterized by the presence of solute crystallization and can occur at a solute concentration higher than 0.8 g g^{-1} (Figure 7).

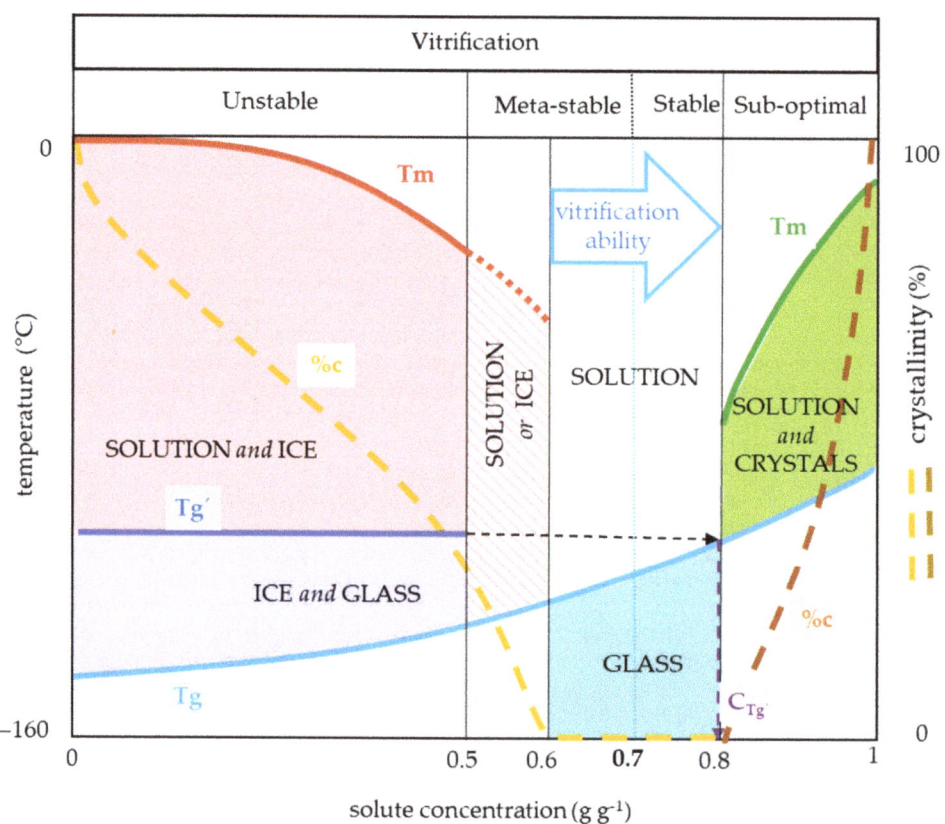

Figure 7. General CPA thermal diagram demonstrating the effect of solute concentration on vitrification ability (VA). Four areas of solute concentration differ in their vitrification ability: (1) low concentration (0–0.5 g g^{-1}) results in crystallization of water during cooling with the exception of an extreme CCR, (2) medium concentration (0.5–0.6 g g^{-1}) carries the risk of crystallization of water during warming, (3) optimal concentration (0.6–0.8 g g^{-1}) for the glass transition, (4) high concentration (0.8–1.0 g g^{-1}) carries the risk of solute crystallization, osmotic damage or CPA toxicity. x-axis—solute concentration (g g^{-1}), left y-axis—temperature (°C), right y-axis—percentage of crystallinity of water or solute (% w/w), Tg—glass transition temperature of the solution, Tg'—glass transition temperature of the freeze-dehydrated solution, Tm—melting point of water (red line) or solute (green line), %c—percentage of crystallinity of water (orange) or solute (brown), C$_g$—concentration of the freeze-concentrated solution. The solute concentration of 0.7 g g^{-1} represents the proposed universal vitrification concentration of CPAs.

Moreover, some transient zones can be detected: the simultaneous occurrence of the water crystallization and the actual glass transition in Gly or Suc solutions during the cooling cycle (Figures 5 and 6); the solute crystallization absence in the fully concentrated Gly (Figure 5). Concerning the vitrification ability at the standard C/W rates of 10 °C min^{-1}, we recommend the third region of the CPA concentrations (0.6–0.8 g g^{-1}) for safe cryopreservation utilizing vitrification methods without any risk of water crystallization at the metastable or stable vitrification conditions with corresponding CCR and CWR. A high solute concentration, close to the C$_{g'}$, is strongly recommended when the C/W rates are limited due to method or sample specificity. The presence of some very effective CPA components as DMSO can shift this region to the lower values. The second region (0.5–0.6 g g^{-1}) is applicable in the case of very high C/W rates. The advantage of this region is an opportunity for the CPA concentration decrease to avoid its toxicity. The first region (0–0.5 g g^{-1}) is recommended mostly for the freezing cryopreservation method and its use for vitrification is possible only under extreme conditions—very small sample size together with ultrarapid C/W rates. The appropriate CCR and CWR should be verified. The last region of the CPA concentrations (0.8–1 g g^{-1}) is inapplicable due to unstable supersaturated solution presence or extreme CPA concentration.

4. Materials and Methods

Osmotic mixtures of two common combined CPAs and their single osmotic components were tested by differential scanning calorimetry. Osmotic components of PVS2 consist of 15% (w/v) ethylene glycol, 15% (w/v) DMSO, 30% (w/v) glycerol, and 13.7% (w/v) sucrose [46]. Nutrients—inorganic salts of low concentration according to Murashige and Skoog [78] were omitted; instead, the osmotics were diluted in water. The above-mentioned composition of the original PVS2 solution was defined as 100% PVS2. Next, six solutions were prepared by dilution of the original solution in water to the following concentrations: 90% (w/v) PVS2, 80% (w/v) PVS2, 70% (w/v) PVS2, 60% (w/v) PVS2, 50% (w/v) PVS2, and 40% (w/v) PVS2.

The second combined CPA, PVS3, consists of 50% (w/v) glycerol and 50% (w/v) sucrose in water [53]. The original concentration of above mentioned osmotics was defined as 100% PVS3. Next, six solutions were prepared by dilution of the original solution in water to the following concentrations: 80% (w/v) PVS3, 60% (w/v) PVS3, 40% (w/v) PVS3, and 20% (w/v) PVS3.

Single components of the above-mentioned complex mixtures were diluted in water to the following concentrations: 0, 10, 20, 30, 40, 50, 60, 70, and 80% (w/w) of ethylene glycol (EG), dimethyl sulfoxide (DMSO), glycerol (Gly), and sucrose (Suc). In the case of liquids (EG, DMSO, Gly), next two concentrations were prepared: 90 and 100% (w/w).

Thermal analysis was performed by differential scanning calorimetry using TA 2920 (TA Instruments, New Castle, DE, USA) and an LN cooling system (LNCS), using helium as a purge gas and hermetically sealed aluminium pans in three repetitions. Standard C/W rates of 10 °C min^{-1} were used in a temperature range from +25 to −140 °C during the cooling cycle of measurement and a range from −140 to 30 °C during the warming cycle of the measurement. Measured thermal events of prepared CPA solutions were evaluated by the Universal Analysis software (TA Instrument, New Castle, DE, USA). The presence of endotherms (crystallization), exotherms (melting), and the glass transitions during the C/W cycles of the measurement were detected. The temperature of glass transition (Tg), the temperature of onset melting peak (Tm) was measured. The amount of crystallinity was calculated based on the measured heat of fusion. Standard heat of fusion of 334, 181, and 173 J g^{-1} K^{-1} was used for water, EG, and DMSO, respectively.

5. Conclusions

The appropriate vitrification ability of two combined (PVS2 and PVS3) and four single (EG, DMSO, Gly, Suc) CPAs usually ranged from solute concentrations of 0.6 to 0.8 g g^{-1} for the CCR and the CWR of 10 °C min^{-1} or higher. Gly was the only single cryoprotectant which possessed vitrification ability up to 100% solute concentration. DMSO was the most effective CPA with an appropriate vitrification ability already starting at a solute concentration of 0.5 g g^{-1}. We demonstrated that the Suc solution can be used as a natural single CPA when dehydrated to a concentration of 0.7 g g^{-1} or higher. In the combined CPAs, the appropriate solute concentration was proved only in 100% PVS2 or 80 and 100% PVS3.

No crystallization in all CPAs during the cooling cycle is associated with the detection of the Tg corresponding to the actual solution concentration. The presence of a glassy state of the freeze-concentrated solution, characterized by Tg′, indicates water crystallization occurrence during the cooling cycle and a possible injury of a sample by ice crystal growth. On the other hand, the Tg' value can help to identify the $C_{g'}$ of the maximally freeze-concentrated solution, which corresponds to the solute concentration typical for the stable vitrification conditions [13]. The solute concentration of 100% PVS3 close to the $C_{g'}$ suggests its higher vitrification ability even at very low CCR and CWR compared to 100% PVS2.

Generally, combined CPAs are recommended to avoid the risk of sample damage due to CPA cytotoxicity [25,79]. The design of combined CPAs can decrease the portion of the individual components below their toxicity threshold [79]. While the higher proportion of DMSO in the combined CPA decreases the total solute concentration with appropriate vitrification ability, the higher proportion of Suc increases the Tg value of the CPA mixture and subsequently increases the safe storage temperature of samples. Knowledge of the CPAs thermal properties can help to design a new combined CPA with the appropriate vitrification ability.

Author Contributions: Conceptualization, M.F., A.B. and J.Z.; methodology, M.F. and J.Z.; investigation, M.F.; resources, J.Z.; data curation, M.F. and A.B.; writing—original draft preparation, M.F.; writing—review and editing, M.F. and J.Z.; project administration, M.F.; funding acquisition, M.F. All authors have read and agreed to the published version of the manuscript.

Funding: This research was funded by the Ministry of Agriculture of the Czech Republic, grant number MZERO0418 and QK1910277.

Institutional Review Board Statement: Not applicable.

Informed Consent Statement: Not applicable.

Data Availability Statement: The data presented in this study are available on request from the corresponding author.

Conflicts of Interest: The authors declare no conflict of interest.

References

1. Clarke, A.; Morris, G.J.; Fonseca, F.; Murray, B.J.; Acton, E.; Price, H.C. A low temperature limit for life on Earth. *PLoS ONE* **2013**, *8*, e66207. [CrossRef]
2. Fahy, G.M.; MacFarlane, D.; Angell, C.A.; Meryman, H. Vitrification as an approach to cryopreservation. *Cryobiology* **1984**, *21*, 407–426. [CrossRef]
3. Pegg, D.E. Principles of cryopreservation. In *Cryopreservation and Freeze-Drying Protocols*; Humana Press: Totowa, NJ, USA, 2007; pp. 39–57.
4. Park, S.; Seawright, A.; Park, S.; Dutton, J.C.; Grinnell, F.; Han, B. Preservation of tissue microstructure and functionality during freezing by modulation of cytoskeletal structure. *J. Mech. Behav. Biomed.* **2015**, *45*, 32–44. [CrossRef] [PubMed]
5. Fuller, B.; Paynter, S. Fundamentals of cryobiology in reproductive medicine. *Reprod. Biomed. Online* **2004**, *9*, 680–691. [CrossRef]
6. Wowk, B. Thermodynamic aspects of vitrification. *Cryobiology* **2010**, *60*, 11–22. [CrossRef]
7. Day, J.G.; Harding, K.C.; Nadarajan, J.; Benson, E.E. Cryopreservation. In *Molecular Biomethods Handbook*; Humana Press: Totowa, NJ, USA, 2008; pp. 917–947.
8. Sakai, A.; Engelmann, F. Vitrification, encapsulation-vitrification and droplet-vitrification: A review. *CryoLetters* **2007**, *28*, 151–172.
9. Gonzalez-Arnao, M.T.; Engelmann, F. Cryopreservation of plant germplasm using the encapsulation-dehydration technique: Review and case study on sugarcane. *CryoLetters* **2006**, *27*, 155–168. [PubMed]
10. Benson, E.E. Cryopreservation of Phytodiversity: A Critical Appraisal of Theory & Practice. *Crit. Rev. Plant Sci.* **2008**, *27*, 141–219. [CrossRef]
11. Engelmann, F. Plant cryopreservation: Progress and prospects. *In Vitro Cell. Dev. Biol. Plant* **2004**, *40*, 427–433. [CrossRef]
12. Fahy, G.M.; Wowk, B. Principles of ice-free cryopreservation by vitrification. In *Cryopreservation and Freeze-Drying Protocols*; Humana Press: Totowa, NJ, USA, 2021; pp. 27–97.
13. Fahy, G.M.; Wowk, B. Principles of cryopreservation by vitrification. *Methods Mol. Biol.* **2015**, *1257*, 21–82. [CrossRef]
14. Fahy, G.M. Cryoprotectant toxicity neutralization. *Cryobiology* **2010**, *60*, S45–S53. [CrossRef]
15. Hubálek, Z. Protectants used in the cryopreservation of microorganisms. *Cryobiology* **2003**, *46*, 205–229. [CrossRef]
16. Elliott, G.D.; Wang, S.; Fuller, B.J. Cryoprotectants: A review of the actions and applications of cryoprotective solutes that modulate cell recovery from ultra-low temperatures. *Cryobiology* **2017**, *76*, 74–91. [CrossRef]
17. Panis, B.; Piette, B.; Swennen, R. Droplet vitrification of apical meristems: A cryopreservation protocol applicable to all Musaceae. *Plant Sci.* **2005**, *168*, 45–55. [CrossRef]

18. Reed, B.M.; Kovalchuk, I.; Kushnarenko, S.; Meier-Dinkel, A.; Schoenweiss, K.; Pluta, S.; Straczynska, K.; Benson, E.E. Evaluation of critical points in technology transfer of cryopreservation protocols to international plant conservation laboratories. *CryoLetters* **2004**, *25*, 341–352.
19. Matsumoto, T. Cryopreservation of plant genetic resources: Conventional and new methods. *Rev. Agric. Sci.* **2017**, *5*, 13–20. [CrossRef]
20. Benson, E.; Harding, K.; Ryan, M.; Petrenko, A.; Petrenko, Y.; Fuller, B. Alginate encapsulation to enhance biopreservation scope and success: A multidisciplinary review of current ideas and applications in cryopreservation and non-freezing storage. *Cryoletters* **2018**, *39*, 14–38. [PubMed]
21. Popova, E.; Shukla, M.; Kim, H.-H.; Saxena, P.K. Root cryobanking: An important tool in plant cryopreservation. *Plant Cell Tissue Organ Cult. PCTOC* **2021**, *144*, 49–66. [CrossRef]
22. Han, Z.; Bischof, J.C. Critical cooling and warming rates as a function of CPA concentration. *CryoLetters* **2020**, *41*, 185–193. [PubMed]
23. Teixeira, A.S.; Gonzalez-Benito, M.E.; Molina-Garcia, A.D. Measurement of Cooling and Warming Rates in Vitrification-Based Plant Cryopreservation Protocols. *Biotechnol. Progr.* **2014**, *30*, 1177–1184. [CrossRef]
24. Paredes, E.; Mazur, P. The survival of mouse oocytes shows little or no correlation with the vitrification or freezing of the external medium, but the ability of the medium to vitrify is affected by its solute concentration and by the cooling rate. *Cryobiology* **2013**, *67*, 386–390. [CrossRef]
25. Best, B.P. Cryoprotectant Toxicity: Facts, Issues, and Questions. *Rejuvenation Res.* **2015**, *18*, 422–436. [CrossRef]
26. Sestak, J. Some Thermodynamic Aspects of the Glassy State. *Thermochim. Acta* **1985**, *95*, 459–471. [CrossRef]
27. Boutron, P.; Kaufmann, A. Stability of Amorphous State in System Water-Glycerol-Dimethylsulfoxide. *Cryobiology* **1978**, *15*, 93–108. [CrossRef]
28. Roos, Y.H.; Karel, M. Phase-Transitions of Amorphous Sucrose and Frozen Sucrose Solutions. *J. Food Sci.* **1991**, *56*, 266–267. [CrossRef]
29. Roos, Y.H. Glass Transition and Re-Crystallization Phenomena of Frozen Materials and Their Effect on Frozen Food Quality. *Foods* **2021**, *10*, 447. [CrossRef] [PubMed]
30. Roos, Y. Melting and glass transitions of low molecular weight carbohydrates. *Carbohyd. Res.* **1993**, *238*, 39–48. [CrossRef]
31. Ren, H.; Wei, Y.; Hua, T.; Zhang, J. Theoretical prediction of vitrification and devitrification tendencies for cryoprotective solutions. *Cryobiology* **1994**, *31*, 47–56. [CrossRef]
32. Block, W. Water status and thermal analysis of alginate beads used in cryopreservation of plant germplasm. *Cryobiology* **2003**, *47*, 59–72. [CrossRef]
33. Dumet, D.; Grapin, A.; Bailly, C.; Dorion, N. Revisiting crucial steps of an encapsulation/desiccation based cryopreservation process: Importance of thawing method in the case of Pelargonium meristems. *Plant Sci.* **2002**, *163*, 1121–1127. [CrossRef]
34. Matsumoto, Y.; Morinaga, Y.; Ujihira, M.; Oka, K.; Tanishita, K. Improvement in the viability of cryopreserved cells by microencapsulation. *Int. J. Ser. C Mech. Syst. Mach. Elem. Manuf.* **2001**, *44*, 937–945. [CrossRef]
35. Wowk, B.; Darwin, M.; Harris, S.B.; Russell, S.R.; Rasch, C.M. Effects of solute methoxylation on glass-forming ability and stability of vitrification solutions. *Cryobiology* **1999**, *39*, 215–227. [CrossRef] [PubMed]
36. Kaczmarczyk, A.; Zanke, C.; Senula, A.; Grube, M.; Keller, E.R.J. Thermal Analyses by Differential Scanning Calorimetry for Cryopreservation of Potato Shoot Tips. *Acta Hortic.* **2011**, *908*, 39–46. [CrossRef]
37. Hammond, S.D.; Faltus, M.; Zámečník, J. Methods of Thermal Analysis as a Tool to Develop Cryopreservation Protocols of Vegetatively Propagated Crops. In *Cryopreservation-Current Advances and Evaluations*; IntechOpen: Rijeka, Croatia, 2019; pp. 161–176.
38. Vozovyk, K.; Bobrova, O.; Prystalov, A.; Shevchenko, N.; Kuleshova, L. Amorphous state stability of plant vitrification solutions. *Biologija* **2020**, *66*, 47–53. [CrossRef]
39. Šesták, J.; Zámečník, J. Can clustering of liquid water and thermal analysis be of assistance for better understanding of biological germplasm exposed to ultra-low temperatures. *J. Therm. Anal. Calorim.* **2007**, *88*, 411–416. [CrossRef]
40. Kim, H.H.; No, N.Y.; Shin, D.J.; Ko, H.C.; Kang, J.H.; Cho, E.G.; Engelmann, F. Development of Alternative Plant Vitrification Solutions to be Used in Droplet-Vitrification Procedures. *Acta Hortic.* **2011**, *908*, 181–186. [CrossRef]
41. Sarkar, D.; Naik, P.S. Cryopreservation of shoot tips of tetraploid potato (Solanum tuberosum L.) clones by vitrification. *Ann. Bot.* **1998**, *82*, 455–461. [CrossRef]
42. Choi, C.-H.; Popova, E.; Lee, H.; Park, S.-U.; Ku, J.; Kang, J.-H.; Kim, H.-H. Cryopreservation of endangered wild species, Aster altaicus var. uchiyamae Kitam, using droplet-vitrification procedure. *CryoLetters* **2019**, *40*, 113–122.
43. Yi, J.Y.; Sylvestre, I.; Colin, M.; Salma, M.; Lee, S.Y.; Kim, H.H.; Park, H.J.; Engelmann, F. Improved Cryopreservation Using Droplet-vitrification and Histological Changes Associated with Cryopreservation of Madder (Rubia akane Nakai). *Korean J. Hortic. Sci.* **2012**, *30*, 79–84. [CrossRef]
44. Benson, E.; Reed, B.; Brennan, R.; Clacher, K.; Ross, D. Use of thermal analysis in the evaluation of cryopreservation protocols for Ribes nigrum L. germplasm. *Cryo-Letters* **1996**, *17*, 347–362.
45. Volk, G.M.; Walters, C. Plant vitrification solution 2 lowers water content and alters freezing behavior in shoot tips during cryoprotection. *Cryobiology* **2006**, *52*, 48–61. [CrossRef]

46. Sakai, A.; Kobayashi, S.; Oiyama, I. Cryopreservation of nucellar cells of navel orange (Citrus sinensis Osb. var. brasiliensis Tanaka) by vitrification. *Plant Cell Rep.* **1990**, *9*, 30–33. [CrossRef]
47. Dumetlll, D.; Block, W.; Worland, R.; ReedJ, B.M.; Benson, E.E. Profiling cryopreserv a tion protocols for ribes cilia rum using differential scanning calorimetry. *CryoLetters* **2000**, *21*, 378.
48. Niino, T.; Sakai, A.; Yakuwa, H.; Nojiri, K. Cryopreservation of in vitro-grown shoot tips of apple and pear by vitrification. *Plant Cell Tissue Organ Cult.* **1992**, *28*, 261–266. [CrossRef]
49. Volk, G.M.; Harris, J.L.; Rotindo, K.E. Survival of mint shoot tips after exposure to cryoprotectant solution components. *Cryobiology* **2006**, *52*, 305–308. [CrossRef] [PubMed]
50. Niino, T.; Yamamoto, S.-I.; Fukui, K.; Martínez, C.R.C.; Arizaga, M.V.; Matsumoto, T.; Engelmann, F. Dehydration improves cryopreservation of mat rush (Juncus decipiens Nakai) basal stem buds on cryo-plates. *CryoLetters* **2013**, *34*, 549–560. [PubMed]
51. Kim, H.-H.; Lee, Y.-G.; Shin, D.-J.; Ko, H.-C.; Gwag, J.-G.; Cho, E.-G.; Engelmann, F. Development of alternative plant vitrification solutions in droplet-vitrification procedures. *CryoLetters* **2009**, *30*, 320–334. [CrossRef] [PubMed]
52. Boutron, P. Glass-forming tendency and stability of the amorphous state in solutions of a 2, 3-butanediol containing mainly the levo and dextro isomers in water, buffer, and Euro-Collins. *Cryobiology* **1993**, *30*, 86–97. [CrossRef]
53. Nishizawa, S.; Sakai, A.; Amano, Y.; Matsuzawa, T. Cryopreservation of Asparagus (Asparagus-Officinalis L) Embryogenic Suspension Cells and Subsequent Plant-Regeneration by Vitrification. *Plant Sci.* **1993**, *91*, 67–73. [CrossRef]
54. Kim, H.H.; Popova, E.V.; Yi, J.Y.; Cho, G.T.; Park, S.U.; Lee, S.C.; Engelmann, F. Cryopreservation of Hairy Roots of Rubia Akane (Nakai) Using a Droplet-Vitrification Procedure. *Cryoletters* **2010**, *31*, 473–484. [PubMed]
55. Teixeira, A.S.; Faltus, M.; Zamecnik, J.; Gonzalez-Benito, M.E.; Molina-Garcia, A.D. Glass transition and heat capacity behaviors of plant vitrification solutions. *Thermochim. Acta* **2014**, *593*, 43–49. [CrossRef]
56. Wang, M.-R.; Zhang, Z.; Zámečník, J.; Bilavčík, A.; Blystad, D.-R.; Haugslien, S.; Wang, Q.-C. Droplet-vitrification for shoot tip cryopreservation of shallot (Allium cepa var. aggregatum): Effects of PVS3 and PVS2 on shoot regrowth. *Plant Cell Tissue Organ Cult. PCTOC* **2020**, *140*, 185–195. [CrossRef]
57. Kim, J.-B.; Kim, H.-H.; Baek, H.-J.; Cho, E.-G.; Kim, Y.-H.; Engelmann, F. Changes in sucrose and glycerol content in garlic shoot tips during freezing using PVS3 solution. *CryoLetters* **2005**, *26*, 103–112. [PubMed]
58. Le, K.C.; Kim, H.H.; Park, S.Y. Modification of the droplet-vitrification method of cryopreservation to enhance survival rates of adventitious roots of Panax ginseng. *Hortic. Environ. Biotechnol.* **2019**, *60*, 501–510. [CrossRef]
59. Ree, J.F.; Guerra, M.P. Exogenous inorganic ions, partial dehydration, and high rewarming temperatures improve peach palm (Bactris gasipaes Kunth) embryogenic cluster post-vitrification regrowth. *Plant Cell Tissue Organ Cult.* **2021**, *144*, 157–169. [CrossRef]
60. Faltus, M.; Bilavčík, A.; Zámečník, J. Thermal analysis of grapevine shoot tips during dehydration and vitrification. *VITIS J. Grapevine Res.* **2015**, *54*, 243–245.
61. Gao, C.; Zhou, G.-Y.; Xu, Y.; Hua, T.-C. Glass transition and enthalpy relaxation of ethylene glycol and its aqueous solution. *Thermochim. Acta* **2005**, *435*, 38–43. [CrossRef]
62. Fahy, G.M.; Levy, D.; Ali, S. Some emerging principles underlying the physical properties, biological actions, and utility of vitrification solutions. *Cryobiology* **1987**, *24*, 196–213. [CrossRef]
63. Roos, Y. Frozen state transitions in relation to freeze drying. *J. Therm. Anal. Calorim.* **1997**, *48*, 535–544. [CrossRef]
64. Sherlock, G.; Block, W.; Benson, E.E. Thermal analysis of the plant encapsulation-dehydration cryopreservation protocol using silica gel as the desiccant. *CryoLetters* **2005**, *26*, 45–54.
65. Dereuddre, J.; Kaminski, M. Applications of Thermal-Analysis in Cryopreservation of Plant-Cells and Organs. *J. Therm. Anal.* **1992**, *38*, 1965–1978. [CrossRef]
66. Lynch, P.T.; Souch, G.R.; Záměník, J.; Harding, K. Optimization of water content for the cryopreservation of Allium sativum in vitro cultures by encapsulation-dehydration. *CryoLetters* **2016**, *37*, 308–317. [PubMed]
67. International Society for Biological and Environmental Repositories. Best practices for repositories I: Collection, storage, and retrieval of human biological materials for research. *Cell Preserv. Technol.* **2005**, *3*, 5–48. [CrossRef]
68. Schafer-Menuhr, A.; Muller, E.; Mix-Wagner, G. Cryopreservation: An alternative for the long-term storage of old potato varieties. *Potato Res.* **1996**, *39*, 507–513. [CrossRef]
69. Kartha, K.; Leung, N.; Mroginski, L. In vitro growth responses and plant regeneration from cryopreserved meristems of cassava (Manihot esculenta Crantz). *Z. Pflanzenphysiol.* **1982**, *107*, 133–140. [CrossRef]
70. Kaczmarczyk, A.; Rutten, T.; Melzer, M.; Keller, E.R.J. Ultrastructural changes associated with cryopreservation of potato (Solanum tuberosum L.) shoot tips. *Cryoletters* **2008**, *29*, 145–156. [PubMed]
71. Weng, L.; Stott, S.L.; Toner, M. Exploring dynamics and structure of biomolecules, cryoprotectants, and water using molecular dynamics simulations: Implications for biostabilization and biopreservation. *Annu. Rev. Biomed. Eng.* **2019**, *21*, 1–31. [CrossRef] [PubMed]
72. Halmagyi, A.; Deliu, C.; Isac, V. Cryopreservation of Malus cultivars: Comparison of two droplet protocols. *Sci. Hortic.* **2010**, *124*, 387–392. [CrossRef]
73. Volk, G.M.; Caspersen, A.M. Cryoprotectants and components induce plasmolytic responses in sweet potato (Ipomoea batatas (L.) Lam.) suspension cells. *In Vitro Cell. Dev. Biol. Plant* **2017**, *53*, 363–371. [CrossRef]

74. Kim, H.-H.; Lee, Y.-G.; Park, S.-U.; Lee, S.-C.; Baek, H.-J.; Cho, E.-G.; Engelmann, F. Development of alternative loading solutions in droplet-vitrification procedures. *CryoLetters* **2009**, *30*, 291–299.
75. Lee, H.; Park, H.; Popova, E.; Lee, Y.-Y.; Park, S.-U.; Kim, H.-H. Ammonium-free medium is critical for regeneration of shoot tips of the endangered species Pogostemon yatabeanus cryopreserved using droplet-vitrification. *CryoLetters* **2021**, *42*, 290–299.
76. Weng, L.; Beauchesne, P.R. Dimethyl sulfoxide-free cryopreservation for cell therapy: A review. *Cryobiology* **2020**, *94*, 9–17. [CrossRef]
77. Kasai, M.; Mukaida, T. Cryopreservation of animal and human embryos by vitrification. *Reprod. BioMed. Online* **2004**, *9*, 164–170. [CrossRef]
78. Murasnige, T.; Skoog, F. A revised medium for rapid growth and bio agsays with tohaoco tissue cultures. *Physiol. Plant* **1962**, *15*, 473–497. [CrossRef]
79. Warner, R.M.; Ampo, E.; Nelson, D.; Benson, J.D.; Eroglu, A.; Higgins, A.Z. Rapid quantification of multi-cryoprotectant toxicity using an automated liquid handling method. *Cryobiology* **2021**, *98*, 219–232. [CrossRef] [PubMed]

Article

Effect of Cryopreservation on Proteins from the Ubiquitous Marine Dinoflagellate *Breviolum* sp. (Family Symbiodiniaceae)

Hsing-Hui Li [1,2,†], Jia-Lin Lu [2,†], Hui-Esther Lo [1], Sujune Tsai [3,*] and Chiahsin Lin [1,2,*]

[1] National Museum of Marine Biology & Aquarium, Pingtung 944, Taiwan; hhli@nmmba.gov.tw (H.-H.L.); estherlo1995@yahoo.ca (H.-E.L.)
[2] Institute of Marine Biology, National Dong Hwa University, Pingtung 944, Taiwan; alice0783@yahoo.com.tw
[3] Department of Post Modern Agriculture, Mingdao University, Peetow, Chang Hua 369, Taiwan
* Correspondence: stsai@mdu.edu.tw (S.T.); chiahsin@nmmba.gov.tw (C.L.); Tel.: +886-925750025 (S.T.); +886-8-8825001 (ext. 1356) (C.L.)
† These authors contributed equally to this work.

Abstract: Coral reefs around the world are exposed to thermal stress from climate change, disrupting the delicate symbiosis between the coral host and its symbionts. Cryopreservation is an indispensable tool for the preservation of species, as well as the establishment of a gene bank. However, the development of cryopreservation techniques for application to symbiotic algae is limited, in addition to the scarceness of related studies on the molecular level impacts post-thawing. Hence, it is essential to set up a suitable freezing protocol for coral symbionts, as well as to analyze its cryo-injury at the molecular level. The objective of this study was to develop a suitable protocol for the coral symbiont *Breviolum* subjected to two-step freezing. The thawed *Breviolum* were then cultured for 3, 7, 14, and 28 days before they were analyzed by Western blot for protein expression, light-harvesting protein (LHP), and red fluorescent protein (RFP) and tested by adenosine triphosphate bioassay for cell viability. The results showed the highest cell viability for thawed *Breviolum* that was treated with 2 M propylene glycol (PG) and 2 M methanol (MeOH) and equilibrated with both cryoprotectants for 30 min and 20 min. Both treatment groups demonstrated a significant increase in cell population after 28 days of culture post-thawing, especially for the MeOH treatment group, whose growth rate was twice of the PG treatment group. Regarding protein expression, the total amounts of each type of protein were significantly affected by cryopreservation. After 28 days of culture, the protein expression for the MeOH treatment group showed no significant difference to that of the control group, whereas the protein expression for the PG treatment group showed a significant difference. *Breviolum* that were frozen with MeOH recovered faster upon thawing than those frozen with PG. LHP was positively and RFP was negatively correlated with Symbiodiniaceae viability and so could serve as health-informing biomarkers. This work represents the first time to document it in Symbiodiniaceae, and this study established a suitable protocol for the cryopreservation of *Breviolum* and further refined the current understanding of the impact of low temperature on its protein expression. By gaining further understanding of the use of cryopreservation as a way to conserve Symbiodiniaceae, we hope to make an effort in the remediation and conservation of the coral reef ecosystem and provide additional methods to rescue coral reefs.

Keywords: cryopreservation; Symbiodiniaceae; two-step freezing; protein expression; coral

1. Introduction

Cryopreservation is a technique that prolongs the viability of structurally intact cells and tissues by freezing under low-temperature treatment at relatively low cost and time [1]. Its procedure can also reduce the expense and work required for the culture of algae, as opposed to the traditional method, and prevent possible contamination and genetic shift during culture [2].

In recent years, temperature change, acidification, and eutrophication [3,4] have caused great impacts on the marine environment, leading to catastrophic consequences on the coral reef ecosystems. Coral reefs, the irreplaceable habitats for many marine organisms, are currently endangered by the disruption of their symbiotic relationship with symbiotic dinoflagellates from the Symbiodiniaceae family due to global environmental change. The disruption of the symbiosis ceases the nutrient supply for corals from Symbiodiniaceae photosynthesis, leading to coral bleaching [5,6]. Corals cannot survive without their symbiotic dinoflagellates, and therefore, conserving the symbionts will be a critical factor in coral reef conservation [7].

Much research is now focused on the application of cryopreservation on Symbiodiniaceae conservation [8,9]. For example, Zhao et al. 2017 [10] tested for the optimal cryopreservation protocol of *Gerakladium* extracted from *Gorgonacea*, using vitrification and slow freezing method. In addition, Chong et al. 2016c [11] made the first successful attempt on the two-step freezing of *Gerakladium*, using 1 M MeOH with 0.4 M sucrose as a cryoprotectant (CPA); the survival rate of the thawed Symbiodiniaceae reached 56.93%. By using Symbiodiniaceae extracted from *Pseudopterogorgia elisabethae*, Santiago-Vázquez et al. 2017 [12] found that, of four different freezing protocols, the two-step freezing resulted in the highest survival rate of thawed algae during culture. Furthermore, Hagedorn and Carter, 2015 [13] attempted cryopreservation of Symbiodiniaceae from different seasons of a year and found that *Cladocopium* quality varies with season. The information accumulated through the preceding studies aids in the establishment of a database for the gene bank of symbiotic dinoflagellates.

Aside from symbiotic dinoflagellates, two-step freezing is also being commonly applied to other algae. For example, in a 1984 study conducted by Van Der Meer and Simpson, the rhodophyta *Gracilaria tikvahiae* was preserved for as long as 4 years using the two-step freezing. Zhang et al. 2008 [14] also performed the same procedure on the economically valuable algae *Laminaria japonica*; of the 43% of algae that survived, there was more of the male gametophyte than female gametophyte, suggesting that two-step freezing could possibly assist in selective breeding in the future. The two-step freezing involves two stages of cooling: the cell is first cooled to between -30 and $-50\ °C$ [15,16], then it is immersed into liquid nitrogen. The purpose of the first cooling stage is to equilibrate the cellular osmotic pressure with the implementation of cryoprotectants (CPAs) to efflux excess water from the cell, thus preventing the formation of intracellular crystals during cryopreservation [2,17].

Current studies on topics regarding the molecular level impact of cryopreserved Symbiodiniaceae only cover as much as mitochondria DNA [18], gene expression [19,20], and DNA integrity [12]. Most research conducted on the protein expression of Symbiodiniaceae focused on how environmental factors influence it, such as the affected protein expression and gene loss caused by different light intensity [21] and warming of seawater [22], as well as the impact of photosynthesis on Symbiodiniaceae protein level and chlorophyll expression, and on the endosymbiotic relationship between algae and coral [23]. However, relevant research on the impact of protein in cryopreserved Symbiodiniaceae is still lacking. Although the role of red fluorescence proteins (RFP) had been extensively studied, its research is still a highly discussed topic [24,25]. Meanwhile, Huang et al. (2017) [26] had explored a high titer antibody for Symbiodiniaceae light-harvesting protein (LHP), which is involved in photosynthesis [27,28]. The objective of the study is to obtain a suitable protocol for the two-step freezing of *Breviolum*, as well as to observe the growth of post-thawed *Breviolum* in culture. The present study also investigated the effect of cryopreservation at the molecular level in *Breviolum* by analyzing proteins (RFP and LHP).

2. Results

2.1. Effect of Equilibration Time and Concentration of CPAs on the Viability of Frozen-Thawed Breviolum

The experiment investigated the viability of thawed *Breviolum* that were treated with four types of CPAs (MeOH, PG, DMSO, and Gly) at three varying concentrations (1, 2, and

3 M) and equilibrated for either 20 or 30 min (Figure 1). ATP content (presented as relative percentages against experimental controls) was used as a proxy for cell viability. Between the MeOH treatment groups equilibrated for 20 and 30 min, the former showed relatively higher *Breviolum* viability after cryopreservation (Figure 1a). The MeOH treatment group equilibrated for 20 min showed significantly higher *Breviolum* viability in the 1 and 2 M groups than at 30 min (pairwise t-tests, $p = 0.021$ and $p = 0.035$, respectively). In the PG treatment group equilibrated for 20 and 30 min, it was found that thawed *Breviolum* treated with 2 M PG and equilibrated for 30 min had the highest cell viability (Figure 1b). Moreover, there were no effects of equilibration time (20 vs. 30 min) at any of the three PG concentrations (Figure 1b). Aside from all that, the results revealed that, in the PG treatment group equilibrated for 20 min, *Breviolum* viability decreased with increasing CPA concentration (one-way ANOVA, $F(3/8) = 39.452$, $p < 0.001$). Lastly, there was no significant difference ($p > 0.05$) in *Breviolum* viability among the DMSO and Gly treatment groups (Figure 1c,d, respectively), regardless of equilibration time or CPA concentration; the only exception to this was the 2 M DMSO treatment group, which showed a significant temporal difference (pairwise t-tests, $p = 0.029$).

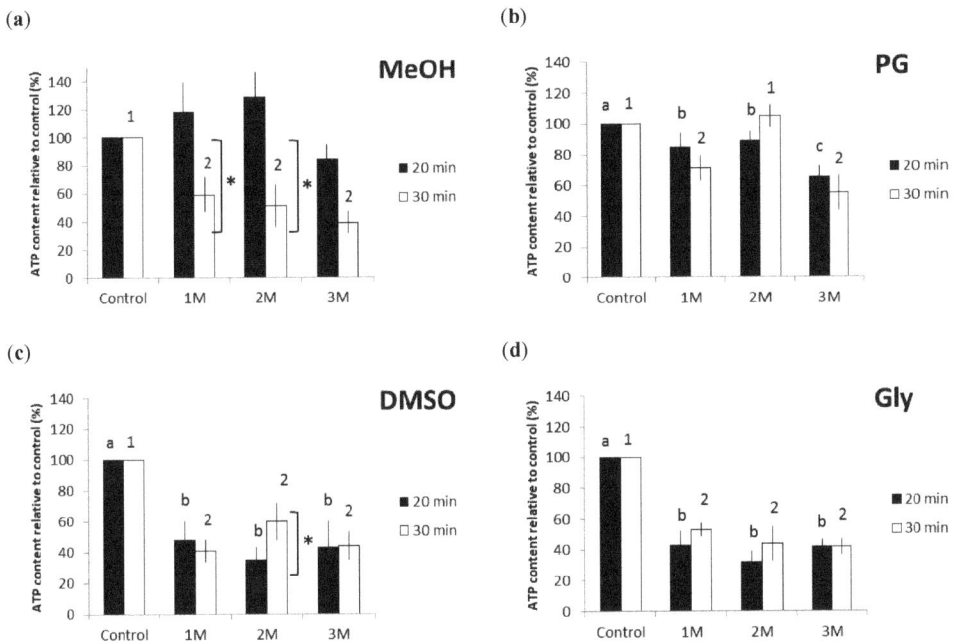

Figure 1. ATP content for thawed *Breviolum* that was cryopreserved with different CPAs: (**a**) MeOH, (**b**) PG, (**c**) DMSO, and (**d**) Gly. *Breviolum* was cryopreserved with 4 different types of CPAs at 3 varying concentrations (1, 2, and 3 M) and equilibrated for either 20 min or 30 min. The control group represents *Breviolum* not subjected to any treatment. Different letters and numbers represent a significant difference ($p < 0.05$) in treatment groups of varying concentrations equilibrated for 20 and 30 min. Asterisks are marked for any significance difference observed under the same CPA concentrations but over different equilibration times. The error bar represents standard deviations.

The experiment yielded the highest *Breviolum* viability for the 2 M MeOH treatment, equilibrated for 20 min, with an increase in ATP response reaching 130%. The next best symbiont viability was obtained with the 2 M PG group equilibrated for 30 min, with an ATP response reaching 100%. Ensuing were the 2 M DMSO and 1 M Gly treatment groups, both equilibrated for 30 min, at 60% and 58%, respectively. The MeOH and PG treatments yielded higher *Breviolum* viability than either of the DMSO and Gly treatment groups. The

Gly treatment group displayed its highest *Breviolum* viability with a 1 M concentration, whereas the other three CPA treatment groups showed their respective highest viability with a 2 M concentration. In terms of equilibration time, only the MeOH treatment group displayed its highest *Breviolum* viability with 20 min, whereas with the other three CPAs, the 30 min equilibration yielded the best *Breviolum* viability.

2.2. Breviolum Growth Rate after Four Culture Periods

Thawed *Breviolum* were cultured independently under different culture durations (3, 7, 14, and 28 days), during which growth was quantified (Figure 2). The number of algal cells was determined using a hemocytometer. The cell culture density for the control groups (enumerated immediately post-thaw) for the MeOH group *Breviolum* was 8.28×10^5 (3-day culture), 9.35×10^5 (7-day culture), 6.16×10^5 (14-day culture), and 7.10×10^5 cells/mL (28-day culture) (Figure 2a). The cell culture density for the control groups for the PG group *Breviolum* was 6.85×10^5 (3-day culture), 4.36×10^5 (7-day culture), 7.87×10^5 (14-day culture), and 6.27×10^5 cells/mL (28-day culture) (Figure 2b). Results showed that *Breviolum* growth commenced after 7 days of culture for the MeOH treatment group (Figure 2a) and after 14 days of culture for the PG treatment group (Figure 2b). Both treatment groups displayed growth by the 28th day of culture, but the cell culture density of the MeOH treatment group was twice as high as that of the PG treatment group. The cell culture density and normalized percentage for MeOH group *Breviolum* after 3, 7, 14, and 28 days of culture were 3.16×10^5, 8.16×10^5, 9.27×10^5, and 1.36×10^7 cells/mL, respectively (Figure 2a), whereas that of PG group *Breviolum* were 2.00×10^4, 3.20×10^4, 1.69×10^5, and 5.54×10^6 cells/mL, respectively (Figure 2b).

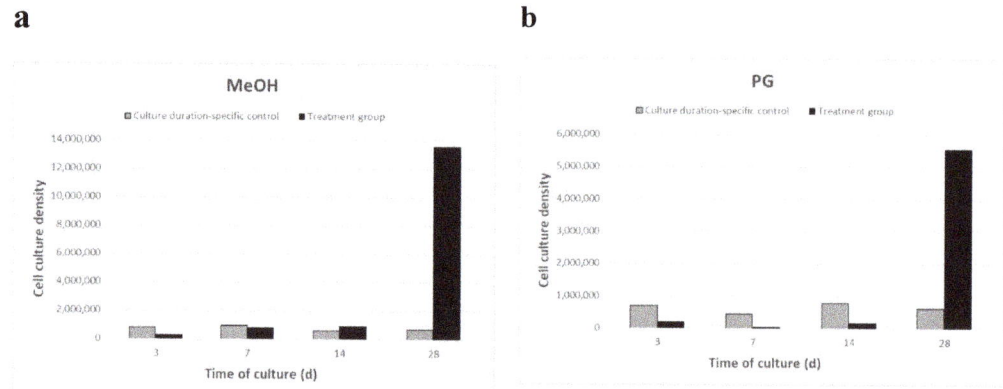

Figure 2. Cell culture density for the MeOH (**a**) and PG (**b**) treatment groups of thawed *Breviolum* after 3, 7, 14, and 28 days of culture; each culture duration trial was conducted independently. Control densities for the MeOH treatment: 3.16×10^5 (3-day culture), 8.16×10^5 (7-day culture), 9.27×10^5 (14-day culture), and 1.36×10^7 cells/mL (28-day culture); PG treatment: 2.00×10^4 (3-day culture), 3.20×10^4 (7-day culture), 1.67×10^5 (14-day culture), and 5.54×10^6 cells/mL (28-day culture). The horizontal axis shows the number of days of culture, at which 0 days represents the cell population immediately after thawing.

2.3. Total Protein Expression Pattern of Breviolum

After extraction of *Breviolum* protein samples, an equal amount of total proteins were loaded in SDS-PAGE, and a SYPRO® Ruby (Invitrogen, Heidelberg, Germany) staining gel confirmed that each protein sample was consistent for follow-up experimentation. Figure 3 showed that the total protein patterns were different for the MeOH (Figure 3a) and PG (Figure 3b) treatment groups of *Breviolum* undergoing two-step freezing. Results also showed that the total protein expression showed different patterns after 3, 7, 14, and 28 days of post-thawed culture.

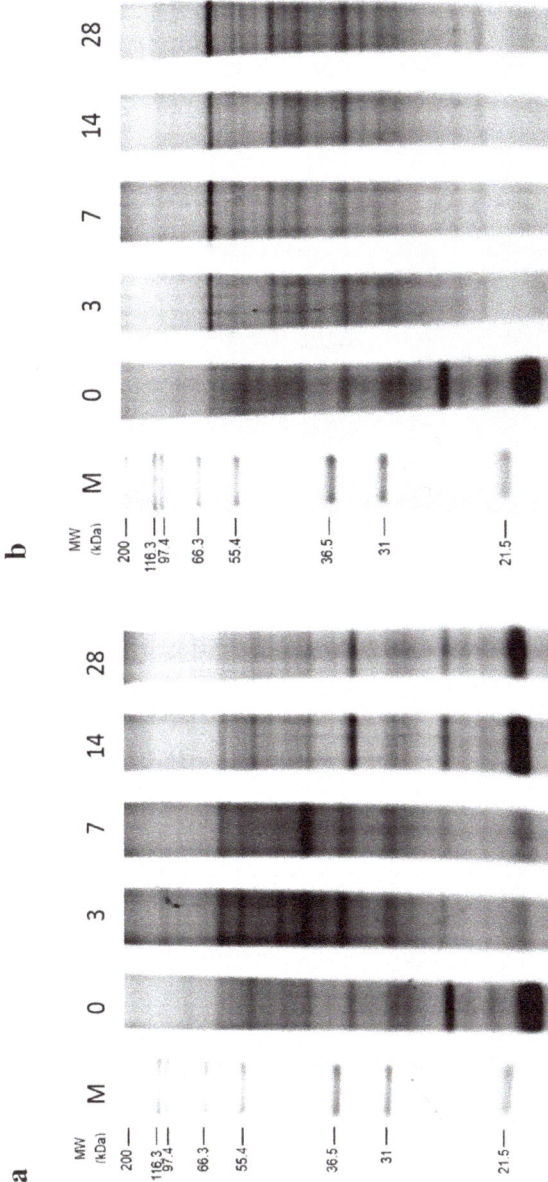

Figure 3. *Breviolum* protein expression for (**a**) MeOH and (**b**) PG treatment groups by the 3rd, 7th, 14th and 28th day of culture after thawing. The horizontal axis shows different days of culture, with M representing the marker and 0 representing *Breviolum* protein expression of untreated control group. The vertical axis indicates molecular size.

2.4. LHP and RFP Protein Expression Pattern of Breviolum

Western blotting was performed using anti-RFP and anti-LHP antibodies to detect the protein expression level of *Breviolum* at different time points during the post-thaw cultures in the MeOH and PG treatments (Figures 4 and 5). For the MeOH treatment

(Figure 4a,c), there was a significant effect of culture day on LHP expression (one-way ANOVA, $F(4/10) = 15.311$, $p < 0.001$). Specifically, no LHP was detected on the 3rd day of culture, though an LHP signal appeared on the 7th day, and expression peaked on the 14th. By the 28th day, its expression was not significantly different from the control group (post-hoc test $p > 0.05$). Meanwhile, for the PG treatment group (Figure 4b,d), LHP was never expressed over the full course of the 28-day culture. The effect of cryopreservation on RFP expression for both MeOH and PG treatment groups of *Breviolum* is shown in Figure 5. It was found that there was no expression of RFP from normal *Breviolum* culture, the control group. Meanwhile, for the MeOH treatment group (Figure 5a,c), the RFP expression was maximum on the 3rd day of culture but gradually reduced on the 7th day and could not be detected during 14–28 days. As for the PG treatment group (Figure 5b,d), RFP expression started right after thawing and continued over the whole course of the 28-day culture.

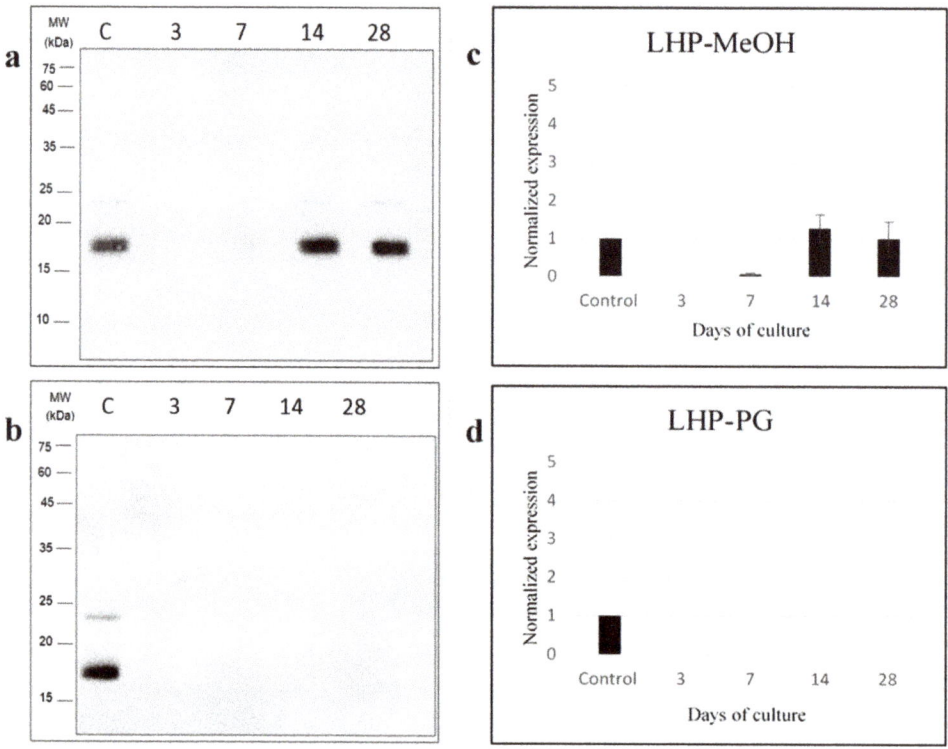

Figure 4. Western blot for *Breviolum* LHP expression for the (**a**) MeOH and (**b**) PG treatment groups after days of culture. The horizontal axis shows C as the control group (fresh *Breviolum* without freezing), and 3, 7, 14 and 28, as days of culture after thawing *Breviolum* LHP expression for the (**c**) MeOH and (**d**) PG treatment groups after same days of culture. The error bars represent standard deviation, and each group was repeated 3 times.

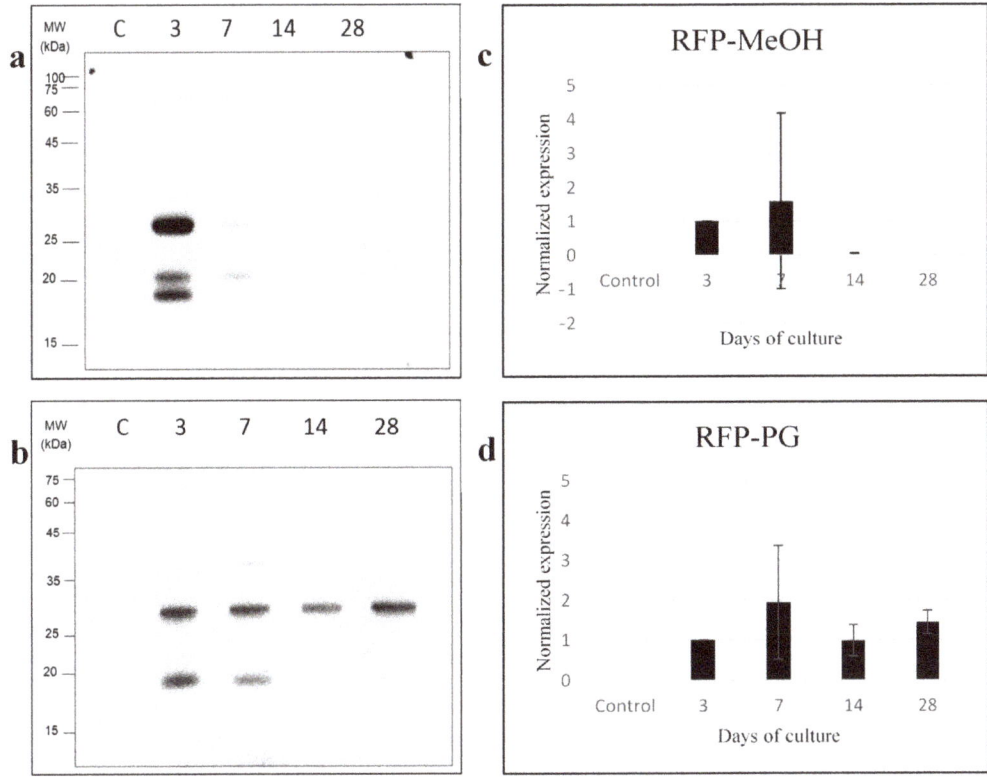

Figure 5. Western blot for *Breviolum* RFP expression for the (**a**) MeOH and (**b**) PG treatment groups after days of culture. The horizontal axis shows C as the control group (fresh *Breviolum* without freezing), and 3, 7, 14 and 28, as days of culture after thawing *Breviolum* RHP expression for the (**c**) MeOH and (**d**) PG treatment groups after same days of culture. The error bars represent standard deviation, and each group was repeated 3 times.

3. Discussion

Cryopreservation is now a common technique used among algae; however, studies of it regarding *Breviolum* and other dinoflagellates are still few [11,29,30]. There are many methods in algae cryopreservation, and the same is true for Symbiodiniaceae, including the two-step freezing [10,11,31] and the single-step direct immersion into liquid nitrogen at −196 °C [32,33]. In 2017, Zhao [10] applied vitrification, a rapid freezing method, to cryopreserve *Gerakladium*; the study found a much higher Symbiodiniaceae viability than programmable freezing that cools at a rate of 1 °C/min. Cooling rates vary among different freezing methods; in fact, it may change depending on the apparatus setting or procedure [8,9]. Our study chose *Breviolum* to freeze using the two-step cryopreservation with a cooling rate of 59.83 °C/min; the same cooling rate is also applicable to *Durusdinium* and *Gerakladium* [11,34]. In 2005, Gwo et al. [35] had attempted to freeze the algae *Nannochloropsis oculata* using the programmable freezing at a rate of 1 °C/min, down to different temperatures, for algae cryopreservation in liquid nitrogen; results found that by controlling the freezing rate, algal survival rates were higher than the single-step freezing. However, no research had yet been conducted on the single-step freezing, by direct liquid nitrogen immersion, for Symbiodiniaceae cryopreservation. Morris [36], in 1981, had mentioned that freezing involving single-step direct immersion of cells into liquid nitrogen did not provide the cell sufficient time to dehydrate and would thus lead to intracellular crystallization, injuring cellular membrane and organelles. On the contrary, if the cooling rate was overly slow, this may cause excessive dehydration, thus reducing the post-thawing recovery ability of the cell itself [12]. Hence, the cooling rate is an important factor in the freezing process. In the present study, the first stage of two-step freezing involved cooling in liquid nitrogen vapor at 59.83 °C/min. This alteration simplified the cooling method and made the overall setup handy, allowing greater mobility and convenience to the operation of cryopreservation.

The common choices of CPA for cryopreserving freshwater algae and marine algae, including dinoflagellates from the Symbiodiniaceae family, are permeating CPAs [12,31,34,36–39], for which DMSO is the most common type [13,32,37,40]. However, the present study showed that the DMSO treatment group during cryopreservation for *Breviolum* did not yield high *Breviolum* viability in comparison with the MeOH treatment group. A similar observation was found in the study of Chong et al. 2016c [11] with *Gerakladium*, in which MeOH added with sucrose gave rise to better viability than DMSO added with sucrose. In the 2007 study by Santiago-Vázquez et al. [12] on *Breviolum*, MeOH treatment also resulted in better viability after freezing than treatment with EtOH. Moreover, Hubálek 2003 [41] also showed that MeOH was the only CPA that successfully cryopreserved the cultured microalga *Euglena*; and in 2006, Rhodes et al. [42] successfully cryopreserved the microalgae *Chaetoceros calcitrans* and *Nitzcschia ovalis* using a low concentration (0.5%–5%) of MeOH. DMSO has been applied to the cryopreservation of many types of algae, but in this present study, DMSO did not yield high viability post-thaw. A possible reason for this is that while DMSO is non-toxic to cells at low temperatures (0–5 °C), its toxicity increases as temperature increased from room temperature [43]. Therefore, we propose that viability for the DMSO treatment group was reduced during equilibration at room temperature and during thawing due to toxicity of the cells, resulting in lower viability than that of the MeOH treatment group. Moreover, MeOH is more permeable than DMSO to cells, meaning MeOH enters cells and replaces intracellular water content at a faster rate, allowing the cells to reach osmotic equilibrium faster [43,44]. A faster equilibrium would be needed with the relatively fast cooling rate used in this study, compared to slower controlled rate cooling of <1 °C/min. This explains the higher viability for the MeOH treatment group than that of the DMSO treatment group when both underwent 20 min of equilibration. Although many studies have found suitable CPA for Symbiodiniaceae, the study of Hagedorn et al. 2010 [29] on *Cladocopium* found that the Symbiodiniaceae CPA tolerance did not change among different subclades within the genus. Our results have also shown that *Breviolum* from the 2 M MeOH treatment group showed higher ATP content than that of the control group. Our observations are in agreement with the studies by Chong et al. 2016c [11] and Lin et al. 2019 [34]. It was speculated that during the course of cryopreservation, Symbiodiniaceae lost its normal function and thus would require additional energy (such as ATP) for cellular repair.

In the present study, *Breviolum* was cultured for 3, 7, 14, and 28 days after cryopreservation. Results showed that *Breviolum* density began to increase at 2 weeks (14 days) after thawing. In 2007, Santiago-Vázquez et al. [12] conducted the first successful cryopreservation on *Breviolum*, isolated from corals, in liquid nitrogen using four different cryopreservation techniques. They pointed out that during culture, the cells first underwent a period of slow growth, then followed by a growth spurt; this can probably be explained by the fact that the cells need to adjust to their new environment. A similar growth trend was also observed in the present study even though the culture medium was kept the same prior to and after cryopreservation; the slow growth rate observed could also be explained by possible cryoinjuries in cells, causing the thawed *Breviolum* to prioritize cellular repair over growth.

The present study initially showed opposite trends in RFP and LHP expression between the cryopreserved symbionts and the control group in the first week of post-thaw culture. The subject of this study, *Breviolum*, is a temperature-sensitive organism. Based on the endosymbiotic relationship between Symbiodiniaceae and corals, it has been suggested that temperature change is often the main cause of disruption of their relationship [6]. The endosymbiotic relationship between Symbiodiniaceae and corals is compromised when the coral host and its endosymbionts are exposed to temperatures higher or lower than what they are acclimated to. Although relevant research regarding the effect of low temperature on algal proteins is still unavailable, similar research to the cryopreservation of fish oocytes and sperm had recorded protein denaturation [45,46]. In the study by Zilli et al. 2014 [47], it had been noted that many protein expressions from frozen-thawed fish sperm were lower than those of the control (non-treated sperm). Changes in protein expression within Symbiodiniaceae would affect its usual function, thus altering its endosymbiotic relationship with corals [48,49]. Hence, the investigation of protein content within Symbiodiniaceae would not only aid in the refinement of its cryopreservation but also in the understanding of the effect protein denaturation may have in the endosymbiosis with corals.

RFP, encoded in the genes, exhibits certain phenotypic trait in the organism that allows it to be used as a genetic marker in biotechnology [50]. RFP expression was associated with the oocyte formation in *Euphyllia ancora*, as it was expressed at the early stage of oogenesis and aided in breaking down hydrogen peroxide (H_2O_2), protecting the oocytes from oxidation [51]. RFP also has a protective function among corals; in Baird et al. 2009 [52] study, its expression increased when environmental conditions deteriorated, suggesting it to be some sort of protective mechanism. It is well documented that RFP is detected in coral host cells, but concentration within Symbiodiniaceae has not been observed. In our study, we found no detection of RFP in the control group, but RFP concentration peaked by the 7th day within cultured *Breviolum* in two CPA (MeOH and PG) treatment groups. This is the first documentation of RFP detection in Symbiodiniaceae. *Breviolum* within these two treatments showed higher growth compared to the control by the 14th day of our experiment (Figure 2). It is possible that the first 7 days were representative of the repair stage and that by the 14th day, the completion of cellular repair resulted in reduced RFP concentration. We propose that RFP may be an indicator of cellular stress or play a protective or repair role in Symbiodiniaceae; however, a robust understanding of the role of RFP in Symbiodiniaceae requires further investigation.

LHP is involved in Symbiodiniaceae photosynthesis; it was shown to be an internal control protein in Symbiodiniaceae [26]. Normally, Symbiodiniaceae should express a large amount of LHP in assistance to photosynthesis, as was observed in the control group *Breviolum* in the present study. The present study also found that LHP expression was affected after freezing; thus, LHP would not be a suitable internal control protein for cryotemperature experiments. After 28 days of culture in the PG treatment group, there was still no LHP expression, whereas by then, it was expressed in the MeOH group. The figure of cell growth for the PG treatment group showed that the *Breviolum* required a longer period of time to initiate cell growth; thus, it could be that LHP expression requires more time for recovery or that some cellular pathways were changed due to altered protein expression in the Symbiodiniaceae. Overall, PG is a more effective CPA than MeOH for cryopreservation of *Breviolum*. Meanwhile, for the MeOH treatment group, LHP expression started after 7 days of culture, meaning that *Breviolum* probably finished repair around the same time. LHP expression peaked by the 14th day, and by the 28th day, LHP expression showed no significant difference to that of the control group. The present authors had previously conducted cryopreservation of the dinoflagellate *Durusdinium* using 2 M PG as CPA [34], and the resulting analysis of LHP expression was similar to the present study. There was no LHP expression on the 3rd and 7th days of culture, and it was not until the 14th day that LHP expression was observed. Based on LHP expression, we suggest that the MeOH treatment group finished intracellular repair by the 28th day of culture, contrary to that of the PG treatment group, which did not finish its repair. Thus, the amount of LHP expression could be an indicator of Symbiodiniaceae repair after cryopreservation.

4. Materials and Methods

This study can be divided into two sections. The first section focused on the testing of different types and concentrations of CPAs and equilibration time for *Breviolum* cryopreservation by measuring their viability after 3, 7, 14, and 28 days of culture post-thawing. The second section made further analyses on the differing protein expression of frozen-thawed *Breviolum* over succeeding days of culture. The control group was not cryopreserved and monitored during the same period as the experimental Symbiodiniaceae. The *Breviolum* used in this study are the free-living algae isolated from the sea anemone *Exaiptasia diaphana* (previously named *Aiptasia pulchella*) [53].

4.1. Symbiodiniaceae Identification

Cladistic identification of Symbiodiniaceae was performed via the larger secondary sequence of the 23S rDNA from chloroplasts. Extraction of the Symbiodiniaceae DNA was performed using ZR Plant/Seed Kit (ZYMO Research, Irvine, CA, USA). Then, 10xAccuPrime PCR buffer solution was added together with 200 nM 23S rDNA primer and 5U/μL AccuPrime Pfx (Invitrogen, CA, USA) into a 50 uL sample. The sequence for the forward primer of the 23S rRNA is 5′-CACGACGTTGTAAAACGACGGCTGTAACTATAACGGTCC-3′, while the sequence for the reverse primer is 5′GGATAACAATTTCACACAGGCCATCGTATTGAACCCAGC3′ [54]. Polymerase chain reaction was performed for DNA amplification, first heating the sample to 94 °C for 5 min, for initial denaturation, then annealment occurred over two 30 s rounds at 94 and 50 °C, respectively, and finally, fragment elongation for 60 s under 72 °C. All steps were repeated for 30 cycles, and one final elongation at 72 °C for 10 min. Electrophoresis was performed for the product of PCR over 1% agarose gel, then dyed with ethidium bromide, followed by purification using Axygen DNA Gel Extraction Kit (Axygen Biosciences, Union City, CA, USA). The sample was then magnified and copied using PCR-Blunt II TOPO cloning kit (Invitrogen, Carlsbad, CA, USA), sequenced with ABI 3100 (Applied Biosystems, CA, USA), and finally, matched via BlastN for interspecies nucleotide sequence, confirming the specimens to belong to the genus *Breviolum*.

4.2. Culture of Breviolum

Breviolum was cultured in a plant culture chamber (LTI-613, Double Eagle, Taiwan), with temperature set to 25 °C, photoperiod of 12 h of light (40 μmol m^{-2} s^{-1}), and 12 h of dark. The culture used autoclaved artificial seawater (A.S.W) (Coralife, scientific grade marine salt dissolved in deionized water, Franklin, WI, USA) together with f/2 (Guillard's f/2 medium, Sigma-Aldrich, Saint Luis, MO, USA) added with penicillin (penicillin 10 units/mL, Sigma-Aldrich, USA). Thawed *Breviolum* was centrifuged (3000 rpm, 3 min, 25 °C), and the supernatant was removed. Then under the same setting, after adding the culture medium (for rinsing purposes), the sample was centrifuged again. Lastly, the sample was added into a T-75 cell culture flask (Nunc, Thermo Fisher Scientific, Waltham, MA, USA) for culture. Half of the culture medium was renewed each week to maintain the nutrient supply for *Breviolum*.

4.3. Preparation of Artificial Sea Water

First, 160 g of sea salt was dissolved into 5 L of pure water (Purelab Ultra, Elga, Antony, France); it was made sure that the salinity of the solution was 32%. This was followed by pouring 970 mL of seawater into each 1 L serum bottle, which was then autoclaved and cooled. Then, 30 mL of f/2 culture medium plus 1 mL of penicillin was added to the solution, mixed thoroughly, and filtered (0.45 μm Flow Filter, Thermo Scientific™ Nalgene™ Rapid-Flow™, IL, USA). Finally, the solution was stored in a refrigerator (Sampo, Kaohsiung, Taiwan) at 4 °C. The solution was warmed back to room temperature before each use.

4.4. Renewal of Culture Medium

Culture medium renewal had to be performed of the dark period during *Breviolum* culture. About 80 mL of the culture solution was poured out; the remaining was then gently swirled and mixed. The remaining medium was centrifuged in a 50 mL centrifuge flask, and the supernatant was removed. The medium was then returned to the serum bottle and filled back to 160 mL with fresh culture medium for continuing culture. After each renewal, 10 and 30 µL of the Symbiodiniaceae solution were removed for Symbiodiniaceae count by hemocytometer and for the adenosine triphosphate (ATP) bioassay, respectively.

4.5. ATP Bioassay

ATP bioassay was conducted prior to and after *Breviolum* cryopreservation to compare the change in *Breviolum* viability throughout the experiment. In normal cells, ATP acts as the energy currency for cellular functions, whereas its concentration reduces when the cell dies. Hence, ATP bioassays can be used as a preliminary assessment of cellular survival rate. This study performed ATP bioassays using ApoSENSOR™ Cell viability assay kit (BioVision, Milpitas, CA, USA). First, 5 µL of the sample was put into a luminescence test tube, followed by adding 100 µL of nucleotide releasing buffer, and was left for 3 min at room temperature for completion of the reaction. This was followed by adding 5 µL of ATP monitoring enzyme, thorough mixing and rest period another 30 s at room temperature. Finally, a measurement was made using a luminometer (Lumat LB 9507, Berthold Technologies GmbH & Co. KG, Germany). The use of nucleotide releasing buffer ruptured the *Breviolum*, allowing the binding of luciferin to ATP, catalyzed by luciferase, and the oxidation of ATP into adenosine monophosphate (AMP) and two phosphate groups, emitting a light blue light. The light was captured and measured by the luminometer for ATP concentration based on the light intensity.

4.6. Two-Step Freezing and Culture of Post-Thawed Breviolum

Three pools of the *Breviolum* for each treatment were analyzed, and the experiments were repeated three times. Therefore a total of nine measurements from three biological pools originating from treatment was obtained. The freezing method used for *Breviolum* in this study was the two-step freezing [11]. Prior to cryopreservation, the *Breviolum* were detached from the sides of the culture flask by gentle swirling and centrifugation (3000 rpm, 3 min, 25 °C) to remove excess culture medium. Then, *Breviolum* were separated by flushing the medium with a syringe (23 G × 1/2″, 0.45 × 13 mm, Top, Kaohsiung, Taiwan) 10 times. Then, the ATP bioassay was conducted after having confirmed with a hemocytometer that the *Breviolum* density was around 1×10^6. After both measurements, centrifugation (3000 rpm, 3 min, 25 °C) was performed again to remove all remaining culture medium. Four cryoprotectants were tested, methanol (MeOH), propylene glycol (PG), dimethyl sulfoxide (DMSO), and glycerol (Gly) at concentrations of 1, 2, or 3 M. ASW was used as a base medium to dissolve the CPAs. CPAs were added directly onto the pelleted sample, which was resuspended by flushing with a syringe another ten times. Equilibration was performed at room temperature for 20 and 30 min for the CPA treatment groups. During equilibration, the sample was transferred into a 500 µL straw, and afterward, placed on the two-step freezing device (Taiwan patent M394447) [11], 5 cm above the surface of the liquid nitrogen (cooling rate at 59.83 °C/min) to cool for 30 min. After the first stage of cooling, the straw was immersed into liquid nitrogen for 2 h. Afterward, the sample was thawed in a 37 °C water bath for 10 s, followed by cell counts and an ATP bioassay analysis. Thawed *Breviolum* were cultured independently across 4 experiments with different culture durations (3, 7, 14, and 28 days). The number of algal cells was determined using a hemocytometer, and the relative cell culture density of *Breviolum* was normalized based on a culture duration-specific control group.

4.7. Protein Extraction

The best post-thaw results were obtained with the CPAs PG (2 M) and MeOH (2 M), and therefore only *Breviolum* that were cryopreserved with these two treatments were selected for the 3-, 7-, 14-, or 28-day cultures and for further experiments. Protein expression in each treatment was quantified three times. After culture, the *Breviolum* were transferred from the culture flask into a 50 mL centrifuge tube for centrifugation to remove excess culture medium. The pellet of *Breviolum* was transferred into a 1.5 mL centrifuge tube, and 500 µL of RIPA buffer [50 mM Tris-base pH 7.4, 0.5% NP-40, 0.125% Na-deoxycholate, 150 nM NaCl and 1 mM EDTA], 20 µL of protease inhibitor (Roche cat. no 11836153001), and micro glass beads (Sigma cat. no G9268) were added. The tube was then sealed with paraffin and placed into the TissueLyser LT (Qiagen Inc., Valencia, CA, USA) machine to homogenize the sample for 30 min at 25 Hz. Then the sample was centrifuged (15,000 rpm, 10 min, 4 °C), and the supernatant was transferred into a new 1.5 mL centrifuge tube. For protein precipitation, an equal volume of 20% trichloroacetic acid was added to the tube and then placed into a -20 °C freezer overnight. After protein precipitation, the solution was centrifuged (15,000 rpm, 10 min, 4 °C), and the supernatant was removed. The precipitated proteins were rinsed with acetone with 0.1% of dithiothreitol (DTT, 17-1318-02; GE Healthcare, USA) twice and then rinsed with pure acetone. Finally, the protein pellet was vacuum dried into a powder. The dry protein powder was stored in a -20 °C freezer and ready to be used.

4.8. SDS-PAGE and Western Blotting

The 1× SDS-PAGE sample buffer (62.5 mM Tris-HCl pH 6.8, 2% SDS, 10% glycerol, 10 mM DTT) was added to dissolve dry protein powder. Then, a Qubit® Protein Assay Kit and a Qubit® 2.0 fluorometer (Thermo Fisher Scientific, Waltham, MA, USA) were used to measure protein concentration. For the SDS-PAGE, 10 µg of each protein sample was separated by 14% polyacrylamide gel and stained with SYPRO® Ruby protein dye (S12000, Invitrogen) according to the manufacturer's instructions. The gel image was scanned using a laser scanning imager (Typhoon FLA 9500, GE Healthcare Life Sciences, Marlborough, MA, USA). For Western blotting, 5 µg of each protein sample were separated by 14% polyacrylamide gel and then blotted onto PVDF membranes (immobilon-PSQ 0.45 mm; Millipore, Darmstadt, Germany). The membranes were incubated in 5% skim milk/TBST buffer (Tris-buffered saline, 0.1% Tween-20) at RT for 1 h, followed by the incubation with the rabbit-anti-RFP antibody (1: 2500 dilution, cat. no ab59457, Abcam, Cambridge, UK) or mouse-anti-LHP antibody (1:10,000 dilution, custom antibody) in TBST buffer at 4 °C overnight. The membranes were then washed 3 times with TBST buffer for 5 min each time and incubated with anti-rabbit or anti-mouse IgG antibodies in TBST buffer for 1 h. The membranes were subsequently washed with TBST buffer and visualized using a SuperSignal West Pico Chemiluminescent substrate kit (cat. no 34080, Thermo Fisher Scientific) according to the manufacturer's recommendations. Finally, the membranes were placed into a luminometer (Vilber Lourmat Fusion FX 7 Fluorescence/Luminescent CCD Image Analyzer, Vilber Lourmat, France), and an image was collected.

4.9. Protein Expression Analysis

Each experiment was repeated in triplicate to collect three images. Image J software was used for protein quantitative analysis. For each image, the position of the protein, based on its molecular size, was marked to show the relative position of each protein on the waveform plot. Then, the background value was zeroed to obtain the relative amount of protein concentration. Lastly, the 3 values from each treatment group were then averaged to obtain protein value. The relative amount was measured in each sample based on staining intensity.

4.10. Statistical Analysis

Statistical analysis for the data of the present study was conducted using the SPSS 17.0 version. Normality was verified with a Kolmogorov–Smirnov test, and homogeneity of variance was verified using Levene's test. One-way analyses of variance (ANOVA) were conducted to detect equilibration time following different concentrations of CPA treatments. Tukey or Games–Howell post-hoc tests were used to verify differences between individual means, and t-tests were used to compare 20 vs. 30 min equilibration times at different CPA concentrations.

5. Conclusions

This study not only demonstrated that the protocols reported here do affect protein expression, but it also provided a suitable protocol for *Breviolum* cryopreservation. This is the first research conducted to investigate the effect of cryopreservation on Symbiodiniaceae protein expression as well as the first documentation of RFP expression in Symbiodiniaceae (RFP was an indicator of cell damage or cell stress). By gaining further understanding of the use of cryopreservation as a way to conserve Symbiodiniaceae, we hope to make an effort in the remediation and conservation of the coral reef ecosystem and provide additional methods to rescue coral reefs.

Author Contributions: C.L. and S.T. conceived the experiment, H.-H.L., H.-E.L. and J.-L.L. conducted the experiment, C.L., S.T., H.-H.L., H.-E.L. and J.-L.L. analyzed the results, J.-L.L., C.L. and S.T. wrote the paper. All authors have read and agreed to the published version of the manuscript.

Funding: Please add: This research was funded by the Ministry of Science and Technology, Taiwan, grant number MOST 109-2635-B-451-001 and MOST 110-2313-B-291-001-MY3.

Informed Consent Statement: Not applicable.

Data Availability Statement: Date is contained within the article.

Conflicts of Interest: The authors declare no conflict of interest.

References

1. Tsai, S.; Lin, C. Advantages and Applications of Cryopreservation in Fisheries Science. *Braz. Arch. Biol. Technol.* **2012**, *55*, 425–434. [CrossRef]
2. Taylor, R.; Fletcher, R.L. Cryopreservation of eukaryotic algae—A review of methodologies. *J. Appl. Phycol.* **1998**, *10*, 481–501. [CrossRef]
3. Ricaurte, M.; Schizas, N.V.; Ciborowski, P.; Boukli, N.M. Proteomic analysis of bleached and unbleached *Acropora palmata*, a threatened coral species of the Caribbean. *Mar. Pollut. Bull.* **2016**, *107*, 224–232. [CrossRef] [PubMed]
4. Baumann, J.H.; Davies, S.W.; Aichelman, H.E.; Castillo, K.D. Coral Symbiodinium Community Composition Across the Belize Mesoamerican Barrier Reef System is Influenced by Host Species and Thermal Variability. *Microb. Ecol.* **2018**, *75*, 903–915. [CrossRef] [PubMed]
5. Gates, R.D.; Baghdasarian, G.; Muscatine, L. Temperature stress causes host cell detachment in symbiotic cnidarians: Implications for coral bleaching. *Biol. Bull.* **1992**, *182*, 324–332. [CrossRef] [PubMed]
6. Weis, V.M. Cellular mechanisms of Cnidarian bleaching: Stress causes the collapse of symbiosis. *J. Exp. Biol.* **2008**, *211*, 3059–3066. [CrossRef]
7. Cirino, L.; Wen, Z.-H.; Hsieh, K.; Huang, C.-L.; Leong, Q.L.; Wang, L.-H.; Chen, C.-S.; Daly, J.; Tsai, S.; Lin, C. First instance of settlement by cryopreserved coral larvae in symbiotic association with dinoflagellates. *Sci. Rep.* **2019**, *9*, 1–8. [CrossRef]
8. Lin, C.; Tsai, S. Fifteen years of coral cryopreservation. *Platax* **2020**, *17*, 53–76.
9. Di Genio, S.; Wang, L.H.; Meng, P.J.; Tsai, S.; Lin, C. Symbio-Cryobank: Towards the development of a cryogenic archive for the coral reef dinoflagellate symbiont Symbiodiniaceae. *Biopreserv. Biobank.* **2020**, *19*, 91–93. [CrossRef]
10. Zhao, Y. Cryopreservation of the Gorgonian Coral Endosymbiont Symbiodinium Using Vitrification and Programmable Freezing. Master's Thesis, National Dong Hwa University, Shoufeng, Taiwan, 2017; pp. 1–70.
11. Chong, G.; Tsai, S.; Wang, L.-H.; Huang, C.-Y.; Lin, C. Cryopreservation of the gorgonian endosymbiont Symbiodinium. *Sci. Rep.* **2016**, *6*, 18816. [CrossRef]
12. Santiago-Vázquez, L.Z.; Newberger, N.C.; Kerr, R.G. Cryopreservation of the dinoflagellate symbiont of the octocoral Pseudopterogorgia elisabethae. *Mar. Biol.* **2007**, *152*, 549–556. [CrossRef]
13. Hagedorn, M.; Carter, V.L. Seasonal Preservation Success of the Marine Dinoflagellate Coral Symbiont, Symbiodinium sp. *PLoS ONE* **2015**, *10*, e0136358. [CrossRef]
14. Zhang, Q.; Cong, Y.; Qu, S.; Luo, S.; Yang, G. Cryopreservation of gametophytes of Laminaria japonica (Phaeophyta) using encapsulation-dehydration with two-step cooling method. *J. Ocean Univ. China* **2008**, *7*, 65–71. [CrossRef]
15. Liu, J.; Woods, E.J.; Agca, Y.; Critser, E.S.; Critser, J.K. Cryobiology of rat embryos II: A theoretical model for the development of interrupted slow freezing procedures. *Biol. Reprod.* **2000**, *63*, 1303–1312. [CrossRef] [PubMed]
16. Viveiros, A.; Lock, E.; Woelders, H.; Komen, J. Influence of Cooling Rates and Plunging Temperatures in an Interrupted Slow-Freezing Procedure for Semen of the African Catfish, Clarias gariepinus. *Cryobiology* **2001**, *43*, 276–287. [CrossRef]
17. Morris, G.J.; Canning, C.E. The Cryopreservation of Euglena gracilis. *J. Gen. Microbiol.* **1978**, *108*, 27–31. [CrossRef]
18. Tsai, S.; Yen, W.; Chavanich, S.; Viyakarn, V.; Lin, C. Development of Cryopreservation Techniques for Gorgonian (Junceella juncea) Oocytes through Vitrification. *PLoS ONE* **2015**, *10*, e0123409. [CrossRef]

19. Chong, G.; Kuo, F.-W.; Tsai, S.; Lin, C. Validation of reference genes for cryopreservation studies with the gorgonian coral endosymbiont Symbiodinium. *Sci. Rep.* **2017**, *7*, 39396. [CrossRef]
20. Liu, C.; Wu, G.; Huang, X.; Liu, S.; Cong, B. Validation of housekeeping genes for gene expression studies in an ice alga Chlamydomonas during freezing acclimation. *Extremophiles* **2012**, *16*, 419–425. [CrossRef] [PubMed]
21. Iglesias-Prieto, R.; Trench, R. Acclimation and adaptation to irradiance in symbiotic dinoflagellates. II. Response of chlorophyll-protein complexes to different photon-flux densities. *Mar. Biol.* **1997**, *130*, 23–33. [CrossRef]
22. Takahashi, S.; Whitney, S.; Itoh, S.; Maruyama, T.; Badger, M. Heat stress causes inhibition of the de novo synthesis of antenna proteins and photobleaching in cultured Symbiodinium. *Proc. Natl. Acad. Sci. USA* **2008**, *105*, 4203–4208. [CrossRef]
23. Prézelin, B.B.; Haxo, F.T. Purification and characterization of peridinin-chlorophyll a-proteins from the marine dinoflag-ellates Glenodinium sp. and Gonyaulax polyedra. *Planta* **1976**, *128*, 133–141. [CrossRef]
24. Gross, L.A.; Baird, G.S.; Hoffman, R.C.; Baldridge, K.K.; Tsien, R.Y. The structure of the chromophore within DsRed, a red fluorescent protein from coral. *Proc. Natl. Acad. Sci. USA* **2000**, *97*, 11990–11995. [CrossRef] [PubMed]
25. Bevis, B.J.; Glick, B.S. Rapidly maturing variants of the Discosoma red fluorescent protein (DsRed). *Nat. Biotechnol.* **2002**, *20*, 83–87. [CrossRef]
26. Huang, K.-J.; Huang, Z.-Y.; Lin, C.-Y.; Wang, L.-H.; Chou, P.-H.; Chen, C.-S.; Li, H.-H. Generation of clade- and symbiont-specific antibodies to characterize marker molecules during Cnidaria-Symbiodinium endosymbiosis. *Sci. Rep.* **2017**, *7*, 1–12. [CrossRef] [PubMed]
27. Stauber, E.J.; Fink, A.; Markert, C.; Kruse, O.; Johanningmeier, U.; Hippler, M. Proteomics of Chlamydomonas reinhardtii Light-Harvesting Proteins. *Eukaryot. Cell* **2003**, *2*, 978–994. [CrossRef] [PubMed]
28. Boldt, L.; Yellowlees, D.; Leggat, W. Hyperdiversity of Genes Encoding Integral Light-Harvesting Proteins in the Dinoflagellate Symbiodinium sp. *PLoS ONE* **2012**, *7*, e47456. [CrossRef] [PubMed]
29. Hagedorn, M.; Carter, V.; Leong, J.; Kleinhans, F. Physiology and cryosensitivity of coral endosymbiotic algae (Symbiodinium). *Cryobiology* **2010**, *60*, 147–158. [CrossRef]
30. Lin, C.; Thongpoo, P.; Juri, C.; Wang, L.-H.; Meng, P.-J.; Kuo, F.-W.; Tsai, S. Cryopreservation of a Thermotolerant Lineage of the Coral Reef Dinoflagellate Symbiodinium. *Biopreserv. Biobank.* **2019**, *17*, 520–529. [CrossRef]
31. Cañavate, J.P.; Lubian, L.M. Tolerance of six marine microalgae to the cryoprotectants dimethyl sulfoxide and methanol. *J. Phycol.* **1994**, *30*, 559–565. [CrossRef]
32. Mitbavkar, S.; Anil, A.C. Cell damage and recovery in cryopreserved microphytobenthic diatoms. *Cryobiology* **2006**, *53*, 143–147. [CrossRef] [PubMed]
33. Tsuru, S. Preservation of marine and fresh water algae by means of freezing and freeze-drying. *Cryobiology* **1973**, *10*, 445–452. [CrossRef]
34. Lin, C.; Chong, G.; Wang, L.; Kuo, F.; Tsai, S. Use of luminometry and flow cytometry for evaluating the effects of cryoprotectants in the gorgonian coral endosymbiont Symbiodinium. *Phycol. Res.* **2019**, *67*, 320–326. [CrossRef]
35. Gwo, J.C.; Chiu, J.Y.; Chou, C.C.; Cheng, H.Y. Cryopreservation of a marine microalga, *Nannochloropsis oculata* (Eustigmatophyceae). *Cryobiology* **2005**, *50*, 338–343. [CrossRef]
36. Morris, G. *Cryopreservation: An Introduction to Cryopreservation in Culture Collections*; Institute of Terrestrial Ecology: Cambridge, UK, 1981; pp. 1–27.
37. Yang, D.; Li, W. Methanol-Promoted Lipid Remodelling during Cooling Sustains Cryopreservation Survival of Chlamydomonas reinhardtii. *PLoS ONE* **2016**, *11*, e0146255. [CrossRef] [PubMed]
38. Morris, G. Cryopreservation of 250 strains of Chlorococcales by the method of two-step cooling. *Br. Phycol. J.* **1978**, *13*, 15–24. [CrossRef]
39. Fenwick, C.; Day, J.G. Cryopreservation of Tetraselmis suecica cultured under different nutrients regimes. *J. Appl. Phycol.* **1992**, *4*, 105–109. [CrossRef]
40. Chong, G.; Tsai, S.; Lin, C. Factors responsible for successful cryopreservation of algae. *J. Fish Soc. Taiwan.* **2016**, *43*, 153–162.
41. Box, J. Cryopreservation of the blue-green alga *Microcystis aeruginosa*. *Eur. J. Phycol.* **1988**, *23*, 385–386. [CrossRef]
42. Hubálek, Z. Protectants used in the cryopreservation of microorganisms. *Cryobiology* **2003**, *46*, 205–229. [CrossRef]
43. Rhodes, L.; Smith, J.; Tervit, R.; Roberts, R.; Adamson, J.; Adams, S.; Decker, M. Cryopreservation of economically valuable marine micro-algae in the classes Bacillariophyceae, Chlorophyceae, Cyanophyceae, Dinophyceae, Haptophyceae, Prasinophyceae, and Rhodophyceae. *Cryobiology* **2006**, *52*, 152–156. [CrossRef] [PubMed]
44. Lin, C.; Tsai, S. The effect of chilling and cryoprotectants on hard coral (Echinopora spp.) oocytes during short-term low temperature preservation. *Theriogenology* **2012**, *77*, 1257–1261. [CrossRef] [PubMed]
45. Zhang, Y.Z.; Zhang, S.C.; Liu, X.Z.; Xu, Y.J.; Hu, J.H.; Xu, Y.Y.; Li, J.; Chen, S.L. Toxicity and protective efficiency of cryo-protectants to flounder (*Paralichthys olivaceus*) embryos. *Theriogenology* **2005**, *63*, 763–773. [CrossRef]
46. Chong, G.; Tsai, S.; Lin, C. Cryopreservation and its molecular impacts on microorganisms. *J. Fish Soc. Taiwan* **2016**, *43*, 263–272.
47. Zilli, L.; Beirão, J.; Schiavone, R.; Herraez, P.; Gnoni, A.; Vilella, S. Comparative Proteome Analysis of Cryopreserved Flagella and Head Plasma Membrane Proteins from Sea Bream Spermatozoa: Effect of Antifreeze Proteins. *PLoS ONE* **2014**, *9*, e99992. [CrossRef] [PubMed]
48. Voolstra, C.R.; Schnetzer, J.; Peshkin, L.; Randall, C.J.; Szmant, A.M.; Medina, M. Effects of temperature on gene expression in embryos of the coral Montastraea faveolata. *BMC Genom.* **2009**, *10*, 627. [CrossRef]

49. Weston, A.J.; Dunlap, W.C.; Shick, J.M.; Klueter, A.; Iglic, K.; Vukelic, A.; Starcevic, A.; Ward, M.; Wells, M.L.; Trick, C.G.; et al. A Profile of an Endosymbiont-enriched Fraction of the Coral Stylophora pistillata Reveals Proteins Relevant to Microbial-Host Interactions. *Mol. Cell. Proteom.* **2012**, *11*. [CrossRef]
50. Shaner, N.; Campbell, R.E.; Steinbach, P.A.; Giepmans, B.; Palmer, A.E.; Tsien, R.Y. Improved monomeric red, orange and yellow fluorescent proteins derived from Discosoma sp. red fluorescent protein. *Nat. Biotechnol.* **2004**, *22*, 1567–1572. [CrossRef]
51. Shikina, S.; Chiu, Y.-L.; Chung, Y.-J.; Chen, C.-J.; Lee, Y.-H.; Chang, C.-F. Oocytes express an endogenous red fluorescent protein in a stony coral, Euphyllia ancora: A potential involvement in coral oogenesis. *Sci. Rep.* **2016**, *6*, 25868. [CrossRef] [PubMed]
52. Baird, A.H.; Bhagooli, R.; Ralph, P.; Takahashi, S. Coral bleaching: The role of the host. *Trends Ecol. Evol.* **2009**, *24*, 16–20. [CrossRef] [PubMed]
53. Grajales, A.; Rodríguez, E. Morphological revision of the genus Aiptasia and the family Aiptasiidae (Cnidaria, Actiniaria, Metridioidea). *Zootaxa* **2014**, *3826*, 55–100. [CrossRef] [PubMed]
54. Santos, S.R.; Taylor, D.J.; Kinzie, R.A., III; Hidaka, M.; Sakai, K.; Coffroth, M.A. Molecular phylogeny of symbiotic di-noflagel-lates inferred from partial chloroplast large subunit (23S)-rDNA sequences. *Mol. Phylogenet. Evol.* **2002**, *23*, 97–111. [CrossRef]

Article

Seed Cryopreservation, Germination, and Micropropagation of Eastern Turkeybeard, *Xerophyllum asphodeloides* (L.) Nutt.: A Threatened Species from the Southeastern United States

Michelle Issac [1], Princy Kuriakose [1], Stacie Leung [1], Alex B. Costa [1], Shannon Johnson [2], Kylie Bucalo [2], Jonathan M. Stober [3], Ron O. Determann [4], Will L. Rogers [5], Jenifer M. Cruse-Sanders [4,5] and Gerald S. Pullman [1,2,*]

[1] School of Biology, Georgia Institute of Technology, Atlanta, GA 30332, USA; mmissac@gmail.com (M.I.); princyk730@gmail.com (P.K.); Stacie.leung@gmail.com (S.L.); alexbcosta87@gmail.com (A.B.C.)
[2] Renewable Bioproducts Institute, Georgia Institute of Technology (Formerly the Institute of Paper Science and Technology), 500 10th Street NW, Atlanta, GA 30332, USA; smjohnson777@yahoo.com (S.J.); kbucalo@ohanainstitute.org (K.B.)
[3] US Forest Service, USDA, Talladega National Forest, Shoal Creek Ranger Dist, Heflin, AL 36264, USA; jonathan.stober@usda.gov
[4] Department of Conservation Research, Atlanta Botanical Garden, 1345 Piedmont Ave., NE, Atlanta, GA 30309, USA; rondetermann@gmail.com (R.O.D.); crusesanders@uga.edu (J.M.C.-S.)
[5] State Botanical Garden of Georgia, 2450 Milledge Avenue, Athens, GA 30605, USA; bonjour@uga.edu
* Correspondence: jerry.pullman@rbi.gatech.edu

Abstract: *Xerophyllum asphodeloides* (Xerophyllaceae), known as eastern turkeybeard, is an herbaceous perennial found in eastern North America. Due to decline and destruction of its habitat, several states rank *X. asphodeloides* as "Imperiled" to "Critically Imperiled". Protocols for seed cryopreservation, in vitro germination, sustainable shoot micropropagation, shoot establishment in soil, and seed germination are presented. Seeds from two tested sources were viable after 20 months of cryopreservation. Germination of isolated embryos in vitro was necessary to overcome strong seed dormancy. Shoot multiplication and elongation occurred on $\frac{1}{2}$ MS medium without PGRs. Shoots rooted in vitro without PGRs or with 0.5 mg/L NAA or after NAA rooting powder treatment and placement in potting mix. When planted in wet, peaty soil mixes, shoots grew for two months and then declined. When planted in a drier planting mix containing aged bark, most plants continued growth. In the field, plant survival was 73% after three growing seasons. Safeguarding this species both ex situ and in situ is possible and offers a successful approach to conservation. Whole seeds germinated after double dormancy was overcome by incubation under warm moist conditions for 12 weeks followed by 12 weeks cold at 4 °C and then warm.

Keywords: conservation; cryopreservation; endangered species; micropropagation; shoot culture; *Xerophyllum asphodeloides*; Eastern Turkey Beard

1. Introduction

Xerophyllum asphodeloides (L.) Nutt. is a monocotyledonous herbaceous perennial species in the Xerophyllaceae, Beargrass Family, ref. [1] within the Liliales [2]. One of two species in the genus, *X. asphodeloides*, also known as eastern turkeybeard, is native to eastern North America where it is found in the Pine Barrens of New Jersey and also in the Appalachian Mountains of Georgia, Carolinas, Virginia, West Virginia, Tennessee, and Alabama and is presumed extirpated in Kentucky and Delaware [3,4]. The two habitats share features of dry, acidic, nutrient poor, and sandy or gravelly soils.

Threats for this species include trampling by hikers, collection as a homeopathic remedy, conversion of Pine Barrens to residential properties, and fire suppression leading to succession of wood canopy [4]. Loss of habitat has threatened the presence of this species and increased its risk of endangerment. *X. asphodeloides* is currently ranked as G4, being "Apparently Secure". However, due to the decline and destruction of its habitat, several states, including West Virginia, Alabama, and Georgia, rank *X. asphodeloides* as S1 (Critically Imperiled) and South Carolina rank it as S2 (imperiled) [4]. In 2001, the USDA reported that *X. asphodeloides* could be found in a total of 41 counties across its range of habitat [4]. In Georgia, 10 populations are known, with 2 on national forest land and 3 on state conservation lands [5]. In Alabama, 16 populations are known to exist in one county [4].

Conservation of this species is important due to its ecological relationships. As to its coexistence with many other plant species, threat to *X. asphodeloides* may have a parallel effect on its habitat. Additionally, it has been significant for medicinal use in homeopathic remedies treating constipation, dysmenorrhea, and skin irritations [6]. Research on its congener, *X. tecta*, indicates that culturally relevant management of the landscape through indigenous fire stewardship and leaf harvest for weaving positively impacts demographic processes and changes in long term practices result in population declines [7,8]. Cryopreservation has been used for seed and shoot tip conservation, providing long-term storage [9–12], and is considered to be the best long-term approach to preserving endangered species [13].

In vitro techniques can be used to multiply rare or endangered plants asexually [12,14–16]. Micropropagation has many advantages, including rapid production of identical plants. Under in vitro conditions, many disease-free plants can be produced in a small space that are often more robust and show accelerated growth compared to similar plants produced by seeds or cuttings. With careful planning to manage genetic source material, plants of endangered species produced through micropropagation can be used to preserve genetic diversity, supplement small populations, or to establish new planting sites for conservation, research, education, or recreational purposes.

A simple method for whole seed germination is required for plant production from cryopreserved seeds. At the time this research began, information on *X. asphodeloides* dormancy was not available. Reports indicated that seeds cannot tolerate long term desiccation and seed germination in the greenhouse and field only occurs after two growing seasons (19,20). In 2014, based on the report of Brumback (19), Baskin and Baskin (29) classified *X. asphodeloides* as having morphophysiological dormancy where the seed contains an undeveloped embryo and germination is initiated after 4 weeks of incubation.

The objectives of this research were to develop reliable protocols for cryogenic storage of seeds and micropropagation to preserve this species for the future using seed to start cultures. A good protocol for germination was required to recover plants from cryopreserved seeds. Through a series of experiments, three approaches to ex situ conservation were tested: (1) in vitro germination of seeds and seed cryopreservation for long term storage, (2) germination of isolated embryos in vitro to produce plants for micropropagation, and (3) results from in vitro experiments led to testing methods to form rooted shoots and establish them in soil.

2. Results

2.1. X. asphodeloides Seed Sterilization and Germination Experiments

Germination did not occur in Sections 4.2.1 and 4.2.3–4.2.10. In the more than 1000 seeds that were tested for germination the only treatment that contained any germinants was with the application of 4 min sulfuric acid as a scarification treatment in Section 4.2.2. Most tests with H_2O_2 surface sterilization or H_2SO_4 scarification showed little to no contamination. Scarification is known to remove dormancy and also to decontaminate the seed surface [17,18]. Tests to repeat and optimize the scarification treatment in Sections 4.2.5–4.2.7, 4.2.9 and 4.2.10 with source 2 seeds showed no germination.

Section 4.2.2. Seed scarification. In the first experiment, seeds scarified for 1 min with sulfuric acid showed the least seed coat damage, while treatment for 10 min caused almost complete removal of the coat. Seeds treated for 4 min showed some damage, and

90% of seeds scarified for 1 min with acid showed fungal contamination compared to no contamination when scarified for 4 or 10 min.

After approximately 30 days, two seeds in the 4 min treatment began to show emergence of an off-white tissue presumed to be the radical (Figure 1A). Two more seeds germinated later, for a total of 4 out of ten. Germinants appeared to form a cluster of tiny bulbs that continued to grow producing a bulbous tissue that after another month began to produce several shoots with strap-like leaves (Figure 1B–D) and roots (Figure 1F).

In the second experiment, the percentage of seeds with fungal contamination was high for short scarification times for source 1: 3 min (45%), 3.5 min (67%), 4 min (27%). Minor contamination was observed for source 2: 2 min (17%), 3 min (0%), 4 min (0%), 5 min (0%). However, no germination occurred, and seeds were scarified for 4 min and planted in greenhouse soil did not germinate.

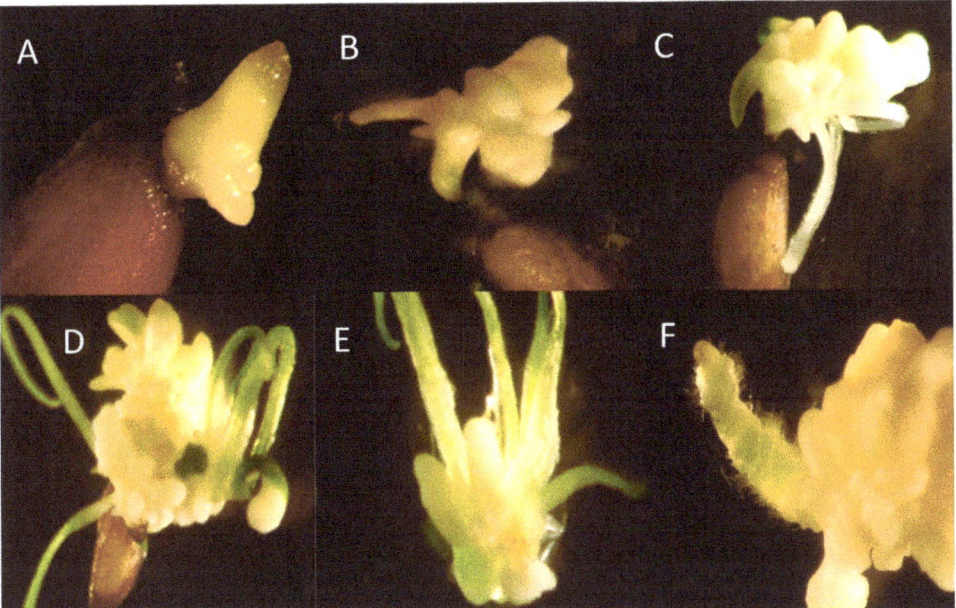

Figure 1. Seed source 1 *Xerophyllum asphodeloides* germinant from a 4-min H_2SO_4 scarification treatment. Days after seed began to germinate. (**A**) Day 1, (**B**) Day 26, (**C**) Day 36, (**D**) Day 92, (**E**) Shoot clump 3 weeks after division, (**F**) Emergence of a root from the growing tissue.

2.2. Tetrazolium Chloride Seed Viability Staining

In the initial experiments, little or no germination occurred. We became concerned that seeds may not be viable. Seeds from sources 1 and 2 stained red after soaking in tetrazolium chloride for 48 h indicating seeds were viable. However, seeds that underwent a 10 min scarification did not stain.

2.3. X. asphodeloides Embryo Isolation and In Vitro Germination Experiments

With little success in germinating *X. asphodeloides* seeds, sporadic germination occurring after scarification and confirmation of viability, we hypothesized that seed coat-imposed dormancy prevented germination, and removal of the embryo from the seed would allow germination to occur.

Seeds contained small oval embryos that were isolated with forceps (Figure 2A). Percent contamination of the embryo cultures grown on $\frac{1}{2}$ MS without PGRs, full MS without PGRs, $\frac{1}{2}$ MS + NAA and full MS + NAA media was recorded: 23%, 14%, 24%, and

18%, respectively. After two weeks growth of the isolated embryo was visible. Embryos on $\frac{1}{2}$ and full-strength MS medium without PGRs developed small roots and formed bulb-like structures that enlarged in size (Figure 2B,C). Embryos placed on media with NAA formed bulb-like tissue with or without pointy green structures. After four weeks presence of roots and/or bulbous tissue were 25%, 20%, 19%, and 3% for $\frac{1}{2}$ M without PGRs S, full M without PGRs S, $\frac{1}{2}$ MS + NAA, and full MS + NAA media, respectively. The best response with the most leaves and the least callus occurred on $\frac{1}{2}$ MS medium without NAA.

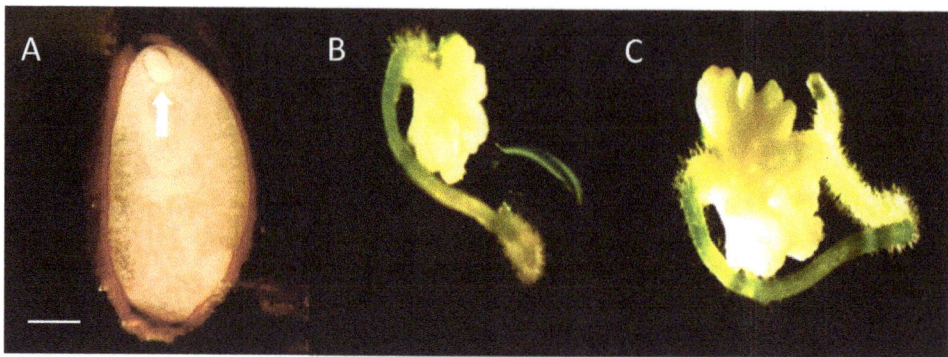

Figure 2. *Xerophyllum asphodeloides* seed with embryo. (**A**) Unstained embryo before isolation (arrow). (**B**) Embryo excised from sterilized seeds growing on $\frac{1}{2}$ MS without PGRs medium 35 days after placement on medium. (**C**) 56 days after placement on medium. Scale bar = 1 mm.

2.4. Embryo Growth on $\frac{1}{2}$ MS Medium with Different Sucrose Concentrations

Embryo germination for sources 2 and 3 increased as sucrose concentration increased with the highest germination percentage at 6%, however, differences were not statistically significant (Table 1). Source 1 embryo germination was similar for 3 and 6% sucrose (Table 1). Embryo germination for the five Alabama sites tested with 3 and 6% sucrose concentrations did not differ significantly, although the interaction between germination and seed source was statistically significant (Table 2).

Table 1. *Xerophyllum asphodeloides* embryo germination percentage for three seed sources (1, 2, and 3) on $\frac{1}{2}$ MS medium without PGRs with different sucrose concentrations after eight weeks.

Seed Source and Medium Sucrose Content	Number Embryos Tested	Embryo Germination (%)
Source 2		
3% sucrose	18	50 a
4% sucrose	18	61 a
5% sucrose	18	61 a
6% sucrose	18	67 a
8% sucrose	18	50 a
10% sucrose	18	33 a
Source 1		
3% sucrose	12	42 a
6% sucrose	12	33 a
Source 3		
3% sucrose	20	35 a
6% sucrose	20	55 a

Values followed by the same letter for a seed source are not statistically different by the multiple range test at $p < 0.05$.

Table 2. *Xerophyllum asphodeloides* embryo germination percentages after seven weeks for different source sites (2017) on $\frac{1}{2}$ MS medium with different sucrose concentrations.

Seed Source Site (2017)	Number of Non-Contaminated Explants for Media Containing 3 and 6% Sucrose [2]	Embryo Germination (%)	
		3% Sucrose	6% Sucrose
2	17, 16	87.5 ± 7.2	65.0 ± 11.9
6	25, 25	56.0 ± 7.5	44.0 ± 11.7
8	20, 20	61.3 ± 9.7	72.5 ± 11.1
9	23, 21	73.3 ± 4.2	58.0 ± 6.6
10	25, 21	64.0 ± 11.7	74.0 ± 9.7
Average [1]		68.6 a	62.7 a

[1] Values followed by the same letter are not statistically different by the multiple range test at $p = 0.05$. Analyses are based on arcsine $\sqrt{\%}$ transformation. [2] Non-contaminated explants did not show visible contamination.

2.5. X. asphodeloides Shoot Multiplication Experiments

Shoots obtained from the four germinants in trial 2 grown on $\frac{1}{2}$ MS without PGRs continued to grow and produce new shoots on all media tested, producing four clones each with multiple shoots. A few of the smallest shoot clumps did not survive. The highest number of shoots grew on $\frac{1}{2}$ MS medium without PGRs, although differences were not statistically significant (Figure 3).

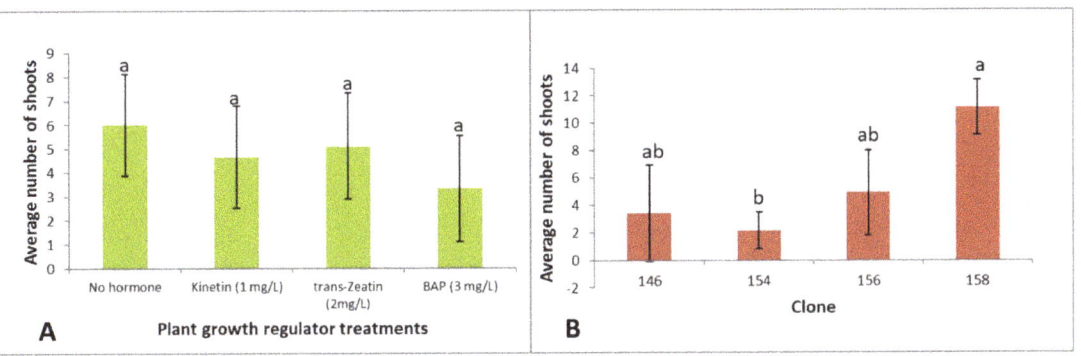

Figure 3. *Xerophyllum asphodeloides* shoot multiplication on $\frac{1}{2}$ MS medium without PGRs after five weeks. Standard errors are shown for each treatment. Treatments with same letter are not statistically different by multiple range test at $p = 0.05$. (**A**) Effect of cytokinins averaged for the four clones. (**B**) Effect of clone averaged for the four media tested.

Thidiazuron had a strong effect on shoot multiplication over a four-month subculture. Low concentrations of thidiazuron (0.1 mg/L) influenced the shoot formation showing the highest number of shoots (Table 3). Differences were statistically significant ($p = 0.05$) with respect to control 1. DMSO, used as control n.2, did not show a statistically significant effect on shoot multiplication.

Thidiazuron again had a strong effect on shoot multiplication over a two-month subculture, in particular on shoots with minor size (Table 4). Shoot multiplication increased as thidiazuron increased with an optimal concentration at 0.15 mg/L and declined as thidiazuron was further increased. Differences were statistically significant for both small and large shoots ($p = 0.05$).

Shoot cultures grown on 0.15 mg/L thidiazuron for two 2–4-month subcultures continued to multiply but showed expanded and deformed shoot bases, suggesting that either thidiazuron concentration was too high or continuous exposure to thidiazuron is detrimental to shoot quality. On the contrary, shoots from all seed sources grown on PGR-free multiplication medium continued to produce new shoots during micropropagation cycles of 2–4 months over several years, suggesting shoot production is sustainable over time. With these results, further shoot multiplication was carried out without PGRs.

Table 3. *Xerophyllum asphodeloides* shoot multiplication over four months for three seed sources (1, 2, and 3) on $\frac{1}{2}$ MS medium without PGRs varying in thidiazuron content.

Thidiazuron and DMSO Content	Number of New Shoots Per Explant
No thidiazuron, no DMSO (control 1)	4.3 a
No thidiazuron, 0.1 mg/L DMSO (control 2)	6.6 ab
Thidiazuron 0.1 mg/L, DMSO 0.1 mg/L	15.0 b
Thidiazuron 0.2 mg/L, DMSO 0.1 mg/L	9.8 ab
Thidiazuron 0.5 mg/L, DMSO 0.1 mg/L	5.6 ab
Thidiazuron 1.0 mg/L, DMSO 0.1 mg/L	8.3 ab

Values followed by the same letter are not statistically different by the multiple range test at $p = 0.05$.

Table 4. *Xerophyllum asphodeloides* shoot multiplication over two months for three seed sources on $\frac{1}{2}$ MS medium varying in thidiazuron content.

Thidiazuron	Average # of Shoots Grown Per Starting Shoot		
	<3 cm	>3 cm	Total
None	0.5 a	1.0 a	1.5 a
0.05 mg/L	1.7 b	0.9 a	2.7 bc
0.1 mg/L	1.4 b	0.7 a	2.3 ab
0.15 mg/L	1.8 b	1.4 b	3.3 c
0.2 mg/L	1.9 b	1.1 ab	2.9 bc

Values within the same column followed by the same letter are not statistically different by the multiple range test at $p = 0.05$.

2.6. X. asphodeloides Root Induction and Acclimation to Soil

Shoots in the preliminary rooting experiment grew for about 3 months and then damped off and died.

In the first experiment, after one month almost all shoots were still alive regardless of presence of roots at the time of planting (Table 5). Source 3 plants with roots at the time of planting appeared to be a bit stronger. Some non-rooted shoots from source 2 showed yellowing and some dead leaves but were still alive. Examined shoots in all treatments showed new root growth (Figure 4). After two months, many of the plants were showing new shoot and root growth (Figure 5). Between two and three months, shoots began to decline. At four months many of the plants had died. Shoots were considered dead if they were completely brown and withered. Almost all of the surviving plants had roots and were larger than when planted, but most were yellowing and showing dead leaves, suggesting continuing decline (Figure 6).

Table 5. *Xerophyllum asphodeloides* shoot performance during rooting and acclimation to soil.

Seed Source and Treatment	Percentage Surviving		
	1 Month	2 Months	4 Months
Source 2 (DF-1) non-rooted	80%	60%	30%
Source 2 (DF-1) rooted	90%	90%	40%
Source 3 (LC-55) non-rooted	100%	100%	20%
Source 3 (LC-55) rooted	100%	92%	67%

Figure 4. Shoots from isolated embryos from seed source 5 germinated in vitro and transferred to greenhouse soil under domes. Until two months plants grew well, but after 3 months they damped off and died.

Figure 5. *Xerophyllum asphodeloides* shoot growth and rooting after 1 month in greenhouse potting mix. (**A**) DF-1 non-rooted. (**B**) DF-1 rooted. (**C**) LC-55 non-rooted. (**D**) LC-55 rooted.

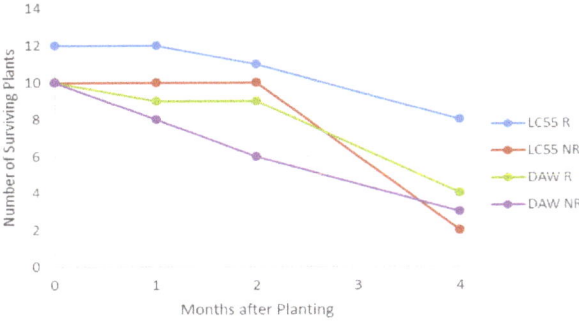

Figure 6. *Xerophyllum asphodeloides* micropropagated shoot survival over time for two clones planted in greenhouse soil. R (shoots with roots at the time of planting), NR (without roots when planted).

In Vitro Root Induction

With repeated observations of plant decline and death after two to three months in the greenhouse, we speculated that even though 80–100% of the plants formed roots, the sparse number of roots formed spontaneously or due to rooting powder treatment may be improved in number and vigor through in vitro rooting treatments with auxin.

Roots in the in vitro rooting experiment began to develop at three weeks and over time root length, number of roots per shoot, and the percentage of shoots forming roots all increased (Figure 7). Maximum rooting percentages (90%) across the four clones (DAW, LC-51, LC-57, and TNF) occurred at nine weeks in the treatment containing 0.5 mg/L NAA (Table 6); differences between treatments were statistically significant at $p < 0.01$. Shoot rooting and health were evaluated prior to planting (Table 6). After nine weeks of in vitro rooting, the largest number of dead shoots occurred in the 1 mg/L NAA treatment suggesting that long term exposure to high NAA concentration reduced shoot health. Clone LC-51 also showed low survival across all treatments (60% died), likely due to presence of bacterial contamination. Shoots were planted in potting mix after nine weeks of in vitro rooting treatment. After planting, shoot survival steadily declined regardless of treatment (Figure 8).

Figure 7. *Xerophyllum asphodeloides* after six weeks in 0.5 mg/L NAA rooting medium. Multiple vigorous roots have formed.

With three prior observations of plant decline and death after two to three months in the greenhouse, we further speculated that once roots form they may require a better drained planting mix to better match the dry environment they grow in naturally to avoid damping off. *X. asphodeloides* normally grows in dry acidic, rocky, loose, well-drained soil with a northwest exposure [4].

The plants placed in SBG1 soil did very well, they survived and continuously grew over nine months. Plants did not exhibit the decline we usually saw after two months in peaty soils that were watered daily. This suggested that the drier environment and possible beneficial microorganisms contained in the aged bark improved growth and survival.

The native soil was very dense and would adhere to roots easily. ABG mix is rocky due to perlite and does not cling to roots like the native soil. SBG mix is more like a traditional soil, as it contains aged pine bark and adheres easily to the roots. All rooted shoots survived and continued to grow. Plant decline was not seen as in prior experiments with ABG potting mixes and daily watering. The reduced watering appeared to be beneficial. Addition of native soil did not improve plant growth or appearance. Plants in SBG1 soil mix exhibited faster growth with a greater number of leaves.

Table 6. *Xerophyllum asphodeloides* shoot rooting percentage after placement in in vitro rooting medium containing $\frac{1}{2}$ MS and varying amounts of NAA and activated carbon.

Treatment and Clone	3 Wks Rooting	6 Wks Rooting	8 Wks Rooting	9 Wks Rooted and Surviving
Control, no PGRs—DAW	0%	0%	20%	0%
Control, no PGRs—LC51	0%	0%	0%	0%
Control, no PGRs—LC57	20%	0%	20%	0%
Control, no PGRs—TNF	40%	60%	80%	40%
Average	15%	15%	30% a	10%
0.5 NAA—DAW	0%	80%	100%	20%
0.5 NAA—LC51	60%	80%	80%	80%
0.5 NAA—LC57	20%	80%	100%	80%
0.5 NAA—TNF	20%	60%	80%	80%
Average	25%	75%	90% c	65%
1.0 NAA—DAW	40%	40%	40%	0%
1.0 NAA—LC51	40%	40%	40%	40%
1.0 NAA—LC57	20%	100%	100%	0%
1.0 NAA—TNF	80%	40%	40%	20%
Average	45%	55%	55% b	15%
1.0 NAA + AC—DAW	40%	20%	0%	0%
1.0 NAA + AC—LC51	20%	0%	0%	0%
1.0 NAA + AC—LC57	20%	0%	20%	40%
1.0 NAA + AC—TNF	20%	40%	20%	0%
Average	25%	15%	10% a	10%

Average values within the 8 week data column are statistically different by the multiple-range test at $p < 0.01$ as indicated by different letters.

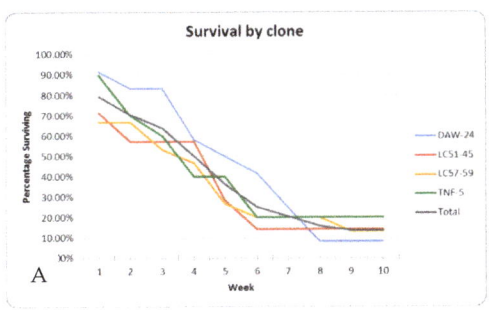
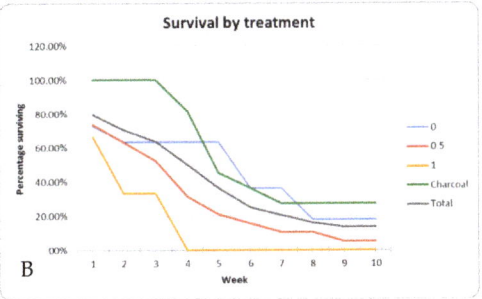

Figure 8. *Xerophyllum asphodeloides* micropropagated shoot survival over time after in vitro rooting treatments and shoots from four clones planted in greenhouse soil. (**A**) Survival by clone. (**B**) Survival by in vitro rooting treatment.

2.7. *X. asphodeloides* Acclimated Greenhouse Plants Planted in the Field

Plants transferred to the field survived and grew normally (Figure 9A). After 21 months of in-field growth, survival per plot was as follows: Plot A = 100%, Plot B = 57%, Plot C = 67%, Plot D = 63%, with an overall 73% survival. After two growing seasons plant size and leaf count were similar across all plots, with plants averaging about 20 cm in height.

Figure 9. (**A**) *Xerophyllum asphodeloides* plants after one year in the field. (**B**) Whole seed germinated after prolonged stratification.

2.8. X. asphodeloides Seed Cryopreservation

Water content of source 3 seeds was determined to be 10.2%.

Whole cryopreserved seeds did not germinate after LN treatment likely due to seed coat-imposed dormancy. For this reason, embryos were excised from cryopreserved seeds.

Twenty percent of embryos (20%) grew from cryopreserved seed, while none of the control isolated embryos grew.

Water content of seed sources 4 and 5 measured 5.63% and 4.61%, respectively. Cryopreserved and control seeds showed similar percentages of isolated embryo germination with averages for the two seed sources of 57% for control seed and 57% for cryopreserved seed (Table 7).

Table 7. *Xerophyllum asphodeloides* isolated embryo germination after 8 weeks for control and seed cryopreserved for one month.

Seed Source and Treatment	# of Starting Seeds	# Contaminated Seeds	Isolated Embryo Germination (%)
		Seed source 4	
Control	25	0	56% a
Cryopreserved	25	0	48% a
		Seed source 5	
Control	23	5	61% a
Cryopreserved	42	5	62% a

Values within a trial followed by the same letter are not statistically different by the multiple range test at $p = 0.05$.

The additional seeds from sources 2, 3, 4, and 5 were evaluated for survival after varying periods of time at room temperature or after cryopreservation (Table 8). Seed sources varied, but seeds appeared to decline in viability with storage longer than 1 year at room temperature. Cryopreservation of viable seeds appeared to halt the decline as seeds from two sources cryostored for nearly two years after 4–5 months at room temperature showed survival similar to fresh non-cryopreserved seeds.

2.9. X. asphodeloides In Vitro and In Vivo Germination with Treatment for Double Dormancy

None of the seeds in medium or soil germinated after 12 weeks at room temperature or after the next period of 12 weeks at 4–5 °C. Seeds began to germinate only after the moist/warm and cold exposures ended and seeds were returned to room temperature. After 8 weeks in the final phase of room temperature, germination was recorded for the in vitro and in vivo treatments. Based on total seeds tested, both treatments showed similar germination percentages (Table 9). Seed source 2017-3 showed 50 and 47% germination for seeds germinated in medium vs. soil. Seed source 2017-5 showed 60 and 53% germination for seeds in medium vs. soil. Microbial contamination in medium was high for both sources with 53 and 40% for sources 2017-3 and 2017-5, respectively. Overall germination percentages in medium or soil did not differ significantly. Since seeds were surface sterilized, it is interesting to note that almost all seeds germinated on medium when not contaminated. Contaminated seeds in soil likely did not germinate and therefore overall germination percentages were similar in both treatments due to the loss of seed viability from microbial contaminants within the seeds.

Table 8. *Xerophyllum asphodeloides* isolated embryo germination in vitro after long term seed storage at room temperature or storage at room temperature followed by cryopreservation.

Seed Source and Treatment	Number Of Starting Seeds	Number of Contaminated Seeds	Number/total (%) of Isolated Embryos Germinating after 3 Months on Medium
	Seed source 2 (DF, collected October 2010)		
3 years at RT	33	0	0%
	Seed source 3 (LC, collected November 2010)		
3 years at RT	32	0	0%
	Seed source 4 (TNF, collected August 2012)		
15 months at RT	10	1	22%
5 mo at RT + 20 months CP	32	24	63%
	Seed source 5 (DF, collected October 2012)		
12 months at RT	10	1	11%
24 months at RT	33	0	6%
4 mo at RT + 20 months CP	33	0	67%

mo = months, RT = room temperature, CP = cryopreserved.

Table 9. *Xerophyllum asphodeloides* whole seed germination on medium in vitro or in greenhouse soil after treatment to remove double dormancy [1].

Seed Source	Contamination In Vitro # of Seeds (%)	Non-Contaminated Seed Germination on Medium # of Seeds (%)	Overall Germination In Vitro # of Seeds (%)	Germination in Greenhouse Mix # of Seeds (%) [2]
2017-3	16/30 (53%)	14/14 (100%)	14/30 (47%) a	15/30 (50%) a
2017-5	12/30 (40%)	16/18 (89%)	16/30 (53%) a	18/30 (60%) a

[1] Double dormancy treatment—12 weeks in moist medium or soil in dark at room temperature followed by 12 weeks of 4–5 °C in dark followed by incubation in a lighted culture room at room temperature. [2] Values followed by the same letter in a row are not significantly different using a *t*-test.

When seeds in moist sphagnum moss were removed from the cold after 16 weeks, surprisingly, three yellow-leafed germinants were present. After planting in SBG1 mix and placement in the lighted greenhouse, the yellow germinants turned green and continued growing.

Seeds placed in bags of sphagnum moss at 4–5 °C in the dark began to germinate after 14–16 weeks of stratification. When moved to grow lights and later planted in 10 cm pots germination continued to occur, with final germination shown in Table 10 and Figure 8B. Note that seeds were stored dry for eight months at room temperature and had probably lost some viability from storage. Germination percent ranged from 0 to 23 and nine of the 11 seed sources showed some germination confirming our prior observations of germination after prolonged stratification.

Table 10. *Xerophyllum asphodeloides* seed germination after 14–16 weeks of stratification at 4–5 °C in wet sphagnum moss.

Seed Source	Number of Seeds Stratified	Number of Seeds Germinating (%)
2017-1	44	23%
2017-2	44	9%
2017-3	44	0%
2017-4	35	11%
2017-5	44	18%
2017-6	44	0
2017-7	44	14%
2017-8	88	23%
2017-9	44	11%
2017-10	50	2%
2017-11	5	20%

3. Discussion

Xerophyllum asphodeloides is an interesting, important, and little-studied species that is threatened by fire suppression and the loss and fragmentation of its habitat. We report here, for the first time, cryopreservation and micropropagation procedures for *X. asphodeloides*. There is little information on the germination of *X. asphodeloides*. Some treatments are shown to overcome coat-imposed seed dormancy both in vitro or in vivo, using an embryo isolation method or whole seeds.

Brumback [19] categorized *X. asphodeloides* as a species that must be sown immediately outdoors upon collection and may take two years to germinate if freshly sown. Cullina [20] classified *X. asphodeloides* as a semi recalcitrant or hydrophilic seed type that does not tolerate dry storage and must be sown immediately upon ripening. A category of seeds intermediate between orthodox and recalcitrant is recognized in which seeds survive desiccation, but are damaged during dry storage at low temperatures [21].

Our research shows that *X. asphodeloides* seeds lose viability over 1–2 years of dry storage. Cullina [20] suggested the seeds are hydrophilic and do not take desiccation very well. Desiccated seeds stored at room temperature may lose viability while seeds kept moist in plastic with cool temperatures or in a soil bank may remain viable for years [22]. This condition is a problem for seed conservation [18,23–25]. Cryopreservation of seed, somatic embryos, plant meristems, or other tissues is an excellent tool for germplasm safekeeping [10–12,26,27]. The first set of cryopreserved seed did not germinate likely due to seed coat-imposed dormancy. The 2nd set of seed was 13 months old and probably lost viability over time. However, 20% survival in the 2nd set of seeds after cryopreservation indicated that *X. asphodeloides* seeds can tolerate cryopreservation. The 3rd cryopreservation experiment with fresher and drier seeds (stored 3–5 months at room temperature with 4.6–5.6% water content) showed similar embryo germination before and after cryoreservation as well as cryostorage prevention of viability loss over 20 months. Optimum moisture content for seed cryopreservation varies from 7 to 14% depending on species and seed lipid content [9]. Our determined seed water contents of 4.6–10.2% allowed successful cryopreservation of *X. asphodeloides* seeds.

Seed dormancy will often prevent germination even if the environment is fine. Seed dormancy provides additional time for a seed to disperse over geographic distances. Two types of dormancy occur: embryo dormancy often due to the ratio of gibberellins to abscisic acid (ABA) and dormancy imposed by the seed coat, endosperm, or pericarp [28]. For the latter, embryos will usually germinate once the seed coat and surrounding tissues have been damaged or removed. To overcome seed-coat dormancy and germinate, seeds often require long-term cold exposure while moist or damage to the outer tissues by long-term microbial activity, mechanical abrasion or contact with strong acids in the digestive system of an animal [29]. Acid or mechanical scarification is often used to speed germination of many seeds with thick coats. The ABA/gibberellin (GA) balance theory suggests dormancy in some species is controlled by antagonistic effects between ABA and GAs. Long-term cold stratification has been shown in some species to increase GAs that counteract germination inhibitors such as ABA [30]. Conditions that occur during moist stratification at 4 °C or chemical scarification can thus break dormancy by altering the ABA: GA ratio and/or thinning the seed coat to increase chemical diffusion and decrease coat strength.

Usually, germination in vitro is simple and involves surface sterilizing seeds in bleach or hydrogen peroxide and the seed germinates on medium within a few weeks. With X. asphodeloides this has been a difficult step because we did not have information about its germination requirements. Several methods to stimulate germination were tested including: MS medium salt strength, scarification, cold exposure, GA and or fluridone exposure, ABA exposure, heating exposure, smoke water exposure, and long times on germination medium. Out of about 1000 seeds, only a few germinated and these occurred when seeds were acid scarified suggesting seed coat-imposed dormancy prevented germination. However, further testing will be necessary with seed acid scarification alone or combined with other dormancy-breaking treatments to obtain high seed germination percentage

After numerous preliminary tests to germinate intact X. asphodeloides seeds in vitro, we decided to apply a different approach sometimes used to overcome interspecific incompatibility for rare lily hybrids [31]. X. asphodeloides seeds were surface-sterilized followed by dissection and removal of the embryo. Because dormancy is often caused by the seed coat and tissues surrounding the embryo, ref. [29] isolation of the embryo followed by placement in vitro on medium can sometimes remove dormancy. This approach worked well for X. asphodeloides and hundreds of embryos germinated in the trials carried out on $\frac{1}{2}$ MS medium without plant growth regulators.

Seedling or shoot tip culture in vitro is commonly used to propagate plants for agricultural, conservation, horticultural, or medicinal purposes. In particular, cells from shoot tips are generally genetically stable and shoot tip culture can produce identical copies of target plants. Propagation with other tissue culture systems involving callus, organogenesis or somatic embryogenesis may induce somaclonal variation [15,32]. Since our goal is to maximize natural variation and minimize in vitro-induced variation, seed and shoot tip culture methods can provide excellent methods for preservation and reintroduction of X. asphodeloides.

X. asphodeloides shoots from germinated seeds and isolated embryos were tested for multiplication with increases of about 6 new shoots on $\frac{1}{2}$ MS PGR-free medium or 15 shoots on medium with 0.1 mg/L thidiazuron over 2-month subculture cycles. Tests over 12 months showed sustainable shoot multiplication on PGR-free medium. Thidiazuron is a diphenyl-substituted urea and is often used for woody plant tissue culture due to its potent cytokinin activity [33]. Low concentrations of thidiazuron can stimulate axillary shoot proliferation while higher concentrations can stimulate axillary and adventitious shoot formation. Thidiazuron has also shown promise for micropropagation of some recalcitrant species in the Liliaceae and other plant families [34,35].

Rooted shoots from in vitro shoot cultures began to grow easily under domes in a greenhouse when watered daily or sparsely every other day. However, plants in soil mixes watered daily grew for several months and then declined, while light watering supported high survival and continued plant growth. We speculate that daily watering causes damping off and root rot while light watering allows the surface soil to dry out creating a more natural environment for root growth. In addition, the use of substrate, the aged pine bark in SBG 1, may harbor beneficial microorganisms that help X. asphodeloides roots to thrive. Abundant research with micropropagated plants supports the use of bio-agents to improve shoot acclimatization, survival and growth [36–38].

Seeds from some species have very specific germination requirements. For example, some genera in the Liliaceae and Melanthiaceae have species with deep simple double morphophysiological dormancy [29]. Baskin and Baskin [29] reported that seeds of *X. asphodeloides* take two winters to germinate during which the embryo grows to a critical size followed by exposure to environmental conditions that break physical dormancy [29].

When *X. asphodeloides* seeds germinated in a greenhouse, germination usually took more than a year before a shoot emerged from the soil [22,39]. Our findings along with observations of Brumback [19] and Cullina [20] suggest that *X. asphodeloides* is an intermediate type seed with morphophysiological dormancy (MPD) and dormancy can be overcome by exposure to warm moist conditions followed by cold stratification followed by continued cold or warm moist again. While nine levels of MPD are recognized (29), more research is needed before we clearly understand the type of MDP present in *X. asphodeloides*. Our additional observations of germination after prolonged stratification show that embryo growth may occur slowly in the cold followed by dormancy release from stratification. Further testing is required to determine times and temperatures needed at each phase of dormancy treatment to break dormancy. Germination occurred in vivo for *X. tenax* when cold stratification for 14–16 weeks was applied with or without smoke-water treatment [40,41]. Fluridone, an ABA synthesis inhibitor, increased and speeded lily seed germination [42]. However, stratification (14 wks), smoke-water and fluridone did not assist *X. asphodeloides* germination in vitro. Our early cold temperature treatments likely did not break dormancy because the seed embryos were not sufficiently developed. The small undeveloped embryo present in fresh seed can be seen in Figure 2A.

The findings in this study provide a starting point for long term seed storage and production of *X. asphodeloides* plants for conservation and safeguarding (Table 11). Further studies are needed to improve, optimize and simplify procedures to germinate *X. asphodeloides* seeds and to grow planting stock. Major needs include: (1) Optimization of temperatures and discovery of minimum times required at each phase of dormancy treatment for maximum germination, (2) Simplification and optimization of methods to maintain planting mix moisture for optimal early plant growth.

Table 11. Recommended steps for *Xerophyllum asphodeloides* seed cryopreservation, germination of isolated embryos or seeds and growing plants.

Seed cryopreservation
(1) Determine seed moisture content and if seeds need to be dried
(2) Place seeds in cryogenic storage vials
(3) Rapidly immerse vials in liquid nitrogen (LN)
(4) Remove vials from LN and re-warm in a 37 °C water bath for 1–2 min
(5) Germinate seeds in vitro or in vivo
Seed germination in vitro (production of initial shoots for micropropagation)
Option A (whole seed germination over 26–28 weeks, low labor requirement)
(1) Surface sterilize fresh of cryopreserved seeds
(2) Place on $\frac{1}{2}$ MS medium without PGRs in sterile containers
(3) Incubate in dark for 12 weeks at room temperature
(4) Incubate in cold (4 °C) in dark for 16 weeks
(5) Move containers to lighted culture room for germination
(6) Remove germinants and use for micropropagation
Option B (embryo isolation, germination in about 2–4 weeks, high labor and skill required)
(1) Surface sterilize fresh of cryopreserved seeds
(2) Remove embryo under sterile conditions
(3) Place embryo on $\frac{1}{2}$ MS medium without PGRs and use for micropropagation

Table 11. *Cont.*

Seed germination in vivo

(1) Place fresh or cryopreserved seeds in plastic bag with moist sphagnum moss at room temperature for 12 weeks
(2) Move bags to 4 °C in dark for stratification for 16 weeks
(3) Move bags to light at room temperature. As germinants occur transplant to SBG1
(4) Cover with humidity domes for 2–4 weeks. Water sparsely with about 3 mL per 10 cm pot daily for 2 weeks. Water with about 4 mL every other day allowing soil surface to dry. Continue for about 6 months and transplant individual plants into larger pots.

Micropropagation

(1) Place contamination-free germinant or shoot on $\frac{1}{2}$ MS medium without PGRs
(2) Subculture and divide shoots as needed until desired number of shoots are obtained
(3) Root shoots in PGR free medium, $\frac{1}{2}$ MS with 0.5 mg/L NAA or dip shoots in rooting powder
(4) Rinse medium from rooted shoots. Insert rooted shoots or rooting powder treated shoots carefully into SBG1. Cover with humidity domes for 2–4 weeks. Water sparsely with about 3 mL per 10 cm pot daily for about 4 weeks. Water with about 4 mL every other day allowing soil surface to dry. Continue for about 6 months and transplant individual plants into larger pots.

With continued habitat destruction, over-collection, disease, and climate change, loss of *X. asphodeloides* populations are expected to continue. Developing methods for long-term seed storage and biodiversity maintenance are therefore critical. Procedures developed here provide tools to preserve germplasm and genetic diversity in *X. asphodeloides* and should be implemented immediately to assist in conservation of these beautiful plants.

4. Materials and Methods

4.1. Plant Materials, Experimental Design, and Evaluation

X. asphodeloides seeds were obtained from 17 sources (Table 12). After receipt, seeds were stored at room temperature in paper bags until use. Treatments were arranged in a completely randomized design. Data were analyzed by t-test or analysis of variance, and significant differences between treatments were determined by the Duncan's multiple range tests using Statgraphics Plus V4.0 (Manugistics, Rockville, MD, USA).

Table 12. Seed sources and sampling times for *X. asphodeloides*.

Seed Source	Collection Location	Harvesting Time	Seed Condition
1	Talladega National Forest (TNF), AL	6 August 2010	Seeds showed visible white fungal hyphae on coat surface
2	Dawson Forest (DF) in Dawson County, GA	October 2010	–
3	Canton, Cherokee County (LC), GA	November 2010	–
4	TNF	27 August 2012	–
5	DF	22 October 2012	–
6	TNF	12 December 2013	–
2017-1 to 11	TNF Shoal Creek Division	Fall 2017	–

4.2. X. asphodeloides Seed Sterilization and In Vitro Germination Experiments

Due to rarity of seeds, unless otherwise indicated, experiments typically consisted of 10–15 seeds from sources 1 or 2 for each treatment tested. Several factors to break seed dormancy were applied in tests for in vitro germination of seeds of *X. asphodeloides* [29]. Seeds were surface-sterilized for 10 min in 20% H_2O_2 as described by Ma et al. [23]. Scarified seeds were not surface-sterilized. After sterilization or scarification, seeds were placed on $\frac{1}{2}$ strength MS salts (Murashige and Skoog 1962) with 3% sucrose, 100 mg/L myo-inositol, 0.50 mg/L thiamine HCl, 0.25 mg/L pyridoxine HCl, 0.25 mg/L nicotinic acid, and 1.0 mg/L glycine with or without an added plant growth regulator (PGR, see individual trials below). The pH of media was adjusted to 5.7 and solidified with 4.5 g/L Phytagel. Media were autoclaved at 121 °C for 20 min. Seeds were placed on 7 mL of medium in 60 × 15 mm Petri dishes. They were placed in a culture room at 24–25 °C under a 16/8-h (day/night) photoperiod with light supplied by cool white fluorescent lamps at an intensity of approximately 30 μmol m^{-2} s^{-1}. Germination responses were examined weekly for two months. The following germination trials were carried out.

4.2.1. Comparison of 1/3 vs. 1/2 MS Media Without PGRs

Seeds from source 1.

4.2.2. Seed Scarification

In a first experiment, seeds not surface-sterilized were scarified in concentrated sulfuric acid for 1, 4, or 10 min followed by pouring seeds and acid into 200 mL of sterile distilled water for 3 min followed by five sterile water rinses for 5 min each. Source 1 seeds were placed on $\frac{1}{2}$ MS medium without PGRs. In a second experiment Source 1 seeds were scarified for 3, 3.5, and 4 min in H_2SO_4. Source 2 seeds were scarified for 2, 3, 4, and 5 min, rinsed and placed on $\frac{1}{2}$ MS medium without PGRs. Thirty seeds of source 2 were scarified for 4 min, rinsed and planted in greenhouse soil mix (1 part pumice: 3 parts perlite: 1 part milled sphagnum moss). Any germinants were transferred to fresh $\frac{1}{2}$ MS medium without PGRs every 3–4 weeks.

4.2.3. Seed Soaking in Gibberellic Acid (GA)

Seeds were surface sterilized and soaked overnight in 5-drop aliquots of filter-sterilized water solutions of 500, 750, or 1000 mg/L GA. Thirty seeds from source 1 placed on $\frac{1}{2}$ MS medium without PGRs.

4.2.4. Hot Water Treatments

Three hot water treatments varying in severity were tested to induce germination of source 1 seeds: (1) Ten seeds in a stainless-steel strainer were dipped for one second into boiling water. (2) Boiling water (100 mL) was poured over a strainer of 10 seeds for about 5 s. (3) Boiling water (100 mL) was removed from the heat and a strainer with 10 seeds was immersed into the water and allowed to cool to room temperature. Seeds were surface sterilized and placed onto $\frac{1}{2}$ MS medium without PGRs.

4.2.5. Seed Scarification Combined with Smoke Water

Source 2 seeds were scarified for 4 min and rinsed. Seeds were further soaked for 24 h in smoke water prepared from a smoke seed primer disc (FineBushPeople, Cape Town, South Africa) according to directions and placed on $\frac{1}{2}$ MS medium without PGRs.

4.2.6. 2 × 2 Factorial: Surface Sterilization or Scarification x ± Fluridone Soak Treatment

Surface-sterilized or scarified for 4 min. Half of the surface-sterilized or scarified seeds were not treated with fluridone. The other half were further soaked for 24 h in a filter-sterilized water solution of 2.4 μM fluridone. Ten to 25 resulting seeds of source 2 for each treatment were then placed onto $\frac{1}{2}$ MS medium without PGRs.

4.2.7. Seed Scarification and Germination in $\frac{1}{2}$ MS vs. Full-Strength MS Salts

Twenty seeds from source 2 were scarified for 4 min. Half of the 20 seeds were placed onto $\frac{1}{2}$ MS medium and half on full-strength MS medium for germination, each medium without PGRs.

4.2.8. Seed Fluridone Treatment ± GA

Seeds from source 2 (15 per treatment) were placed on $\frac{1}{2}$ MS medium without PGRs, $\frac{1}{2}$ MS medium with 10 µM fluridone, or $\frac{1}{2}$ MS medium with 10 µM fluridone and 100 µM GA.

4.2.9. Seed Cold Treatment ± Scarification

Two sets of 60 seeds from source 2 were prepared for cold treatment. One set was surface sterilized and the other set was scarified for 4 min. All seeds were placed on $\frac{1}{2}$ MS medium without PGRs and incubated in the dark at 4 °C. Ten plates, containing one seed each, were removed from the cold at 4, 6, 8, 10, 12, and 14 weeks for each set and incubated in the light.

4.2.10. Seed Cold Treatment and Homemade Smoke Water with Surface Sterilization or 4 min Scarification

To prepare smoke water, plant debris collected from the base of wild *X. asphodeloides* plants was burned in a barbecue next to a bowl of distilled water. Thirty source 2 seeds were surface sterilized and 30 seeds were scarified for 4 min. All seeds were soaked in filter-sterilized smoke water for 24 h, placed on $\frac{1}{2}$ MS medium without PGRs and incubated at 4 °C for 10, 12, or 14 weeks. Ten seeds from each treatment were removed at each cold treatment time and incubated in the light.

4.3. Tetrazolium Chloride Seed Viability Staining

We became concerned that seeds may not be viable. Tetrazolium chloride is commonly used to assay seed viability [43]. Source 1 and 2 seeds were soaked overnight in 1% tetrazolium chloride in distilled water. Seeds were sliced open and examined for red coloration indicating viability. Seeds scarified for 10 min were also tested for viability.

4.4. X. asphodeloides Embryo Isolation and In Vitro Germination Experiments

Seeds were soaked in distilled water for 48 h to soften the seeds and then surface sterilized. Embryos were dissected out of sterilized seeds in sterile conditions under a dissecting microscope. The seed was held base-up with forceps, while a scalpel was used to make a horizontal cut 2/3 across the seed without cutting either of the ends. The halves were then pulled apart to reveal a small oval shaped embryo. The embryo was carefully removed with forceps.

4.4.1. Embryo Growth on $\frac{1}{2}$ vs. Full Strength MS Medium ± 0.1 mg/L NAA

Isolated embryos were placed onto four test media: (i) $\frac{1}{2}$ MS medium, (ii) $\frac{1}{2}$ MS medium + 0.1 mg/L 1-Napthalenacetic acid (NAA), (iii) full strength MS medium without PGRs, (iv) full-strength MS medium + 0.1 mg/L NAA. Approximately 25 embryos from seed source 2 were isolated per treatment.

4.4.2. Embryo Growth on $\frac{1}{2}$ MS Medium with Different Sucrose Concentrations

Embryos were isolated from surface sterilized seeds from source 2 and placed onto media with different sugar concentrations. Approximately 18 isolated embryos per treatment were placed on $\frac{1}{2}$ MS medium without PGRs varying in sucrose concentrations: 3%, 4%, 5%, 6%, 8%, and 10%.

Additional tests were run isolating embryos from seeds sources 1 (12 embryos per treatment) and 3 (20 embryos per treatment) with placement on $\frac{1}{2}$ MS medium without PGRs containing 3 or 6% sucrose.

A larger trial was run comparing 3 and 6% sucrose using seed collected from five sites in Alabama (sources 2017-2, 6, 8, 9, and 10). Tests compared four or five replications of five isolated embryos per treatment and seed source.

4.5. X. asphodeloides Shoot Multiplication Experiments

Germinated seeds from trial 2 were transferred to $\frac{1}{2}$ MS medium without PGRs monthly. The bulbous tissue and strap-like leaves that formed during germination continued to grow. After three months these masses were cut into two to four pieces and again placed on $\frac{1}{2}$ MS medium without PGRs.

In the first shoot multiplication experiment, shoot cultures from the four source 1 germinants were divided into quarters (shoot clumps) and randomly placed onto four different media: (i) $\frac{1}{2}$ MS without PGRs, (ii) $\frac{1}{2}$ MS + 1 mg/L Kinetin, (iii) $\frac{1}{2}$ MS + 2 mg/L trans-zeatin, (iv) $\frac{1}{2}$ MS + 3 mg/L 6-benylaminopurine (BAP). Single shoot clumps were placed on 20 mL media in Magenta boxes (Magenta, Chicago, IL). Each test medium had nine replicates. All treatments were placed in the same lighted culture room used for germination. The number of shoots after five weeks was evaluated.

In the second experiment shoots produced from the isolated embryos derived from sources 1, 2 and 3 were grown on $\frac{1}{2}$ MS medium without PGRs. Ten shoots approximately 2 cm in size from each clone were placed onto each of six media containing 0 (control 1), 0.1, 0.2, 0.5 or 1.0 mg/L thidiazuron. To account for the possible effect of DMSO used in dissolving thidiazuron, a second control was added with 0.1 mg/L DMSO (control 2, the maximum amount of DMSO contained in the 1 mg/L thidiazuron treatment). All shoots were incubated under fluorescent light (cool white, light intensity at 30 μmoles/m^2/s) at 24–25 °C and evaluated after four months without subculture.

An additional experiment (third experiment) was developed to optimize thidiazuron concentration. Ten shoots from each of four clones from seed sources 1, 2, and 3, each grown on $\frac{1}{2}$ MS medium without PGRs, were each placed onto $\frac{1}{2}$ MS medium containing 0, 0.05, 0.1, 0.15, or 0.2 mg/L thidiazuron. Shoots were grown for as described above and then evaluated after two months for the number of shoots smaller or larger than 3 cm.

Shoot cultures were maintained with 2–4 month subcultures over time on PGR-free medium or medium containing 0.15 mg/L thidiazuron.

4.6. X. asphodeloides Root Induction and Acclimation to Soil

Large shoots about 3 cm in length in PGR-free multiplication medium often formed roots in vitro spontaneously. In a preliminary test several spontaneously rooted shoots were rinsed to remove excess medium from the roots and planted into Atlanta Botanical Garden (ABG) new cutting mix (1 part milled New Zealand sphagnum moss: 1 part perlite: 1 part pumice).

Two trials were carried out to establish shoots in greenhouse potting mix. In the first trial shoots from two clones from seed sources 2 and 3 (DF-1 and LC-55) grown on medium with $\frac{1}{2}$ MS medium without PGRs were separated into 10 (DF-1) or 12 (LC-55) shoots with roots and 10 shoots (DF-1 and LC-55) without roots. All shoots were dipped into Schultz Take Root rooting powder containing 0.1% indole-3-butyric acid (IBA) and planted about 6 mm deep into trays containing approximately 6 mm of horticultural grade charcoal and filled with a modified ABG new cutting mix (1 part milled sphagnum moss: 3 parts perlite: 4 parts pumice). Trays with plants were placed under transparent plastic domes, transferred to a greenhouse and watered daily. The survival rates were recorded after one and two months, and a few shoots were removed from containers to observe root development.

A second rooting trial was carried out. Shoots grown on $\frac{1}{2}$ MS medium without PGRs were dipped into rooting powder with 0.1% IBA and planted into trays with horticultural grade charcoal and ABG new cutting mix. Four clones (TNF from source 1, DAW from source 2, and LC-51 and LC-57 from source 3) with 40 shoots each were tested.

In Vitro Root Induction

Another experiment was developed with four clones containing 20 shoots each. Shoots were split among four in vitro rooting treatments: (i) $\frac{1}{2}$ MS without PGRs, (ii) $\frac{1}{2}$ MS with 0.5 mg/L NAA, (iii) $\frac{1}{2}$ MS with 1.0 mg/L NAA, (iv) $\frac{1}{2}$ MS with 1.0 mg/L NAA plus

100 mg/L activated carbon (Sigma-Aldrich C-4386). Rooting percentages were evaluated at three, six, and eight weeks, and surviving plants with roots were evaluated at nine weeks.

Single shoots, 2–3 cm in size, were inserted about 5 mm into 20 mL test medium in Magenta boxes and incubated under fluorescent light (cool white, light intensity at 30 µmoles/m^2/s) at room temperature (24–25 °C). In vitro rooting was evaluated after 1–2 months. Nine weeks after placement in medium, shoots were removed from the medium, rinsed with water, and planted with or without roots into 5 cm x 5 cm pots with approximately 1 cm charcoal and potting mix containing 3 parts modified cutting mix (1 part sphagnum moss: 1 part pumice: 4 parts perlite) plus 1 part modified ABG Carnivorous plant mix (5 parts peat moss: 1 part sphagnum moss: 1 part builder's sand). Potting mix was tightly packed and shoots were inserted about 1 cm into the mix. Trays with plants were placed under transparent plastic domes and placed in a greenhouse in early summer, watered daily, and observed weekly.

To evaluate different types of soil substrate we planted five rooted shoots in a local mix from the State Botanical Garden of Georgia (SBG1) that contained pine bark. SBG1 potting mix contained 1 part orchid bark mix, 1 part builder's sand, and 1 part SBG2 pine bark mix. SBG2 pine bark mix contained 125 gallons fine grade pine bark, 8 cubic ft coarse vermiculite, 8 cups dolomitic limestone, 2 cups superphosphate, 1 cup each of gypsum, potassium nitrate, calcium nitrate, and Micromax granular micronutrient fertilizer. Individual plants were placed in 4 cm pots with a bottom layer of horticulture grade charcoal. Medium was washed from the roots and plants were carefully planted in the soil and covered with a humidity dome for two weeks. To keep soil drier instead of drenching pots daily, individual pots were watered with 15–20 mL water every other day and provided with an extra 10 mL prior to weekends. After about four weeks water was further reduced to 10 mL every other day.

To determine if beneficial organisms from native *X. asphodeloides* soil help rooted shoots survive, we compared rooted shoot survival in two soil mixes without or with 10% amendment of native soil collected from the base of wild *X. asphodeloides* plants. Soil cores were collected to a depth of 12 inches adjacent to healthy plants from four AL sites. Large rocks, roots, and debris were removed and equal amounts of soil were mixed together from each site. ABG modified new cutting mix (1 part sphagnum moss, 3 parts perlite, 4 parts pumice) or SBG 1 mix were left as is or amended with 10% native *X. asphodeloides* soil forming four treatments. For each treatment five rooted shoots of clone TNF-7 and five shoots of clone TNF-115 were each planted in 4 cm pots with a bottom layer of horticultural grade charcoal and filled with the soil mix to be tested.

4.7. X. asphodeloides Acclimated Greenhouse Plants Planted in the Field

Thirty plants from rooted shoots grown in the greenhouse for six months or longer, resulting from the trials to evaluate soil types above, were planted at field site 11 in the Talladega National Forest. A hilltop near site 11 and resembling natural *X. asphodeloides* habitat was located and divided into four quadrants. Six to nine plants were planted into each quadrant in February, 2019.

4.8. X. asphodeloides Seed Cryopreservation

Twenty seeds from source 2 were used for the first cryopreservation experiment. Seeds were placed 10 seeds per vial into 2 mL Nalgene cryogenic storage vials (Thermo Scientific, Waltham, MA, USA) and rapidly immersed in liquid nitrogen (LN). After 48 h the vials were removed from LN and re-warmed in a 37 °C water bath for 1–2 min (Lynch et al. 2013 [24]). Whole seeds were surface sterilized and embryos were isolated and placed on $\frac{1}{2}$ MS medium without PGRs.

Five seeds from source 3 after 13 months of storage at room temperature were tested for water content. Seeds were placed in small pre-weighed glass vials, covered with aluminum foil to prevent water exchange with the air, weighed, dried for 24 h at 105 °C, weighed again, and water contents were calculated.

Fifteen seeds from source 3 were placed into each of two cryovials. One vial was rapidly immersed into LN and the other was held at room temperature. After 48 h, the vial was removed from LN and re-warmed in a 37 °C warm water. Both sets of 15 seeds were soaked for 24 h in sterile water at 4 °C. Seeds were surface sterilized and embryos were isolated and placed on $\frac{1}{2}$ MS medium without PGRs.

Another cryopreservation experiment was carried out with seed sources 4 and 5, they were stored at room temperature for 3–5 months in paper bags until tested for water content and cryopreservation using isolated embryos. Water content measurements were determined for five seeds from each source. About 350 seeds were used, inserting 30 seeds per cryovial. They were rapidly immersed in LN. A similar number of seeds remained at room temperature for the control.

Additional seeds from sources 2, 3, 4, and 5 were evaluated to determine survival after varying periods of time at room temperature or storage at room temperature followed by cryopreservation (Table 8). Experiments evaluated 32–33 seeds per treatment and storage time using the embryo isolation method.

4.9. X. asphodeloides Germination of Whole Seeds with Treatment for Double Dormancy

To simplify germination from cryopreserved or non-cryopreserved seeds and to produce tissue more easily for micropropagation, we returned to whole seed germination. Prior greenhouse trials for *X. asphodeloides* showed that slow seed germination occurred over two years [19,22,39]. We hypothesized that *X. asphodeloides* seeds have double dormancy and require a warm moist exposure to cause embryo development followed by a cold period to break dormancy followed by a moist, warm environment for root and shoot growth. To test this hypothesis 30 non-cryopreserved seeds from sources 2017-3 and 2017-5 were each tested in two treatments: (1) whole sterilized seeds were placed in individual Petri dishes with $\frac{1}{2}$ MS medium containing 3% sucrose without PGRs; and (2) whole sterilized seeds were placed into individual 14 × 14 cm trays containing soil mix composed of one part perlite and one part aged Monterey pine bark. All seeds were incubated at room temperature in darkness for 12 weeks followed by 12 weeks of incubation at about 4–5 °C. After this period, they were transferred to the lighted culture room as described above for eight weeks. Soil mix trays were misted every few days to maintain soil moisture. Petri plates were inspected weekly and seeds showing visible contamination were recorded and discarded.

Knowing that dry seeds lose viability over several years, 10 leftover seeds from a mix of source 2017 sites were stored in moist sphagnum moss at 4–5 °C for $3\frac{1}{2}$ months.

To evaluate the viability after long term storage at room temperature, seeds from sources 2017-1–11 stored dry at room temperature for 8 months were placed in wet sphagnum moss in Ziplock bags and kept dark at 4–5 °C for 16 weeks. Germinants were planted into SBG1 soil in 10 cm pots with up to ten per pot and placed on grow shelves lighted with cool white fluorescent light and under humidity domes for the next 2–3 weeks. Humidity domes were removed and pots were watered daily with about 3 mL per pot for the next two weeks. Pots were then watered every other day with about 4 mL per pot to allow the soil surface to dry. This watering regime was continued for the next six months until plants were transplanted into individual containers.

Author Contributions: Authors provided the following contributions: Conceptualization, G.S.P.; methodology, G.S.P.; software, G.S.P., K.B. and S.J.; investigation, M.I., P.K., S.L., A.B.C., K.B., S.J., J.M.S., W.L.R. and R.O.D.; resources, G.S.P. and J.M.C.-S.; writing—original draft preparation, M.I., P.K., S.L. and A.B.C.; writing—review and editing, G.S.P., J.M.C.-S., S.J., A.B.C. and J.M.S.; supervision, G.S.P. and J.M.C.-S.; project administration, G.S.P., K.B., S.J. and J.M.C.-S.; funding acquisition, G.S.P. and J.M.C.-S. All authors have read and agreed to the published version of the manuscript.

Funding: We thank the Georgia Institute of Technology for providing materials and supplies for Undergraduate Research in support of this project. Partial funding for the project was provided by the USDA Forest Service to Pete Bettinger, Hargreaves Distinguished Professor in Forest Management Warnell School of Forestry and Natural Resources, University of Georgia. Additional partial support was provided by National Science Foundation NSF# 1655732 to Jennifer Cruse-Sanders.

Institutional Review Board Statement: Not applicable.

Informed Consent Statement: Not applicable.

Data Availability Statement: Not applicable.

Acknowledgments: We thank Jan Midgeley, James Van Horne, Lisa Kruse and the Georgia DNR for providing seeds. The authors gratefully acknowledge the help of Radhika Admin, Michael Boyd, Lesley Hodge, Kate Mullen, William Pentecost, Bill Brumback, and Emily Slater.

Conflicts of Interest: The authors declare no conflict of interest.

References

1. Tajhtajan, A. *Diversity and Classification of Flowering Plants*; Cambridge University Press: Cambridge, UK, 1994; ISBN 10:0512420881/13:9780521420884.
2. Weakley, A.S. *Flora of the Southeastern United States—University of North Carolina Herbarium*; North Carolina Botanical Garden: Chapel Hill, NC, USA, 2020.
3. Bourg, N.A.; McShea, W.J.; Gill, D.E. Putting a Cart before the Search: Successful Habitat Prediction for a Rare Forest Herb. *Ecology* **2005**, *86*, 2793–2804. [CrossRef]
4. NatureServe. Xerophyllum Asphodeloides | NatureServe Explorer 2.0. NatureServe Explorer: An Online Encyclopedia of Life [Web Application]. Available online: https://explorer.natureserve.org/ (accessed on 24 April 2021).
5. Georgia Biodiversity Conservation Data Portal. Xerophyllum_asphodeloides (Gakrakow.Github.Io). Available online: https://georgiabiodiversity.org/natels/profile?es_id=17528 (accessed on 24 April 2021).
6. Xerophyllum Asphodeloides—ABC Homeopathy. Available online: https://abchomeopathy.com/r.php/Xero-a (accessed on 24 April 2021).
7. Hart-Freeluces, G.; Ticktin, T. Fire, Leaf Harvest, and Abiotic Factors Drive Demography of an Ecologically and Culturally Important Understory Plant. *Ecosphere* **2019**, *10*, e02813. [CrossRef]
8. Hart-Freeluces, G.M.; Ticktin, T.; Lake, F.K. Simulated Indigenous Fire Stewardship Increases the Population Growth Rate of an Understorey Herb. *J. Ecol.* **2021**, *109*, 1133–1147. [CrossRef]
9. Pritchard, H.W. Cryopreservation of Desiccation-Tolerant Seeds. In *Cryopreservation and Freeze-Drying Protocols*, 2nd ed.; Day, J.G., Stacey, G.N., Eds.; Methods in Molecular Biology™; Humana Press: Totowa, NJ, USA, 2007; pp. 185–201. [CrossRef]
10. Engelmann, F. Cryopreservation of Embryos: An Overview. In *Plant Embryo Culture. Methods in Molecular Biology (Methods and Protocols)*; Thorp, T., Yeung, E., Eds.; Humana Press: New York, NY, USA, 2000; Volume 710.
11. Engelmann, F. Use of Biotechnologies for the Conservation of Plant Biodiversity. *In Vitro Cell. Dev. Biol. Plant* **2011**, *47*, 5–16. [CrossRef]
12. Reed, B.M.; Sarasan, V.; Kane, M.; Bunn, E.; Pence, V.C. Biodiversity Conservation and Conservation Biotechnology Tools. *In Vitro Cell. Dev. Biol. Plant* **2011**, *47*, 1–4. [CrossRef]
13. Engelmann, F. Cryopreservation of Tropical Plant Germplasm. In *Current Research Progress and Application*; Japan International Research Center for Agriculture Sciences and International Plant Genetic Resources Institte: Rome, Italy, 2000; pp. 8–20.
14. Fay, M.F. Conservation of Rare and Endangered Plants Using in Vitro Methods. *In Vitro Cell. Dev. Biol. Plant* **1992**, *28*, 1–4. [CrossRef]
15. Sarasan, V.; Cripps, R.; Ramsay, M.M.; Atherton, C.; McMichen, M.; Prendergast, G.; Rowntree, J.K. Conservation In Vitro of Threatened Plants—Progress in the Past Decade. *In Vitro Cell. Dev. Biol. Plant* **2006**, *42*, 206–214. [CrossRef]
16. Jaskowiak, M.A. Reviews of Science for Science Librarians: The Conservation of Endangered Plants Using Micropropagation. *Sci. Technol. Libr.* **2014**, *33*, 58–70. [CrossRef]
17. Northcutt, C.; Davies, D.; Gagliardo, R.; Bucalo, K.; Determann, R.O.; Cruse-Sanders, J.M.; Pullman, G.S. Germination In Vitro, Micropropagation, and Cryogenic Storage for Three Rare Pitcher Plants: Sarracenia Oreophila (Kearney) Wherry (Federally Endangered), *S. leucophylla* Raf., and *S. purpurea* Spp. Venosa (Raf.) Wherry. *HortScience* **2012**, *47*, 74–80. [CrossRef]
18. Khanna, S.; Jenkins, H.; Bucalo, K.; Determann, R.O.; Cruse-Sanders, J.M.; Pullman, G.S. Effects of Seed Cryopreservation, Stratification and Scarification on Germination for Five Rare Species of Pitcher Plants. *Cryoletters* **2014**, *35*, 29–39.
19. Brumback, W.E. Propagation of Wildflowers. *Comb. Proc. Int. Plant Propagators Soc.* **1985**, *35*, 542–548.
20. Cullina, B. *Germination Requirements of Seeds*; Georgia Native Plant Society Symposium Handout; Georgia Native Plant Society, 2005; Available online: https://gnps.org/conservation/propagation-germination-requirements-of-seeds/ (accessed on 12 April 2021).

21. Bewley, J.D.; Black, M. *Seeds: Physiology of Development and Germination*; Springer Science & Business Media: Berlin/Heidelberg, Germany, 1994.
22. Brumback, W.E.; (New England Wild Flower Society, Director of Conservation, Boston, MA, USA). Personal Communication, 2014.
23. Ma, X.; Bucalo, K.; Determann, R.O.; Cruse-Sanders, J.M.; Pullman, G.S. Somatic Embryogenesis, Plant Regeneration, and Cryopreservation for Torreya Taxifolia, a Highly Endangered Coniferous Species. *In Vitro Cell. Dev. Biol. Plant* **2012**, *48*, 324–334. [CrossRef]
24. Lynch, S.; Johnston, R.K.; Determann, R.O.; Cruse-Sanders, J.M.; Pullman, G.S. Seed Cryostorage and Micropropagation of Georgia Aster, Symphyotrichum Georgianum (Alexander) Nesom: A Threatened Species from the Southeastern United States. *HortScience* **2013**, *48*, 750–755. [CrossRef]
25. Perullo, N.; Determann, R.O.; Cruse-Sanders, J.M.; Pullman, G.S. Seed Cryopreservation and Micropropagation of the Critically Endangered Species Swamp Pink (*Helonias bullata* L.). *In Vitro Cell. Dev. Biol. Plant* **2015**, *51*, 284–293. [CrossRef]
26. Pence, V.C. Cryopreservation of Seeds of Ohio Native Plants and Related Species. *Cryopreserv. Seeds Ohio Nativ. Plants Relat. Species* **1991**, *19*, 235–251.
27. Reed, B.M. Implementing Cryogenic Storage of Clonally Propagated Plants. *Cryo Lett.* **2001**, *22*, 97–104.
28. Bewley, J.D.; Black, M.; Halmer, P. *The Encyclopedia of Seeds: Science, Technology and Uses*; CABI: Wallingford, UK; Cambrige, MA, USA, 2006.
29. Baskin, C.; Baskin, J. *Seeds: Ecology, Biogeography, and Evolution of Dormancy and Germination*, 2nd ed.; Academic Press: New York, NY, USA, 2014.
30. Yamauchi, Y.; Ogawa, M.; Kuwahara, A.; Hanada, A.; Kamiya, Y.; Yamaguchi, S. Activation of Gibberellin Biosynthesis and Response Pathways by Low Temperature during Imbibition of Arabidopsis Thaliana Seeds. *Plant Cell* **2004**, *16*, 367–378. [CrossRef]
31. Okazaki, K.; Asano, Y.; Oosawa, K. Interspecific Hybrids between Lilium "Oriental" Hybrid L. "Asiatic" Hybrid Produced by Embryo Culture with Revised Media. *Breed. Sci.* **1994**, *44*, 59–64. [CrossRef]
32. Kaeppler, S.M.; Kaeppler, H.F.; Rhee, Y. Epigenetic Aspects of Somaclonal Variation in Plants. In *Plant Gene Silencing*; Matzke, M.A., Matzke, A.J.M., Eds.; Springer: Dordrecht, The Netherlands, 2000; pp. 59–68. [CrossRef]
33. Huetteman, C.A.; Preece, J.E. Thidiazuron: A Potent Cytokinin for Woody Plant Tissue Culture. *Plant Cell Tissue Organ Cult.* **1993**, *33*, 105–119. [CrossRef]
34. Arzate-Fernández, A.; Kazuo, O.; Makoto, M.; Tetsushi, Y.; Tomohide, S. In Vitro Propagation of Miyamasukashi-Yuri (Lilium Maculatum Thunb. Var. Bukosanense), an Endangered Plant Species. *Rev. Fitotec. Mex.* **2007**, *30*, 373–379.
35. Babaei, N.; Psyquay Abdullah, N.A.; Saleh, G.; Lee Abdullah, T. An Efficient In Vitro Plantlet Regeneration from Shoot Tip Cultures of Curculigo Latifolia, a Medicinal Plant. *Sci. World J.* **2014**, *2014*, e275028. [CrossRef]
36. Taylor, J.; Harrier, L.A. Beneficial Influences of Arbuscular Mycorrhizal (AM) Fungi on the Micropropagation of Woody and Fruit Trees. In *Micropropagation of Woody Trees and Fruits*; Jain, S.M., Ishii, K., Eds.; Forestry Sciences; Springer: Dordrecht, The Netherlands, 2003; pp. 129–150. [CrossRef]
37. Niemi, K.; Scagel, C.; Häggman, H. Application of Ectomycorrhizal Fungi in Vegetative Propagation of Conifers. *Plant Cell Tissue Organ Cult.* **2004**, *78*, 83–91. [CrossRef]
38. Hazarika, B.N. Use of Bio-Agents in Acclimatizing Micropropagated Plants—A Review-Indian Journals. *Agric. Rev.* **2006**, *27*, 152–156.
39. Determann, R.O.; (Atlanta Botanical Gardens, Conservatory Director, Atlanta, GA, USA). Personal Communication, 2014.
40. Smart, A.W.; Minore, D. Germination of Beargrass (Xerophyllum Tenax [Pursh] Nutt.). *Plant Propagator* **1977**, *23*, 13–15.
41. Shebitz, D.J.; Ewing, K.; Gutierrez, J. Preliminary Observations of Using Smoke-Water to Increase Low-Elevation Beargrass Xerophyllum Tenax Germination. *Nativ. Plants J.* **2009**, *10*, 13–20. [CrossRef]
42. Pelkonen, V.-P. *Biotechnological Approaches in Lily (Lilium) Production*; Faculty of Science, Department of Biology, University of Oulu: Oulu, Finland, 2005.
43. Grano, C.X. Tetrazolium Chloride to Test Loblolly Pine Seed Viability. *For. Sci.* **1958**, *4*, 50–53. [CrossRef]

Article

Successful Cryopreservation of Dormant Buds of Blackcurrant (*Ribes nigrum* L.) by Using Greenhouse-Grown Plants and In Vitro Recovery

Saija Rantala [1,2,*], Janne Kaseva [3], Anna Nukari [4], Jaana Laamanen [5], Merja Veteläinen [6], Hely Häggman [2] and Saila Karhu [7]

1. Natural Resources Institute Finland (Luke), Production Systems, Survontie 9 A, FI-40500 Jyväskylä, Finland
2. Ecology and Genetics Unit, Faculty of Science, University of Oulu, P.O. Box 3000, FI-90014 Oulu, Finland; hely.haggman@oulu.fi
3. Natural Resources Institute Finland (Luke), Natural Resources, Tietotie 4, FI-31600 Jokioinen, Finland; janne.kaseva@luke.fi
4. Natural Resources Institute Finland (Luke), Production Systems, Latokartanonkaari 9, FI-00790 Helsinki, Finland; anna.nukari@luke.fi
5. Natural Resources Institute Finland (Luke), Natural Resources, Survontie 9 A, FI-40500 Jyväskylä, Finland; jaana.laamanen@luke.fi
6. Boreal Plant Breeding Ltd., Myllytie 10, FI-31600 Jokioinen, Finland; merja.vetelainen@boreal.fi
7. Natural Resources Institute Finland (Luke), Production Systems, Itäinen Pitkäkatu 4 A, FI-20520 Turku, Finland; saila.karhu@luke.fi
* Correspondence: saija.rantala1@luke.fi

Citation: Rantala, S.; Kaseva, J.; Nukari, A.; Laamanen, J.; Veteläinen, M.; Häggman, H.; Karhu, S. Successful Cryopreservation of Dormant Buds of Blackcurrant (*Ribes nigrum* L.) by Using Greenhouse-Grown Plants and In Vitro Recovery. *Plants* 2021, 10, 1414. https://doi.org/10.3390/plants10071414

Academic Editor: Carla Benelli

Received: 1 June 2021
Accepted: 6 July 2021
Published: 10 July 2021

Publisher's Note: MDPI stays neutral with regard to jurisdictional claims in published maps and institutional affiliations.

Copyright: © 2021 by the authors. Licensee MDPI, Basel, Switzerland. This article is an open access article distributed under the terms and conditions of the Creative Commons Attribution (CC BY) license (https://creativecommons.org/licenses/by/4.0/).

Abstract: The cryopreservation of dormant buds can be a feasible method for preserving germplasm of cold-tolerant woody plants. In the present study, we evaluated the effects of pre-desiccation, thawing method, and the rehydration of bud sections on the post-cryopreservation recovery of dormant blackcurrant buds in vitro. The estimated recovery of small- and medium-sized buds was 80.1 and 62.7% respectively for desiccated buds and 67.8 and 72.3% respectively for non-desiccated buds. The pre-desiccation of bud sections enhanced the number of the shoots regenerated from vegetative buds (2.3 vs. 4.7). The estimated recovery of fast-thawed buds was better after 14-day than after 7-day rehydration (85 vs. 59%). In slowly thawed buds the difference between 14-day and 7-day rehydration was not significant (73 vs. 62%). The estimated recovery of vegetative and flower buds was 77.7 and 41.1% respectively after 7-day rehydration, and 95.2 and 43.6% respectively after a 14-day rehydration period. The rehydration of bud sections was not necessary for the in vitro recovery of non-desiccated, fast-thawed buds. Of the 23 blackcurrant cultivars cryopreserved using non-desiccated dormant buds collected from a greenhouse, the estimated recovery of 22 cultivars ranged between 42 and 90%.

Keywords: currants; cryobanking; plant genebanks; plant genetic resources; germplasm collections; liquid nitrogen; long-term preservation; safety back-ups

1. Introduction

The preservation of the genetic resources of agricultural crop plants is important for future plant breeding and food security. The germplasm of vegetatively propagated crops can be maintained as plants in the field or in greenhouses or as in vitro cultures [1,2]. The maintenance of collections solely in the field is a risk, due to the fact that diseases, pests, and adverse weather conditions can compromise the preservation of germplasm. A duplicate field collection or a safety backup by an alternative conservation method is therefore recommended by FAO [1]. Cryopreservation, i.e., the preservation of biological material at ultralow temperatures (in liquid nitrogen and/or its vapor phase at temperatures ca. −196 °C to −140 °C [3]), is a useful and cost-effective option to secure the long-term preservation of plant germplasm [4–6]. For example, in the case of clonally propagated fruit trees that are traditionally maintained in clonal orchards, the utilization of cryopreservation can greatly improve the conservation of germplasm [7]. However, to utilize cryogenic preservation, cryoprotocols suitable for the species or even genotypes in question are needed.

For the long-term preservation of cold-tolerant woody plant species, cryopreservation of dormant buds may be an applicable method [8,9]. The success of dormant bud cryopreservation is affected by the quality of the source material, the steps of the protocol, and recovery practices [8]. Cryopreservation of dormant buds was reported in the 1960s when Sakai discovered that the twigs of cold-hardy poplar (*Populus sieboldii* Mig.) and willow (*Salix koriyanagi* Kimura) pre-frozen at −30 °C were able to survive without fatal intracellular freezing even when immersed in liquid nitrogen [10]. After that, the cryopreservation of dormant buds was studied with species such as apple (*Malus domestica*) [11,12] and mulberry (*Morus bombycis* Koidz.) [13], and to date, many cryoprotocols utilizing dormant buds with different recovery practices have been developed [8,9]. In the case of *Malus*, the recovery of buds is often done by grafting [14], but the recovery of buds via in vitro can also be used as in the case of species such as *Ulmus* [15], *Diospyros kaki* Thunb. [16], *Morus* [13,17], *Betula pendula* Roth, and *Populus tremula* L × *P. tremuloides* Michx. [18]. The cryopreserved twigs of *Salix* can be recovered via direct rooting by placing thawed twigs in moist, sterilized soil and keeping them in high humidity until they have rooted [19,20]. In some temperate fruit trees, thawed twig segments can be forced, after which sprouted shoots can be excised and introduced into the tissue culture [21]. Shoot tips excised from dormant buds prior to cryopreservation may also be utilized for cryopreservation with cryoprotectants [22]. The success of dormant bud cryopreservation varies depending on the protocol and species, e.g., for *Malus* spp., recovery ranges of accessions from 16 to 100% was reported [23].

The pre-desiccation of bud sections has been proven to enhance the post-cryopreservation recovery of dormant buds [11,12,24] by decreasing cells' water content and preventing intracellular ice crystallization of cryopreserved material during cooling and thawing [25]. Therefore, many dormant bud cryoprotocols include the pre-desiccation of bud sections, e.g., in the protocol developed for *Malus* species [23], bud sections are desiccated to a moisture content of 25–30% before slow cooling to a range between −30 and −40 °C, followed by a transfer to liquid nitrogen or its vapor phase. This protocol is used or slightly modified also for some species, e.g., for *Fraxinus* [26] and *Vaccinium* [27].

However, careful monitoring is often needed for evaluating the progress of desiccation, which is usually laborious and needs time. Cryopreservation of non-desiccated dormant buds recovered via sprouting or grafting was reported for *Salix* [28] and for *Malus* [29] but the pre-desiccation of buds is usually omitted from protocols in which the recovery of buds is achieved via in vitro culture, as in the case of *Ulmus* [15] and *Betula pendula* Roth [18].

Blackcurrant *Ribes nigrum* L. is a cold-tolerant woody shrub, and it is cultivated for juicy berries both commercially and in home gardens. According to the FAOSTAT database, the production of currants (mainly blackcurrant) was 647,815 tonnes in 2019 [30]. Many old blackcurrant varieties or local strains are no longer used in commercial berry production, but they may be valuable source for future plant breeding. However, plants preserved in the open field are often exposed to many pests and diseases. Pests such as eriophyid mites, spider mites, moths, gall midges, and aphids are common in currant cultivation in Finland. Fungal diseases, pathogens causing leaf spots, powdery mildew, and rusts may also occur, but these do not always cause severe problems in blackcurrants in Finland [31]. Certain pests may also act as vectors for viruses, and several virus diseases may occur in blackcurrants [32]. Blackcurrant reversion virus (BRV), transmitted by the gall mite (*Cecidophyopsis ribis*), is the most significant virus in blackcurrants, causing disease symptoms to the leaves and flowers, proliferation, and ultimately yield losses [33,34]. Pest and disease infections can be prevented in certified plant production by maintaining pre-basic mother plants (stock material) in insect-proof greenhouse [35].

The Finnish national core collection of blackcurrant includes a total of 27 cultivars and landraces, all called cultivars in this study. The core collection was selected as part of the multinational RIBESCO project in 2007–2011 [36,37]. It is managed by the Finnish National Genetic Resources Programme for Agriculture, Forestry, and Fishery and maintained by the Natural Resources Institute Finland (Luke). A new field collection of the blackcurrant genebank was established in Kaarina, Finland between 2011 and 2016 because of the symptoms of blackcurrant reversion virus were detected in the old field collection. The renewal of field collection was conducted using plants produced via micropropagation. In the context of the renewal of new field collection, the need to create a cryopreserved backup collection was identified. Blackcurrant can be cryopreserved by using explants excised from tissue cultures [38–41] or by using dormant buds [42–44]. Cryopreservation success of blackcurrant varies depending on cultivar. According to our previous study, the estimated post-cryopreservation recovery of in vitro-derived shoot tips ranged by cultivars from 17 to 94% [41]. Previously, we also studied the cryopreservation of dormant blackcurrant buds using greenhouse and field-maintained source plants of the cultivar Mortti, and we reported the estimated post-cryopreservation recovery of buds in vitro from 66 to 86% [43]. In eleven blackcurrant cultivars, a post-cryopreservation viability of dormant buds from $58.9 \pm 1.1\%$ to $73.5 \pm 1.9\%$ in field conditions was reported [44].

The aim of the present study was to evaluate the suitability of dormant bud cryopreservation for the preservation of the blackcurrant germplasm collection using dormant buds derived from greenhouse-maintained plants. Dormant buds of cvs. Mikael, Marski, and Vilma were cryopreserved according to an experiment setup to confirm the utility of a selected cryoprotocol and to evaluate the necessity of pre-desiccation and rehydration of bud sections, and the effect of the thawing method (fast or slow) on the in vitro recovery of cryopreserved buds. Finally, the post-cryopreservation viability of non-desiccated, fast-thawed dormant buds of 24 cultivars was estimated.

2. Results
2.1. Protocol Experiments

The effect of desiccation, thawing method, and rehydration of bud sections was tested according to the experimental set-up with cvs. Mikael, Marski, and Vilma. The actual recovery percentages of cryopreserved buds varied from 20 to 80%, depending on the cultivar and treatment combination (Table 1). The estimated recovery of cryopreserved buds over all treatment combinations was 60, 83, and 67% for cvs. Mikael, Marski, and Vilma, respectively. Cryopreserved buds were mainly floral; out of 160 buds per cultivar, the numbers of vegetative buds were 21, 37, and 52 for cvs. Mikael, Marski, and Vilma, respectively. The average moisture content of non-desiccated twig samples ranged at 55–57%, 54–59%, and 53–55%, whereas the average moisture content of desiccated twig samples ranged at 22–32%, 31–34%, and 30–33% for cvs. Mikael, Marski, and Vilma, respectively. In all three cultivars, the recovery of non-cryopreserved control buds ranged from 90 to 100% for both non-desiccated buds and buds that were desiccated and thereafter rehydrated.

2.1.1. Preliminary Quality Evaluation

When the in vitro cultures were initiated from cryopreserved buds, the percentages of healthy green buds, i.e., buds without blackening or paleness of shoot tip, flower primordia, or leaves, were 71, 68, and 36 for cvs. Mikael, Marski, and Vilma, respectively. According to the results of all three cultivars, the percentages of healthy-looking green buds were higher (i) if buds were desiccated as opposed to when they were not (70 vs. 48%; $p < 0.001$), (ii) if buds were thawed slowly instead of fast thawing (76 vs. 40%; $p < 0.001$), (iii) if buds were small instead of medium-sized buds (67 vs. 54%; $p = 0.011$), and iv) if buds were rehydrated for 7 days instead of 14 days (63 vs. 54%; $p = 0.052$). The frequency of buds without visual damage did not differ between flower buds (58%) and vegetative buds (60%). However, almost all buds (28 out of 30) with their outermost leaves blackened were flower buds of cv. Marski. Leaves with no turgor (i.e., not fully rehydrated) occurred only in buds that were desiccated and then rehydrated after thawing for 7 days. Of 40 buds per cultivar desiccated and rehydrated for only 7 days, the number of not fully rehydrated buds was 18 for cv. Mikael and was 36 for cvs. Marski and Vilma.

Table 1. The number of flower buds out of thawed buds and the actual recovery percentages (%) of the cryopreserved buds of blackcurrant cvs. Marski, Mikael, and Vilma per treatment combination. In each treatment combination, 20 bud sections were cryopreserved and thawed either slowly at 2 °C or for 3 min in water bath at 38 °C and rehydrated for 7 or 14 days (d). The actual recovery percentage of buds in vitro was calculated based on the number of regenerated buds related to the number of uncontaminated ones.

Treatment Combination	Mikael		Marski		Vilma	
	Number of Floral Buds	Recovery of Thawed Buds (%)	Number of Floral Buds	Recovery of Thawed Buds (%)	Number of Floral Buds	Recovery of Thawed Buds (%)
Desiccated bud sections						
Fast thawing + rehydration 7 d	18/20	30	18/20	40	14/19 *	47
Fast thawing + rehydration 14 d	18/20	30	12/20	80	16/20	60
Slow thawing + rehydration 7 d	16/20	50	16/20	50	11/20	60
Slow thawing + rehydration 14 d	18/20	35	17/20	65	12/20	55
Non-desiccated bud sections						
Fast thawing + rehydration 7 d	18/20	50	16/20	70	18/20	25
Fast thawing + rehydration 14 d	17/20	50	17/20	80	9/20	70
Slow thawing + rehydration 7 d	18/20	40	10/20	70	13/20	55
Slow thawing + rehydration 14 d	18/20	20	17/20	60	10/13 **	46

* 1 bud rejected due to contamination; ** 7 buds rejected due to contamination; d: number of days in rehydration.

Notably, the presence of these visual damage of buds did not influence the later recovery percentage of the in vitro cultures ($p = 0.576$). Of the 244 buds that recovered in vitro, 105 (43%) had visual symptoms of damage when the in vitro cultures were initiated. Furthermore, of the 228 buds that did not recover, 136 (60%) did not have visual symptoms of damage at the time of initiation of in vitro cultures.

2.1.2. The Effect of Desiccation on the Recovery of Thawed Buds In Vitro

According to the statistical analysis of the data of the protocol experiment (n = 472 buds), the desiccation of bud sections of the cvs. Mikael, Marski, and Vilma had no significant main effect on the estimated recovery percentage of cryopreserved buds. An interaction was found between the pre-treatment of bud sections (desiccated or not) and the size of the buds ($p = 0.022$) (Figure 1a): desiccation improved the estimated recovery percentage of small buds slightly, but for medium-sized buds the effect was the opposite. However, the difference between desiccated and non-desiccated buds was not significant in either case. In the case of desiccated buds, the estimated recovery was better for small than for medium-sized buds ($p = 0.037$).

The number of proliferated shoots per regenerated bud evaluated 7 weeks after the initiation of in vitro cultures showed an interaction between the desiccation of bud sections (desiccated or not) and bud type (flower or vegetative) ($p < 0.001$). The estimated shoot number per bud was higher for desiccated vegetative buds than for non-desiccated vegetative buds (4.7 vs. 2.3; $p < 0.001$). In the case of flower buds, the difference between desiccated and non-desiccated buds was not significant (2.2 vs. 1.7; $p = 0.121$). In addition, the estimated shoot number per bud was higher for desiccated vegetative buds than for desiccated flower buds (4.7 vs. 2.2; $p < 0.001$), and evidence for the difference of means was found for non-desiccated vegetative and flower buds (2.3 vs. 1.7; $p = 0.094$).

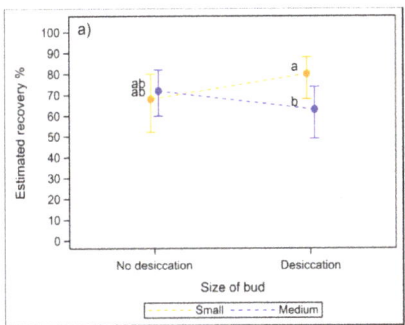 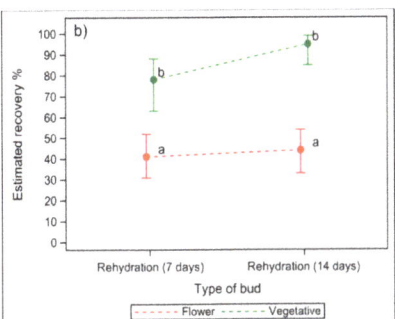

Figure 1. The estimated mean recovery of cryopreserved buds of blackcurrant cvs. Marski, Mikael, and Vilma according to (**a**) the pre-treatment of bud sections (desiccated or not) and the bud length, and (**b**) the bud type and rehydration time of bud sections. The effects of cultivar, pre-treatment, thawing method, rehydration, type of bud, size of bud, and all their 2-way and 3-way interactions were analysed using generalised linear mixed models. The length of the bars indicates the sizes of 95% confidence intervals. Letters a and b indicate significant differences ($p < 0.05$) in estimated recovery rate in vitro between treatments.

The estimated number of shoots was higher for desiccated than for non-desiccated buds for cvs. Mikael (3.3. vs. 1.7; $p < 0.001$) and Marski, (3.8 vs. 2.0; $p < 0.001$) respectively, but in cv. Vilma the difference, although parallel, was not found to be statistically significant (3.1 vs. 2.3; $p = 0.092$). The difference due to bud size was rather small: the estimated number of shoots per regenerated bud was 2.9 for small buds and 2.5 for medium-sized buds ($p = 0.051$).

2.1.3. The Effect of Thawing Speed and Rehydration on the Recovery of Buds In Vitro

According to the results of the protocol experiment, thawing speed had no main effect on the estimated recovery percentage of buds of cvs. Mikael, Marski, and Vilma. However, the interaction between thawing method (slow or fast) and the rehydration time of bud sections was close to significant ($p = 0.053$). The estimated recovery percentage of fast-thawed buds was better after 14-day than after 7-day rehydration (85 vs. 59%; $p = 0.013$), but in slowly thawed buds, there was no significant difference between 14-day rehydration and 7-day rehydration (73 vs. 62%; $p = 0.637$). Vegetative buds had a better estimated recovery than flower buds after both rehydration periods ($p < 0.001$). An interaction between the duration of rehydration treatment and bud type was found ($p = 0.031$). Within the bud type, the estimated recovery of buds did not differ between 7-day and 14-day rehydration treatment, although a longer rehydration time seemed to give some benefit to vegetative buds (Figure 1b). Moreover, an interaction was found between rehydration time and cultivar ($p = 0.027$). In all three cultivars, the estimated recovery of buds was better after 14-day than 7-day rehydration, but the difference was significant only for cv. Marski (92 vs. 68%; $p = 0.016$).

In cvs. Mikael, Marski, and Vilma, the thawing method and the duration of the rehydration treatment did not have significant main effects on the shoot number of regenerated buds. However, an interaction between the rehydration time and cultivar showed evidence of difference ($p = 0.094$), but in pairwise comparisons statistically significant differences could not be found.

2.1.4. The Necessity of Rehydration

To evaluate the necessity of rehydration for non-desiccated, fast-thawed cryopreserved buds, 84 bud sections of cv. Brödtorp were thawed for rehydration test. The average moisture content of bud sections of cv. Brödtorp was 54%, and an average bud length was 4.5 mm (ranged from 2 to 6 mm). Ten weeks after the initiation of the in vitro culture, the estimated recovery percentage did not differ significantly between rehydrated (71 (50–86)) and non-rehydrated buds (90 (70–97), $p = 0.087$). At the time of initiation of in vitro cultures, all non-rehydrated buds were scored as "healthy green", but only 7 of 42 rehydrated ones were "healthy green". However, after two weeks of in vitro culture, blackening and paleness was also observed in shoot tips excised from non-rehydrated buds. In both rehydrated and non-rehydrated treatments, the first shoots started to grow two weeks after the excision of shoot tips (Figure 2). The estimated recovery of buds was again better for vegetative buds than for flower buds (94 vs. 56%; $p = 0.003$), but because of the low number of flower buds (only 6 of 84) and non-recovered buds, the test result may not be accurate.

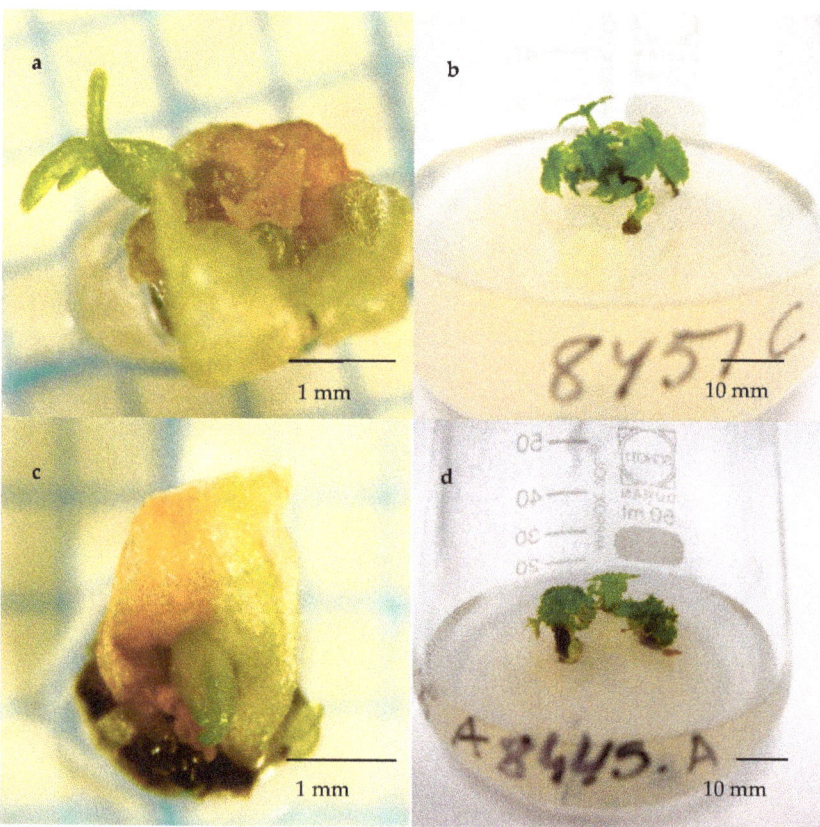

Figure 2. The recovery of cryopreserved buds of cv. Brödtorp in vitro. Top row: the propagule excised from a non-rehydrated bud (**a**) 2 weeks after initiation of in vitro culture (photo Dr. Mauritz Vestberg) and (**b**) 10 weeks after initiation. Bottom row: the propagule excised from a bud rehydrated 11 days (**c**) 2 weeks after initiation of in vitro culture (photo Dr. Mauritz Vestberg) and (**d**) 10 weeks after initiation. Bud sections of cv. Brödtorp were cryopreserved without pre-desiccation and revived from cryostorage via fast thawing. Images a and c were taken on graph paper through a stereomicroscope.

When the shoot number of the non-desiccated, fast-thawed buds of cv. Brödtorp were analysed, no significant differences in the pairwise comparisons between rehydrated and non-rehydrated treatments could be found. However, a significant interaction between bud length (2–3 mm or 4–6 mm) and rehydration (rehydrated or not) was found ($p = 0.047$). In small buds, the estimated number of shoots was higher for rehydrated buds (5.6) compared to that of the non-rehydrated ones (2.5). The estimated shoot number for medium-sized buds did not differ between rehydrated buds (3.5) and non-rehydrated buds (3.3).

2.2. Viability Testing When Cryobanking a Collection of Cultivars

The results from viability assessments of 23 cultivars cryopreserved for long-term cryopreservation are shown in Table 2. All buds were cryopreserved without pre-desiccation and recovered via fast thawing and without rehydration. The estimated recovery of buds otherwise ranged from 42 to 90%, but the estimated recovery of one exceptional cv., Jänkisjärvi, was only 9%, and the difference between cultivars was not found statistically significant ($p = 0.189$). Despite a non-significant p-value of the F-test, cv. Jänkisjärvi obviously differed statistically significantly from a few cultivars. Other statistically significant differences between the estimated recovery rate of cultivars were not found because of relatively wide confidence limits.

Table 2. The measured and estimated recovery of buds and measured and estimated number of regenerated shoots per bud of 23 blackcurrant cultivars. The recovery of buds was evaluated after ten weeks of in vitro culture. Results are based on 20 or 21 thawed buds per cultivar, but contaminated initiations were rejected from evaluations. The estimated values take into count the effect of bud type (vegetative or floral) and the size of the bud. CI, 95% confidence intervals.

Cultivar	Actual Recovery of Thawed Buds%	Estimated Recovery of Thawed Buds % (CI)	Actual Number of Regenerated Shoots Per Recovered Bud Mean (CI)	Estimated Number of Regenerated Shoots Per Recovered Bud Mean (CI)
Karila	95	90 (52–99)	8.4 (5.9–10.8)	5.5 (3.9–7.6)
Vilma *	95	90 (52–99)	3.6 (2.9–4.2)	2.7 (1.9–3.7)
Ri 289 *	90	85 (54–96)	5.9 (4.6–7.2)	4.8 (3.5–6.5)
Suvi-7	95	81 (48–95)	7.7 (5.5–9.9)	5.2 (3.8–7.2)
Hedda	75	78 (51–92)	5.1 ((3.9–6.3)	3.9 (2.8–5.5)
Venny *	65	74 (49–90)	2.6 (1.6–3.6)	2.2 (1.6–3.1)
Marski	85	72 (40–90)	4.2 (3.0–5.4)	3.0 (2.1–4.1)
Öjebyn	75	70 (42–88)	3.7 (2.2–5.2)	2.4 (1.7–3.3)
Mortti	80	68 (38–89)	5.1 (4.2–6.0)	3.9 (2.6–5.6)
Vertti *	55	68 (44–85)	3.4 (1.6–5.1)	3.0 (2.0–4.4)
Mikael	70	66 (39–85)	3.7 (3.0–4.5)	2.9 (2.0–4.1)
Ola	70	65 (38–86)	2.1 (1.6–2.6)	1.6 (1.1–2.2)
Pyhtilän Musta	85	63 (26–89)	1.7 (1.2–2.1)	1.1 (0.7–1.6)
Osmola	75	63 (35–84)	5.9 (3.7–8.1)	3.5 (2.5–5.0)
Nikkala	80	60 (29–85)	3.9 (2.3–5.5)	2.4 (1.7–3.5)
Åström	75	59 (32–82)	3.3 (2.0–4.5)	2.1 (1.5–2.9)
Kangosfors	80	59 (29–83)	4.6 (3.1–6.1)	2.9 (2.0–4.1)
Osmolan musta	75	56 (30–78)	5.0 (2.7–2.6)	3.0 (2.1–4.2)
Kuoksan Musta	80	55 (23–84)	2.0 (1.4–2.6)	1.3 (0.9–2.0)
Gerby	70	55 (29–78)	3.0 (2.1–3.9)	2.0 (1.4–2.8)
Matkakoski	65	54 (29–77)	1.5 (1.1–1.9)	1.2 (0.8–1.7)
Lepaan Musta	63	42 (16–73)	3.5 (2.5–4.5)	2.2 (1.6–3.7)
Jänkisjärvi	15	9 (3–29)	1.3 (0–2.8)	1.1 (0.6–2.3)

* Green fruited.

The estimated recovery of cryopreserved buds was again better for vegetative buds than for floral bud (83 vs. 43%; $p < 0.001$). Of all the buds thawed for viability assessments, the percentage of floral buds was 6% for young donor plants and 28% for donor plants maintained in a greenhouse for several years (i.e., pre-basic mother plants or older gene bank plants). The number of flower buds varied between cultivars, with Vertti (15) having the highest number, followed by cvs. Venny (12), Jänkisjärvi (8), Mikael (6), and Hedda (6). The length of the thawed buds varied from 1–2 mm to 7 mm. Only 3 buds out of 462 were rejected due to contamination.

The estimated number of proliferated shoots per regenerated bud varied by cultivar ($p < 0.001$) and ranged between 1.1 (Pyhtilän Musta) and 5.5 (Karila) (Table 1). The estimated number of shoots per bud was higher for vegetative buds than for flower buds (3.2 vs. 2.0; $p = 0.002$).

3. Discussion

In the present study, dormant buds of blackcurrant were cryopreserved using a two-step freezing method. The success of cryopreservation was evaluated by in vitro recovery and the shoot formation of thawed buds.

Desiccation of buds prior to cooling is considered an essential step for the successful recovery of buds in many dormant bud cryoprotocols [45]. In the present study, the results of the protocol experiment indicated that the pre-desiccation of blackcurrant bud sections was not necessary for the post-thaw recovery of buds via in vitro culture. Results from viability assessments of a genebank collection supported this conclusion: the estimated recovery of 22 blackcurrant cultivars out of 23 that were cryopreserved without desiccation had a success of more than 40%. The result is consistent with our previous study [43] with cryopreserved blackcurrant cv. Mortti, for which the estimated recovery for non-desiccated outdoor and greenhouse-collected buds in vitro was 86 and 66%, respectively. In a previous study, the recovery of the winter buds of nine blackcurrant cultivars rehydrated 7 days before plunging into liquid nitrogen was successful via in vitro but not by grafting [42]. However, recovery by budding was reported to be successful for blackcurrant cuttings desiccated with a moisture content of 28–32% at −4 °C prior to the two-step cryopreservation [44].

According to the results of our protocol experiment, the pre-desiccation of bud sections decreased the number of buds with visual damage, but blackening and paleness were also detected in the desiccated buds. It is possible that the duration and conditions of the desiccation process were not optimal for cryopreservation, and the full benefit of pre-desiccation of bud sections was therefore not realised. It might also explain why desiccation was more effective for small buds than for medium-sized buds. The bud sections of cvs. Mikael, Marski, and Vilma were desiccated for four days at 2 °C, which is quite a short desiccation time compared to that in some other studies. For example, desiccation of 3.5 cm long stem segments of apple *Malus domestica* at −4 °C to a water content of ca. 30% of fresh weight took 11 to 14 days [46], and desiccation of 7 to 10 cm long apple sections to 28–32% moisture took 4 to 6 weeks [23]. However, dormant buds of persimmon (*Diospyros kaki* Thunb.) were desiccated at room temperature for 3 h before stepwise freezing to −30 °C in five days followed preservation at −150 °C [16]. In the case of *Diospyros kaki*, the recovery of buds was successful via in vitro but not by grafting.

The effect of thawing method on the recovery of cryopreserved buds has been studied previously with both grafted and in vitro recovered buds. The in vitro recovery of dormant buds of *Diospyros kaki*, which were dehydrated at 25 °C for 3 h before slow cooling and cryopreservation, was better after thawing in the air at 25 °C for 24 h than after thawing at −1 °C or after thawing at 40 °C in a water bath for 15 min, plus holding at 25 °C for 24 h [16]. In vegetative buds of *Morus bombycis* Koidz., the survival rate of cryopreserved buds in vitro depended on both pre-freezing and thawing temperatures [13] When segments were slowly pre-frozen to −10 °C, rapid thawing at 37 °C for 5 min in water gave good survival rates, but slow thawing at 0 °C for at least 3 h in the air did not. When shoot segments were slowly frozen to −20 °C or −30 °C, the survival of meristems was almost 100%, regardless of the thawing method. However, the shoot formation percentage was about half that of survival, and for segments that were cooled to −30 °C, slow warming gave a better result [13]. In *Malus domestica*, the rapid warming by placing the tubes in a water bath at 30 °C for 3 min did not support the bud burst of grafted buds [47]. In the present study, the thawing method, either fast in a water bath or slowly in a cold room, did not have a statistically significant effect on the recovery of buds in vitro. We therefore concluded that the cryovials containing blackcurrant sections can be thawed in a water bath according to the protocol that was also used for *Betula pendula* and *Populus tremula* L × *P. tremuloides* Michx [18].

The results of cvs. Mikael, Marski, and Vilma indicated that the duration of the rehydration treatment (7 day or 14 day) was not significant for the recovery of the buds, although a 14-day rehydration seemed beneficial for fast-thawed vegetative buds. In addition, according to the results of cv. Brödtorp, the rehydration treatment was not necessary for non-desiccated buds. The rehydration of bud sections in moist cotton prior to the initiation of in vitro culture increases the risk of contaminations. The rehydration of bud sections was therefore not adopted, although it enhanced the regeneration of shoots.

In the present study, the bud type, i.e., floral or vegetative, had a strong effect on the success of recovery, with vegetative buds giving a better result than floral buds. For dormant bud cryopreservation, twigs from the previous season's growth with vegetative buds are usually recommended [9,14]. Moreover, the cold hardiness and cold acclimation state of source plants is considered the most important issue affecting the success of dormant bud cryopreservation [8]. In the present study, dormant buds collected from insect-proof greenhouse-maintained donor plants were used because of their known health status and because these plants had a lower contamination risk compared to outdoor plants in the initiation of in vitro culture [48–50]. The exchange of the vegetative material includes the great risk of disease transfer [7]. The good health status of the source plants is beneficial for the future utilization of the cryopreserved germplasm. Cryopreserved buds of certified pre-basic mother plants may easily be used even for healthy plant production as well as for replacing old field collections after years, without new pest and disease indexing. Bud material that is examined to be free of black currant reversion virus can also be utilized later without renewing the testing of the virus.

The pre-basic mother plants for certified plant production were pruned annually, and the twigs collected for cryopreservation were from the previous season's growth, but the prevalence of floral buds was still quite high compared to young plants. In blackcurrant, floral buds are also formed in young shoots after the first summer [51]. Both vegetative and floral buds of greenhouse-grown blackcurrant plants were used for cryopreservation because the type of intact bud is difficult to define.

The number of flower buds among cryopreserved buds still in a cryotank cannot be known, but the possibility that a proportion of the cryopreserved buds would be flower buds was considered in statistical testing when the post-thaw recovery of cryopreserved buds was estimated. The estimated means therefore differed from the actual measured ones, and the confidence intervals of the estimated means were quite wide.

When appropriate preservation methods are selected for germplasm preservation, the cost-effectiveness of the conservation methods is also important criteria [7]. For some fruit trees such as *Malus* spp. and *Diospyros* spp., either the cryoprotocols utilizing in vitro-derived shoot tips or dormant buds can be used [7]. Previously, a cost-benefit analysis of PVS2-vitrification and dormant bud techniques used for cryopreservation of ancient apple cultivars showed that the dormant bud method was most effective in terms of time and labor [52].

We previously reported a procedure for the cryopreservation of blackcurrant cultivars by using the excised shoot tips of in vitro cultured shoots for freezing procedures [41]. Dormant bud cryopreservation can be a time-saving method in genebanking, even if the recovery of buds is done in vitro, because the initiation and the multiplication of in vitro cultures prior to cryopreservation can be omitted. Moreover, a dormant bud protocol may be easier to implement compared to in vitro-based protocols, which may need considerable optimization before they can be applied in practice [39,53,54]. If the recovery of cryopreserved buds is possible by grafting or direct rooting, the whole cryopreservation process can be done without laboratory facilities [9,20]. However, if laboratory facilities are available, the recovery via in vitro offers an opportunity to revive cryopreserved material throughout the year and makes it possible to multiply plantlets via micropropagation [13]. In our experiments, the recovery process in vitro was shown to be highly beneficial: only a small propagule (a shoot tip in the bud) was excised for the initiation of in vitro culture, and when a new shoot started to sprout, it could be excised from the propagule, even if blackening of the basal part of the propagule occurred.

4. Materials and Methods

4.1. Plant Material

In the present study, the dormant buds of a total of 24 blackcurrant cultivars, i.e., 23 listed in Table 2 and cv. Brödtorp, were cryopreserved in 2010–2015 to reinforce the cryobanking of blackcurrant collection. In addition, two black fruited cultivars, Marski and Mikael, and one green fruited cultivar, Vilma, were cryopreserved for a protocol experiment in 2013. In cv. Brödtorp, a larger amount of bud sections compared to the other cultivars were cryopreserved and thawed to perform a rehydration test. All the cryopreserved cultivars, except cv. Hedda, were included in the Finnish national core collection of blackcurrant (*Ribes nigrum* L.). For cryopreservation, dormant buds were collected from the pre-basic mother plants maintained for certified plant production or from genebank plants produced to establish a new field germplasm collection of blackcurrant (Figure 3.). All source plants were maintained in insect-proof greenhouse at the Laukaa in Central Finland (62°19′13″ N, 25°59′36″ E), where the temperature was kept above 4 °C during the winter months. At the time of collection of the buds, the temperature in greenhouse was ranged from 4 to 7 °C, and no bud burst was detected. All source plants were propagated via micropropagation from plants heat-treated to eradicate virus infections. The pre-basic mother plants grown in tubs were tested to be free from pests and diseases regarding to the legislation demands of certified plant production, and they were pruned annually. Genebank plants were tested to be free from blackcurrant reversion virus.

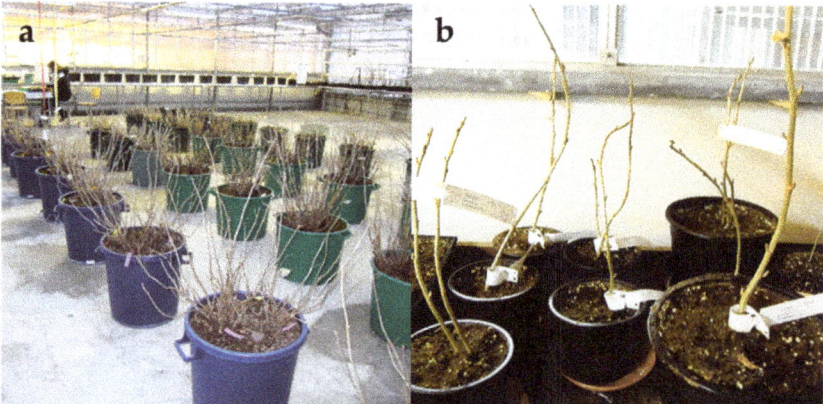

Figure 3. Source plants used for cryopreservation. (**a**) The pre-basic mother plants maintained for plant production. (**b**) Young genebank plants produced for establishment of the new field collection.

4.2. Collection and Handling of Bud Sections

4.2.1. Protocol Experiment

Dormant buds of cultivars Marski, Mikael, and Vilma were collected from greenhouse-maintained pre-basic mother plants used for certified healthy plant production in January 2013. The collected branches were cut into approximately 1.5–2 cm long stem sections, each containing a single bud in the middle of the segment (bud sections). The basal part of stems and the uppermost part of the stems were not used. The length of buds was measured with a calliper. Stem sections containing 3–5 mm long buds were selected for the experiment because they were the most abundant. All bud sections were packed in plastic bags and stored in a cold room at 2 °C for four days.

For the protocol experiment, the bud sections of cvs. Mikael, Marski, and Vilma were cryopreserved and thawed according to the experimental design shown in Figure 4. The bud sections were divided into two groups by cultivar, i.e., those to be cryopreserved after desiccation, and those to be cryopreserved without desiccation. Eighty non-desiccated bud sections per cultivar were sealed in cryotubes, i.e., two per one 1.8 mL cryotube (Sarstedt) and placed in cryoboxes (Sarstedt) which were kept in the cold room overnight and cryopreserved the next day. For desiccation, eighty bud sections per cultivar were spread on plastic containers and held unsealed at 2 °C for four days (Figure 5a). The bud sections were then packed in plastic bags for two days before they were sealed in cryotubes and cryopreserved by the same protocol as the non-desiccated bud sections.

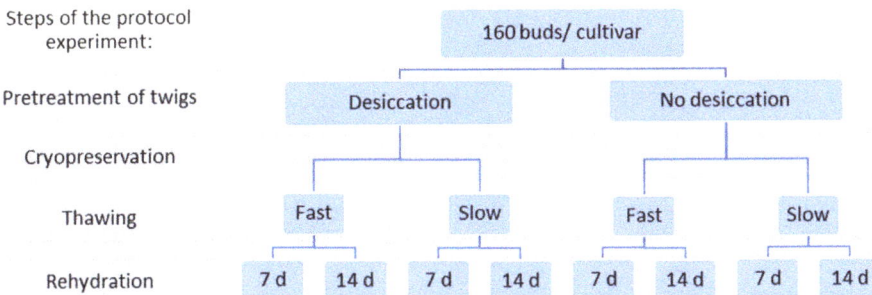

Figure 4. The experimental set-up of the protocol experiment for blackcurrant cvs. Marski, Mikael, and Vilma. Of 160 bud sections of each cultivar, 80 bud section were dehydrated before cryopreservation, and 80 bud sections were cryopreserved without dehydration. Cryopreserved bud sections were thawed either slowly at 2 °C or fast in a water bath at 38 °C, and thereafter rehydrated for 7 days or 14 days. Each treatment combination included 20 bud sections.

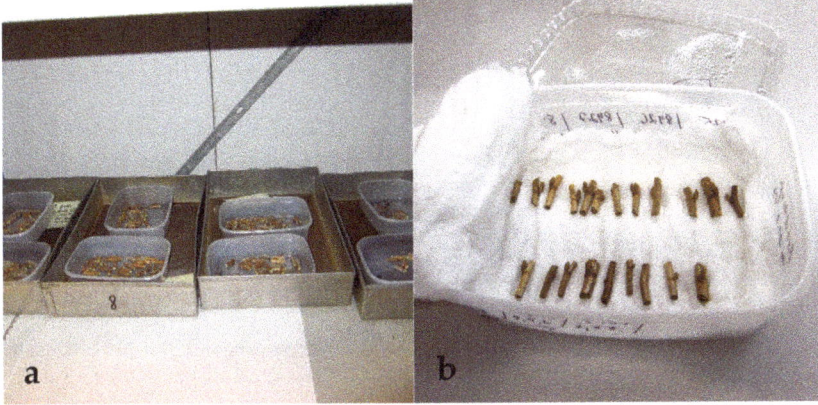

Figure 5. (**a**) Dehydration of bud sections in open freezer containers in cold room at 2 °C. (**b**) Rehydration of bud sections in plastic freezing container in moist cotton wool.

To evaluate the moisture content of bud sections prior to cryopreservation, samples of non-desiccated and desiccated bud sections from each pre-basic mother plant were weighed, oven-dried at 82–84 °C for one day and then reweighed. The moisture content of bud sections was determined using the formula (fresh weight − dry weight)/fresh weight × 100 [55]. To evaluate the progress of desiccation, samples of desiccated bud sections were weighed daily, and their moisture content was determined [55].

To test the in vitro culture success of cvs. Marski, Mikael, and Vilma, an additional 10 non-desiccated and 10 desiccated non-cryopreserved buds per cultivar were initiated. The non-desiccated control buds were initiated five days after the buds were collected and then kept in a cold room at 2 °C. Desiccated buds were initiated after 15 days of rehydration. In vitro culture of these control buds was conducted using the same media as described above for cryopreserved buds.

To test the effect of cooling, thawing, and rehydration on the recovery of the non-cryopreserved buds, an additional 24 non-desiccated and 24 desiccated bud sections of cv. Mikael were cooled to −38 °C and thereafter thawed either fast in a water bath or slowly in a cold room. Thawed bud sections (6 per treatment combination) were rehydrated for two weeks or recovered without rehydration.

4.2.2. Cryobanking of Dormant Buds

To perform the cryobanking of blackcurrant collection, dormant buds were collected from a cool greenhouse at January or at the beginning of February, except for the cultivar Ri 289, which was collected and cryopreserved in mid-December. For long-term cryopreservation, only non-desiccated bud sections were cryopreserved. Measured from the sample bud sections of the source plants, the moisture content of bud sections used ranged between 50 and 59%. After cutting the branches, the bud sections were sealed in cryotubes (1 to 3 bud sections per cryotube) and kept at 2 to 4 °C overnight.

4.3. Cooling and Cryopreservation of Bud Sections

Pre-cooling and cryopreservation of bud sections was conducted according to the protocol developed for dormant buds of silver birch and aspen [18]. Cryoboxes were placed in the chamber of the programmable freezer (Kryo 10–16 series II with programming unit Kryo 10–22 or Kryo 560-16 with MVR controller, Planer PLC, Sunbury-On-Thames, UK) without lids, and gaps of about 2 cm were left between the boxes using wooden pins. The cryotubes were cooled at 0.17 °C min^{-1} from 0 °C to −38 °C and held at −38 °C for about 30 to 50 min. The cryoboxes were then immersed in liquid nitrogen one by one until the bubbling of liquid nitrogen settled. During immersion, the cryotubes were held in place with a grid. After immersion, the cryoboxes were stored at the gas phase of liquid nitrogen in a cryotank (MVE 1520 Eterne).

4.4. Thawing and Rehydration of Bud Sections

In the protocol experiment, bud sections of cvs. Mikael, Marski, and Vilma were thawed either quickly in a water bath or slowly in a cold room. For slow thawing, the bud sections were transferred from the cryotank to a cold room at 2 °C overnight. For fast thawing, the cryovials were placed in a water bath at 37 °C for 3 min. The thawed bud sections were rehydrated in plastic freezing containers inside moist cotton wool at 2–4 °C for 7 or 14 days. Two different rehydration times were used because the results from non-cryopreserved buds of cv. Mikael cooled to −38 °C indicated that the recovery of desiccated buds was very poor without rehydration (data not shown).

To assess the post-cryopreservation viability of 23 cultivars cryopreserved for long term preservation, approximately twenty bud sections per cultivar were thawed after 2 to 42 months of cryostorage. The buds were thawed in a water bath at 37 °C for 3 min and recovered without rehydration. However, a total of 84 bud sections of cv. Brödtorp were thawed to test the necessity of rehydration for the post-cryopreservation recovery of non-desiccated buds. All the buds of cv. Brödtorp were fast-thawed in a water bath, but half the bud sections (42) were rehydrated for 11 days in moist cotton at ca. 4 °C (Figure 5b) before initiation for in vitro cultures, and the remaining 42 buds were initiated without rehydration.

4.5. Recovery of Buds In Vitro

The bud sections cryopreserved for the protocol experiment were thawed and recovered in vitro by cultivar so that buds rehydrated for 7 days and 14 days were cultured on the same schedule. The buds of cv. Mikael were thawed after 6 to 7 weeks of cryostorage, the buds of cv. Marski after 16 to 17 weeks, and the buds of cv. Vilma after 31 to 32 weeks of cryostorage. To initiate the in vitro cultures, the rehydrated bud sections were sterilized with 70% ethanol for ca. 20 s, followed by dipping in pure ethanol. The length of the bud was measured on graph paper under the stereomicroscope. The scales and leaves of the bud were removed, and the shoot tip of buds with two- or three-leaf primordia was dissected. The bud type (floral or vegetative), the turgor of primordial leaves (rehydrated or wizened), and the colour of the bud (entirely healthy green or with visible blackening or paleness of leaves, floral primordia, or shoot tip of the bud) were observed and recorded.

The excised propagules were placed in culture tubes containing a WPM [56] culture medium, supplemented as described in Rantala et al. [43]. The culture tubes were transferred to a growth room at 22 °C and kept covered with gauze or foil for 3 days before they were fully exposed to a 16/8 h light/dark photoperiod under two fluorescent tubes (Osram L 36 W/830 Lumilux warm white, Osram, Munich, Germany), with an average photosynthetically activated radiation of 40 to 60 µmol m^{-2} s^{-1}. After two weeks, the explants were transferred into Erlenmeyer bottles (25 mL) using G basal medium [57] as described in Rantala et al. [43]. The explants were transferred to a fresh medium after two weeks, and the recovery of the buds was evaluated 7 weeks after the initiation of the in vitro culture (Figure 6). Buds that had produced at least one viable shoot were defined as recovered, and the number and quality of shoots produced by a bud was recorded. In cv. Vilma, a total of 8 buds was rejected from the study because of contaminations, but in the case of cvs. Marski and Mikael, no contaminations were detected.

Figure 6. Microplantlets of cv. Vilma after seven weeks in vitro culture in the protocol experiment. (**a**) Shoots regenerated from non-desiccated and non-cryopreserved control buds. (**b**) Shoots regenerated from buds cryopreserved without desiccation and rehydrated 14 day after fast thawing.

The buds that were thawed for viability assessments were cultured in vitro using the same culture medias as in the protocol experiment, but the recovery of buds was evaluated 10 weeks after the initiation of cultures.

4.6. Statistical Analyses

In the protocol experiment, the recovery percentage of healthy green buds in different categories of treatments or characteristics was studied using contingency tables. The tested effects were pre-treatment (desiccation, no desiccation), thawing method (fast, slow), rehydration (7 days, 14 days), type of bud (flower, vegetative), length of bud (±3 mm, 4–5 mm), and cultivar (Mikael, Marski, Vilma). Fisher's exact test was used for dichotomous variables, and the Cochran–Mantel–Haenszel (CMH) test for the comparison of the cultivars [58].

The estimated recovery percentages of buds and the number of shoots per regenerated bud after 7 weeks were analysed using the generalised linear mixed model (GLMM). The effects of the cultivar, pre-treatment, thawing method, rehydration, type of bud, size of bud, and all their 2-way and 3-way interactions were used as fixed effects. Statistically non-significant effects were omitted from the final model, and all significant results were reported. Bud sections from the same pre-basic mother plant tub were used as random effects to account for the sampling structure. Binary distribution with logit link was used to analyse the recovery rate.

GLMMs with the assumptions of binary and lognormal distribution were used for the estimated recovery percentage and for the number of regenerated shoots, respectively, for the rehydration experiment with cv. Brödtorp. Time length of rehydration (7 days, 14 days) and type of bud (flower, vegetative) were used as fixed effects. The interaction of these effects was not included in the previous model because there were only six flower buds, and overall, only eight buds did not survive. In terms of the number of regenerated shoots, the length of bud (2-3 mm or 4-6 mm) and its interaction with rehydration—but not with the type of bud—were included in the model. In both cases, samples from the same cryovial were used as random effects to account for their possible correlation.

In the viability assessments of cryopreserved cultivars, GLMMs were also used to estimate the recovery of the buds and the number of shoots per regenerated bud. The effects of type of bud (flower, vegetative) and size of bud (small, medium) were added to the models as fixed effects to standardise the comparison of cultivars. Binomial distribution with a logit link and lognormal distribution with an identity link were used to analyse the recovery rate and the number of shoots respectively. Cryovials from the same cryopreserved set were used as random effects.

The estimated means were transformed back to the means of the original scale in all models, but median estimates were used in the case of lognormal distribution. Restricted maximum likelihood (REML) or residual pseudo likelihood (RSPL) estimation methods were used, and the degrees of freedom were calculated using the Kenward–Roger method [59]. Tukey's method was used for a pairwise comparisons of means [60], with a significance level of $\alpha = 0.05$. The analyses were performed using the GLIMMIX procedure of the SAS Enterprise Guide 7.15 (SAS Institute Inc., Cary, NC, USA).

5. Conclusions

According to our study, the cryopreservation of non-desiccated dormant buds is an applicable method for the long-term preservation of blackcurrant cultivars. Greenhouse-maintained blackcurrant plants are feasible for bud material in their dormant state. The best results were obtained with vegetative buds. The use of young plants should therefore be preferred, or if older plants are used, cultivation practices that keep the plants in their vegetative state should be used before the start of cryopreservation. Recovery via in vitro culture was useful for the regeneration of cryopreserved buds, and visual symptoms of post-cryo damage detected in buds when in vitro cultures were initiated did not predict that the recovery of the buds might fail. The advantage of the introduced protocol for germplasm preservation is that the cryopreservation process takes only a few days. However, in vitro facilities and a programmable freezer, in addition to the cryopreservation devices, are needed. A methodology to distinguish flower buds from vegetative buds in their early stage might further improve the success of the cryopreservation protocol. In addition, variability in the response of different cultivars in recovering and producing shoots in vitro after cryopreservation should always be considered.

Author Contributions: Conceptualisation S.R., A.N. and J.L.; design of the research, S.R.; statistical analyses, J.K.; writing—original draft preparation, S.R.; writing—review and editing, S.R., S.K., H.H., A.N., J.L., M.V. and J.K.; visualisation, J.K. and S.R.; J.L. was responsible for the plant health testing of the stock plants. All authors have read and agreed to the published version of the manuscript.

Funding: This research was partly funded by the European Commission, Directorate-General for Agriculture and Rural Development, under Council Regulation (EC) No. 870/2004 through Action 071 AGRI GEN RES 870/2004 (RIBESCO), by the Ministry of Agriculture and Forestry, and by the foundations of Maiju and Yrjö Rikalan Puutarhasäätiö and Oiva Kuusisto Säätiö.

Institutional Review Board Statement: Not applicable.

Informed Consent Statement: Not applicable.

Data Availability Statement: The data presented in this study will be retained according to the policy of the Natural Resources Institute Finland (Luke) and it will available on request from the corresponding author.

Acknowledgments: The authors are grateful to Hannu Tiainen, Virpi Tiainen, Virpi Lahtonen, Riitta Toivakka, and Satu-Marja Virtanen for technical support in laboratory and greenhouse work. Our sincere appreciation is also expressed to Marjatta Uosukainen for advice during the various stages of the practical work, to Mauritz Vestberg who took some of the photos and to all persons who helped to process plant material for cryopreservation.

Conflicts of Interest: The authors declare no conflict of interest. The funders had no role in the design of the study; in the collection, analyses, or interpretation of data; in the writing of the manuscript; or in the decision to publish the results.

References

1. FAO. *Genebank Standards for Plant Genetic Resources for Food and Agriculture*; FAO: Rome, Italy, 2013.
2. Panis, B.; Nagel, M.; Van den Houwe, I. Challenges and prospects for the conservation of crop Genetic resources in field genebanks, in in vitro collections and/or in liquid nitrogen. *Plants* **2020**, *9*, 1634. [CrossRef] [PubMed]
3. Benson, E.E. Cryopreservation of Phytodiversity: A Critical Appraisal of Theory & Practice. *Crit. Rev. Plant Sci.* **2008**, *27*, 141–219.
4. Forsline, P.L. Procedures for collection, conservation, evaluation and documentation of Malus germplasm. *Acta Hort.* **2000**, *522*, 223–234. [CrossRef]
5. Popova, E.; Shukla, M.; Kim, H.H.; Saxena, P.K. Plant Cryopreservation for Biotechnology and Breeding. In *Advances in Plant Breeding Strategies: Breeding, Biotechnology and Molecular Tools*; Al-Khayri, J., Jain, S., Johnson, D., Eds.; Springer: Cham, Switzerland, 2015; pp. 66–93. [CrossRef]
6. Pence, V.C.; Ballesteros, D.; Walters, C.; Reed, B.M.; Philpott, M.; Kingsley, W.; Dixon, W.; Pritchard, H.W.; Culley, T.M.; Vanhove, A.-C. Cryobiotechnologies: Tools for expanding long-term ex situ conservation to all plant species. *Biol. Conserv.* **2020**, *250*, 108736. [CrossRef]
7. Benelli, C.; De Carlo, A.; Engelman, F. Recent advances in the cryopreservation of shoot-derived germplasm of economically important fruit trees of *Actinidia, Diospyros, Malus, Olea, Prunus, Pyrus* and *Vitis*. *Biotechhnol. Adv.* **2013**, *31*, 175–185. [CrossRef]
8. Towill, L.E.; Ellis, D.D. Cryopreservation of Dormant Buds. In *Plant Cryopreservation: A Practical Guide*; Reed, B.M., Ed.; Springer: New York, NY, USA, 2008; pp. 421–441.
9. Tanner, J.D.; Chen, K.Y.-C.; Bonnart, R.M.; Minas, I.S.; Volk, G.M. Considerations for large-scale implementation of dormant budwood cryopreservation. *Plant Cell Tissue Organ Cult.* **2020**. [CrossRef]
10. Sakai, A. Survival of the twig of woody plants at −196 °C. *Nature* **1960**, *185*, 392–394. [CrossRef]
11. Tyler, N.J.; Stushnoff, C. The effects of prefreezing and controlled dehydration on cryopreservation of dormant vegetative apple buds. *Can. J. Plant Sci.* **1988**, *68*, 1163–1167. [CrossRef]
12. Tyler, N.; Stushnoff, C. Dehydration of dormant apple buds at different stages of cold acclimation to induce cryopreservability in different cultivars. *Can. J. Plant Sci.* **1988**, *68*, 1169–1176. [CrossRef]
13. Yakuwa, H.; Oka, S. Plant regeneration through meristem culture from vegetative buds of mulberry (Morus bombycis Koidz.) stored in liquid nitrogen. *Ann. Bot.* **1988**, *62*, 79–82. [CrossRef]
14. Jenderek, M.M.; Tanner, J.D.; Chao, C.T.; Blackburn, H. How applicable are dormant buds in cryopreservation of horticultural woody plant crops? The Malus case. In *Proceedings of the III International Symposium on Plant Cryopreservation*; International Society for Horticultural Science: Leuven, Belgium, 2019; pp. 317–322.
15. Harvengt, L.; Meier-Dinkel, A.; Dumas, E.; Collin, E. Establishment of a a cryopreserved gene bank of European elms. *Can. J. For. Res.* **2004**, *34*, 43–55. [CrossRef]
16. Matsumoto, T.; Niino, T.; Shirata, K.; Kurahashi, T.; Matsumoto, S.; Maki, S.; Itamura, H. Long term conservation of Diospyros germplasm using dormant buds by a prefreezing method. *Plant Biotechnol.* **2004**, *21*, 229–232. [CrossRef]
17. Atmakuri, A.R.; Chaudhury, R.; Malik, S.K.; Kumar, D.; Ramachandran, R.; Qadri, S.M.H. Mulberry biodiversity conservation through cryopreservation In vitro Cell. *Dev. Biol. Plant* **2009**, *45*, 639–649. [CrossRef]
18. Ryynänen, L.; Jokipii, S.; Häggman, H. Controlled rate cooling of silver birch and aspen dormant buds. In *Plant Cryopreservation: A Practical Guide*; Reed, B.M., Ed.; Springer: New York, NY, USA, 2008; pp. 432–435.
19. Towill, L.; Volk, G.; Waddel, J.; Bonnart, R.; Widrlechner, M. *Cryopreservation of Willow (Salix) Dormant Buds in Plant Cryopreservation: A Practical Guide*; Reed, B.M., Ed.; Springer: New York, NY, USA, 2008; pp. 436–437.
20. Jenderek, M.M.; Ambruzs, B.D.; Holman, G.E.; Carstens, J.D.; Ellis, D.D. Salix dormant bud cryotolerance varies by taxon, harvest year, and stemsegment length. *Crop Sci.* **2020**, *60*, 1965–1973. [CrossRef]
21. Tanner, J.D.; Minas, I.S.; Chen, K.Y.; Jenderek, M.M.; Wallner, S.J. Antimicrobial forcing solution improves recovery of cryopreserved temperate fruit tree dormant buds. *Cryobiology* **2020**, *92*, 241–247. [CrossRef] [PubMed]

22. Matsumoto, T.; Yamamoto, S.; Fukui, K.; Rafique, T.; Engelman, F.; Niino, T. Cryopreservation of persimmon shoot tips from dormant buds using the D cryo-plate technique. *Hortic. J.* **2015**, *84*, 106–110. [CrossRef]
23. Forsline, P.L.; Towill, L.E.; Waddel, J.W.; Stushnoff, C.; Lamboy, W.F.; McFerson, J.R. Recovery and longevity of cryopreserved dormant apple buds. *J. Am. Soc. Hort. Sci.* **1998**, *123*, 365–370. [CrossRef]
24. Volk, G.M.; Waddell, J.; Bonnart, R.; Towill, L.; Ellis, D.; Luffman, M. High viability of dormant Malus buds after 10 years of storage in liquid nitrogen vapour. *CryoLetters* **2008**, *29*, 89–94.
25. Benson, E.E. Cryopreservation Theory. In *Plant Cryopreservation A Practical Guide*; Reed, B.M., Ed.; Springer: New York, NY, USA, 2008; pp. 15–32.
26. Volk, G.M.; Bonnart, J.; Waddel, J.; Widrlechner, M.P. Cryopreservation of dormant buds from diverse Fraxinus species. *CryoLetters* **2009**, *30*, 262–267.
27. Jenderek, M.M.; Tanner, J.; Ambruzs, B.D.; West, M.; Postman, J.D.; Hummer, K.E. Twig pre-harvest temperature significantly influences effective cryopreservation of Vaccinium dormant buds. *Cryobiology* **2017**, *74*, 154–159. [CrossRef]
28. Towill, L.E.; Widrlechner, M. Cryopreservation of Salix species using sections from winter vegetative scions. *CryoLetters* **2004**, *27*, 71–80.
29. Towill, L.E.; Bonnart, R. Cryopreservation of apple using non-desiccated sections from winter-collected scions. *CryoLetters* **2005**, *26*, 323–332. [PubMed]
30. Food and Agriculture Organization of the United Nations. *FAOSTAT Database*; FAO: Rome, Italy, 2021. Available online: http://www.fao.org/faostat/en/#home (accessed on 15 June 2021).
31. Tuovinen, T.; Parikka, P.; Lemmetty, A. Plant protection in currant production in Finland. *Acta Hortic.* **2008**, *777*, 333–338. [CrossRef]
32. Jones, A.T. Important virus diseases of Ribes, their diagnosis, detection and control. *Acta Hortic.* **2002**, *585*, 279–285. [CrossRef]
33. Jones, A.T. Black currant reversion disease—The probable causal agent, eriophyid mite vectors, epidemiology and prospects for control. *Virus Res.* **2000**, *71*, 71–84. [CrossRef]
34. Lemmetty-Kaukoranta, A. Isolation and Identification of the Causal Agent of Black Currant Reversion Disease. Ph.D. Thesis, University of Turku, Turku, Finland, 30 March 2001; ISBN 951-29-1883-8.
35. Rajamäki, M.-L.; Lemmetty, A.; Laamanen, J.; Roininen, E.; Vishwakarma, A.; Streng, J.; Latvala, S.; Valkonen, J.P.T. Small-RNA analysis of pre-basic mother plants and conserved accessions of plant genetic resources for the presence of viruses. *PLoS ONE* **2019**, *14*, e0220621. [CrossRef]
36. Antonius, K.; Karhu, S.; Kaldmäe, H.; Lacis, G.; Rugenius, R.; Baniulis, D.; Sasnauskas, A.; Schulte, E.; Kuras, A.; Korbin, M.; et al. Development of the Northern European Ribes core collection based on a microsatellite (SSR) marker diversity analysis. *Plant Genet. Resour. Charact. Util.* **2012**, *10*, 70–73. [CrossRef]
37. Karhu, S.; Antonius, K.; Rantala, S.; Pluta, S.; Ryliskis, D.; Schulte, E.; Toldam-Andersen, T.B.; Kaldmäe, H.; Rumpunen, K.; Sasnauskas, A.; et al. A multinational approach for conserving the European genetic resources of currants and gooseberry. In *Proceedings of the XXVIII International Horticultural Congress on Science and Horticulture for People (IHC2010): International Symposium on Berries: From Genomics to Sustainable Production, Quality and Health*; Mezzetti, B., Brás de Oliveira, P., Eds.; International Society for Horticultural Science: Leuven, Belgium, 2012; Volume 926, pp. 27–32.
38. Benson, E.E.; Reed, B.M.; Brennan, R.M.; Clacher, K.A.; Ross, D.A. Use of thermal analysis in the evaluation of cryopreservation protocols for Ribes nigrum L. germplasm. *CryoLetters* **1996**, *17*, 347–362.
39. Reed, B.M.; Dumet, D.; Denoma, J.M.; Benson, E.E. Validation of cryopreservation protocols for plant germplasm conservation: A pilot study using Ribes L. *Biodivers. Conserv.* **2001**, *10*, 939–949. [CrossRef]
40. Reed, B.M.; Schumacher, L.; Dumet, D.; Benson, E.E. Evaluation of a modified encapsulation-dehydration procedure incorporating sucrose pretreatments for the cryopreservation of Ribes germplasm. *In Vitro Cell. Dev. Biol. Plant* **2005**, *41*, 431–436. [CrossRef]
41. Rantala, S.; Kaseva, J.; Nukari, A.; Laamanen, J.; Karhu, S.; Veteläinen, M.; Häggman, H. Droplet vitrification technique for cryopreservation of a large diversity of blackcurrant (*Ribes nigrum* L.) cultivars. *Plant Cell Tissue Organ Cult. PCTOC* **2021**, *144*, 79–90. [CrossRef]
42. Green, J.; Grout, B. Direct cryopreservation of winter buds of nine cultivars of blackcurrant (*Ribes nigrum* L.). *CryoLetters* **2010**, *31*, 341–346.
43. Rantala, S.; Kaseva, J.; Karhu, S.; Veteläinen, M.; Uosukainen, M.; Häggman, H. Cryopreservation of *Ribes nigrum* (L.) dormant buds: Recovery via in vitro culture to the field. *Plant Cell Tissue Organ Cult.* **2019**, *138*, 109–119. [CrossRef]
44. Verzhuk, V.; Pavlov, A.; Novikova, L.; Filipenko, G. Viability of Red currant (*Ribes rubrum* L.) and Black currant (*Ribes nigrum* L.) Cuttings in Field Conditions after Cryopreservation in Vapors of Liquid Nitrogen. *Agriculture* **2020**, *10*, 476. [CrossRef]
45. Stushnoff, C. Cryophysiology of woody plant dormant buds. In *Proceedings of the II International Symposium on Plant Cryopreservation*; Reed, B.M., Ed.; International Society for Horticultural Science: Leuven, Belgium, 2014; Volume 1039, pp. 63–72.
46. Vogiatzi, C.; Grout, B.W.W.; Wetten, A.; Toldam-Andersen, B.T. Cryopreservation of winter-dormant apple buds: II tissue water status after desiccation at −4 °C and before further cooling. *CryoLetters* **2011**, *32*, 367–376. [PubMed]
47. Vogiatzi, C.; Grout, B.W.W.; Wetten, A.; Orididge, M.; Clausen, S.K. Cryopreservation of winter-dormant apple buds IV: Critical temperature variation that can compromise survival. *CryoLetters* **2018**, *39*, 245–250. [PubMed]
48. Niedz, R.P.; Bausher, M.G. Control of in vitro contamination of explants from greenhouse- and field-grown trees. *In Vitro Cell. Dev. Biol. Plant* **2002**, *38*, 468–471. [CrossRef]

49. Dziedzic, E.; Jagla, J. Micropropagation of Rubus and Ribes spp. In *Protocols for Micropropagation of Selected Economically-Important Horticultural Plants*; Lambardi, M., Ed.; Springer Science + Business Media: New York, NY, USA, 2013; Methods in molecular Biology; Volume 11013, pp. 149–160.
50. Aronen, T.; Ryynänen, L. Cryopreservation of dormant in vivo-buds of hybrid aspen: Timing as critical factor. *CryoLetters* **2014**, *35*, 385–394. [CrossRef]
51. Nasr, T.A.A.; Warein, P.F. Studies on flower initiation in black currant 1. Some internal factors affecting flowering. *J. Hortic. Sci.* **1961**, *36*, 1–10. [CrossRef]
52. Lambardi, M.; Benelli, C.; De Carlo, A.; Ozudogru, E.A.; Previati, A.; Ellis, D. Cryopreservation of ancient apple cultivars of Vento: A comparison between PVS2-vitrification and dormant-bud techniques. In *Proceedings of the First International Symposium on Cryopreservation in Horticultural Species*; Panis, P., Lynch, P., Eds.; ISHS: Leuven, Belgium, 2011; pp. 191–198.
53. Reed, B.M.; Kovalchuk, I.; Kushnarenko, S.; Meier-Dinkel, A.; Schoenweiss, K.; Pluta, S.; Straczynska, K.; Benson, E.E. Evaluation of critical points in technology transfer of cryopreservation protocols to international plant conservation laboratories. *CryoLetters* **2004**, *25*, 341–352.
54. Bettoni, C.J.; Bonnart, R.; Volk, G.M. Challenges in implementing plant shoot tip cryopreservation technologies. *Plant Cell Tissue Organ Cult. PCTOC* **2021**, *144*, 21–34. [CrossRef]
55. Towill, L.E. Cryopreservation of apple (Malus domestica) dormant Buds. In *Plant Cryopreservation: A Practical Guide*; Reed, B.M., Ed.; Springer: New York, NY, USA, 2008; pp. 427–429.
56. Lloyd, G.; McCown, B. Commercially-feasible micropropagation of mountain laurel, Kalmia latifolia, by use of shoot-tip culture. *Proc. Int. Plant Propag. Soc.* **1980**, *30*, 421–427.
57. Uosukainen, M. Rooting and weaning of apple rootstock YP. *Agronomie* **1992**, *12*, 803–806. [CrossRef]
58. Agresti, A. *Categorical Data Analysis*; John Wiley & Sons, Inc.: Hoboken, NJ, USA, 2002; pp. 231–232, ISBN 0-471-36093-7.
59. Kenward, M.G.; Roger, J.H. An improved approximation to the precision of fixed effects from restricted maximum likelihood. *Comput. Stat. Data Anal.* **2009**, *53*, 2583–2595. [CrossRef]
60. Westfall, P.; Tobias, R.D.; Wolfinger, R.D. *Multiple Comparisons and Multiple Tests Using SAS*; SAS Publishing: Cary, NC, USA, 2011.

Communication

Non-Uniform Distribution of Cryoprotecting Agents in Rice Culture Cells Measured by CARS Microscopy

Fionna M. D. Samuels [1,*], Dominik G. Stich [2], Remi Bonnart [3], Gayle M. Volk [3] and Nancy E. Levinger [1,4,*]

1. Department of Chemistry, Colorado State University, Fort Collins, CO 80523, USA
2. Advanced Light Microscopy Core, NeuroTechnology Center, University of Colorado School of Medicine, Anschutz Medical Campus, Aurora, CO 80045, USA; DOMINIK.STICH@CUANSCHUTZ.EDU
3. USDA-ARS National Laboratory for Genetic Resources Preservation, 1111 S. Mason St., Fort Collins, CO 80521, USA; remi.bonnart@usda.gov (R.B.); gayle.volk@usda.gov (G.M.V.)
4. Department of Electrical and Computer Engineering, Colorado State University, Fort Collins, CO 80523, USA
* Correspondence: fionna.samuels@colostate.edu (F.M.D.S.); nancy.levinger@colostate.edu (N.E.L.)

Abstract: Cryoprotectants allow cells to be frozen in liquid nitrogen and cryopreserved for years by minimizing the damage that occurs in cooling and warming processes. Unfortunately, how the specific cryoprotectants keep the cells viable through the cryopreservation process is not entirely evident. This contributes to the arduous process of optimizing cryoprotectant formulations for each new cell line or species that is conserved. Coherent anti-Stokes Raman scattering microscopy facilitates the visualization of deuterated cryoprotectants within living cells. Using this technique, we directly imaged the location of fully deuterated dimethyl sulfoxide (d_6-DMSO), the deuterated form of a commonly used cryoprotectant, DMSO, within rice suspension cells. This work showed that d_6-DMSO does not uniformly distribute throughout the cells, rather it enters the cell and sequesters within organelles, changing our understanding of how DMSO concentration varies within the cellular compartments. Variations in cryoprotectant concentration within different cells and tissues will likely lead to differing protection from liquid nitrogen exposure. Expanding this work to include different cryoprotectants and mixtures of cryoprotectants is vital to create a robust understanding of how the distributions of these molecules change when different cryoprotectants are used.

Keywords: cryopreservation; cryoprotectant distribution; raman microscopy

1. Introduction

Preserving cells and tissues for later use is vital in fields from human in vitro fertilization and organ transfers to the preservation of agricultural crops or animals and endangered plant and animal species [1–3]. The importance of plant conservation is recognized by The Convention on Biological Diversity (2002), which had targets focused on conserving 75% of threatened plant species ex situ as well as the conservation of 70% of the genetic diversity of crops, wild relatives, and other economically relevant species by 2020 [4]. In the review of the program twelve years later it is clear that these goals were overly ambitious [5]. Falling short of the targets set in 2002 can be attributed to the recalcitrance of some species to traditional seed banking, such as those that have non-orthodox seeds or that are clonally maintained for which collections must be grown as plants in the field, greenhouse, or in vitro for preservation. These collections are particularly susceptible to environmental threats as well as pests, pathogens, and diseases. Having a secure backup of these collections is critical to their long-term sustainability. When possible, cryopreservation, or long-term storage in liquid nitrogen, provides a secure, efficient backup for plant genebank collections.

Cryopreservation has been utilized in plant genebanks for over thirty years. Cryoreservation methods have been developed to maximize the number of cells that survive storage in liquid nitrogen. These methods are effective at preserving biological materials, e.g., cells and tissues, stored in the liquid or vapor phase of liquid nitrogen [1,2,6]. Methods based on cryoprotectant solutions rely on bathing the materials in mixtures of molecules called cryoprotecting agents (CPAs) that protect cells and tissues from the mechanical and osmotic stresses associated with cooling and rewarming [6]. Examples in plant cryopreservation include Plant Vitrification Solution 2 (PVS2; 30% glycerol, 15% dimethyl sulfoxide, 15% ethylene glycol, 0.4 M sucrose; [7]) and Plant Vitrification Solution 3 (PVS3; 50% glycerol, 50% sucrose; [8]), both shown in the 1990s to be highly effective for cryopreserving cells and plant shoot tips. However, these methods are not universally protective, and each treatment must be optimized for each new plant species conserved. Optimizing these methods for individual plant species takes time and resources that may not be available for endangered species. Determining how they interact will establish how different CPAs protect specific parts of cells. Establishing this fundamental knowledge will guide practices and minimize the resources needed in the traditionally empirical approach to optimization. This work aimed to add to the growing body of literature concerned with CPA–cell interactions by demonstrating that coherent Raman microscopy enables direct visualization of deuterated CPAs within living plant cells.

2. Results and Discussion

When plant cells are exposed to CPAs, they can go through a process of plasmolysis and deplasmolysis [9–11], as shown in Figure 1A. Plasmolysis is attributed to the change in osmotic pressure when a cell is exposed to a cryoprotectant solution, observable when the plasma membrane shrinks away from the cell wall. This is seen in Figure 1, Panel B, in the image acquired 70 s after exposure to 15% dimethyl sulfoxide (DMSO), a commonly used CPA in plant and animal cryopreservation. Following plasmolysis, the cell swells until the cell membrane reaches the cell wall in a process called deplasmolysis. Deplasmolysis is commonly attributed to CPAs entering the cell [9], as it is generally seen when cells are exposed to cell-permeating cryoprotectants, like DMSO. However, simple attribution of CPAs entering the cell to cause deplasmolysis does not explain why some cells, like the cell indicated by the purple arrow in Figure 1B, do not appear to respond to DMSO exposure with a plasmolysis/deplasmolysis cycle, while others, like the cell indicated by the red arrow, appear to respond as expected, plasmolyzing within 70 s of exposure and deplasmolyzing after 210 s of exposure, at room temperature. Both cells had movement within their cytoplasm, leading us to believe that they were both alive. This difference in response was observed in many different exposure experiments and with cells in clusters ranging from only a few cells to hundreds of cells (see Supplementary Information).

Prior to exposure, at 0 s, the two highlighted cells appear similar, both appearing populated with small organelles, which move within the cell. Thus, we might expect similar responses to the 15% DMSO solution. The question remains, did DMSO enter the unresponsive cell, or is it preferentially concentrated in the responsive cell? Although the macroscopic behavior of the cell is easily identifiable with bright field microscopy, the definitive localization of DMSO within these cells remains elusive. A primary objective of the current work was to directly observe the accumulation of DMSO within cells, as previous works have almost entirely relied on observing cellular responses to CPA exposure to understand the protective behavior of CPAs.

There has been some work to characterize the cellular responses to explore how CPAs work to protect or destroy cells [7–16]. Bright field microscopy techniques have shown how cells respond to CPA exposure in real-time [9]. Fluorescence microscopy has identified how fluorescently-labelled organelles move and change upon CPA exposure [12], as well as the histological changes that occur [11]. Electron microscopy has been used in conjunction with fluorescence microscopy to observe ultrastructural changes within cells that occur with CPA exposure and freezing [15,16]. Toxicity studies have demonstrated the applicability of standard CPA formulations to new cell lines [7,8], and various protein assays have determined DNA or RNA damage by CPA exposure [13,14]. All these studies inform how cells respond to and are damaged by CPA exposure, but they do not identify or determine CPA location or translocation within cells. Detailed information on CPA location and translocation within living cells, specifically which organelles are being most impacted by CPA exposure, will enable the development of highly robust and specific cryopreservation protocols that provide improved tolerance to cooling and low temperature storage.

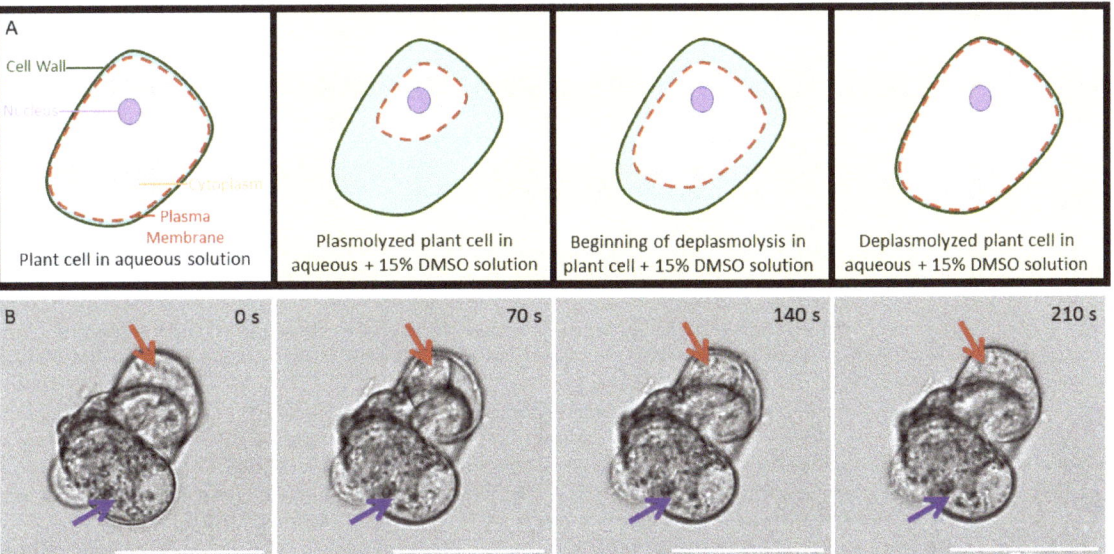

Figure 1. Panel A: Cartoon representation of a cell undergoing plasmolysis (the pulling away of the red dashed plasma membrane from the green cell wall) and deplasmolysis (the moving of the cell membrane back to cell wall) as it is exposed to a 15% DMSO in water solution (yellow background). Panel B: Brightfield microscopy images showing a small cluster of rice suspension cells 0, 70, 140, and 210 s after exposure to 15% aqueous DMSO. Red arrow: a cell that completely plasmolyzed after 70 s and deplasmolyzed after 210 s. Purple arrow: a cell that did not appear to respond to the 15% DMSO in water solution.

Most CPA formulations are mixtures of small molecular components, commonly DMSO, glycerol, ethylene glycol, and sugars. These molecules cannot be directly imaged by bright field microscopy and attaching a dye molecule to any of these molecules dramatically changes their diffusion behavior, making bright field and fluorescence microscopies ill-suited for determining their exact location within cells. However, these small molecules can be imaged using vibrational microscopy, which enlists unique vibrations intrinsic to the molecules of interest to image the sample. Although IR microscopy may be used as a label-free imaging technique, its intrinsically low spatial resolution (due to long IR wavelengths) and water's high IR absorptivity makes it challenging to image biological samples. Recent advances in IR microscopy that enlist both a mid-IR and ultraviolet laser can generate photoacoustic data with resolutions comparable to those collected with fluorescence microscopy [17]. Raman microscopy enables the imaging of live samples with wavelengths in the near-IR and visible region via the vibrations associated with specific molecules [18]. For example, the hydrocarbon in lipid molecules is highly effective for imaging cell membranes [18]. Although spontaneous Raman microscopy has been used to image live cells and tissues, it suffers from low sensitivity and contamination from endogenous sample fluorescence.

Various nonlinear coherent Raman scattering methods demonstrate dramatic enhancement in sensitivity over traditional Raman microscopy [19–22]. For example, coherent anti-Stokes Raman scattering (CARS), used in this work, boasts five to six orders of magnitude higher sensitivity to spontaneous Raman measurements [18,20]. Additionally, coherent Raman microscopy techniques avoid interference by endogenous fluorescence through optical filtering and also offer high spatial resolution, with optimal resolutions ≤300 nm, and high temporal resolution, <1 s, to measure location and translocation of molecules of interest [18,22]. These dramatic improvements enable real-time acquisition in live, biological samples [23].

All coherent Raman processes require two short laser pulses, usually in the picosecond range, aligned in space and time so that both beams simultaneously impinge on the sample. When the frequency difference between the pump and Stokes beams matches a molecular vibration in the sample, a CARS signal is generated (see SI) [20]. A CARS signal is quadratically proportional to the concentration of the molecule of interest [20], so CARS microscopy is most frequently used to image prevalent biological components in cells. Although Raman techniques have been used to image intra-cellular lipid responses to CPAs [24], freezing behavior of water in and around live cells exposed to CPAs [25–27], and bulk distributions of CPAs in frozen mixtures [28], they have yet to be applied to directly image CPAs distribution within living plant cells. In the work reported here, we used CARS microscopy to image deuterated dimethyl sulfoxide (d_6-DMSO) interacting with live rice suspension cells.

As most CPAs are organic molecules composed largely of carbon, hydrogen, and oxygen, their vibrational modes often fall in the same frequency range as the biological sample itself. In this work, we enabled selective CPA detection by deuteration. Figure 2A shows how deuteration in d_6-DMSO shifts the C-H stretch of DMSO (~2900 cm^{-1}) away from the broad C-H stretch region in the plant cells to the relatively quiet C-D stretch region of the spectrum (~2120 cm^{-1}). Targeting the 2120 cm^{-1} stretch ensures that the CARS signal detected reflects d_6-DMSO while cell features like the cell membranes and walls block that signal. The CARS image shown in Figure 2B was collected after the cell was exposed to d_6-DMSO for approximately 3 min, which was the amount of time it took to prepare the sample and begin collecting data. The d_6-DMSO signal appears as yellow, while blue features indicate places where no d_6-DMSO is present or components of the cell or other cells block the d_6-DMSO signal. Figure 2C shows an expanded view of the cell highlighted by the light blue square in Figure 2B. In this cell, the d_6-DMSO signal appears to pool in subcellular organelles. That is, rather than remaining dispersed throughout the cytoplasm, the d_6-DMSO preferentially accumulates in specific organelles. The blue outlines around these pools and in other parts of the image arise from areas where cellular components, such as lipid membranes and cell walls, block the d_6-DMSO signal. Figure 2D contrasts the pixel intensity along the pink and dark blue lines drawn in Figure 2C. The pink line intercepts three obvious d_6-DMSO-rich organelles while the blue line goes through a relatively uniform background part of the cell, contrasting the inside of the organelles to the surrounding cytoplasm. It is clear that there is a higher concentration of d_6-DMSO inside those organelles than in the surrounding cytoplasm. This demonstrates that d_6-DMSO is not uniform in the cell interior, preferentially pooling inside organelles within the cell. Other than a ~8% increase in mass, the properties of d_6-DMSO differ little from those of H_6-DMSO. Thus, we expect the same effect when cells are exposed to H_6-DMSO, as the molar mass is not significantly changed by deuteration.

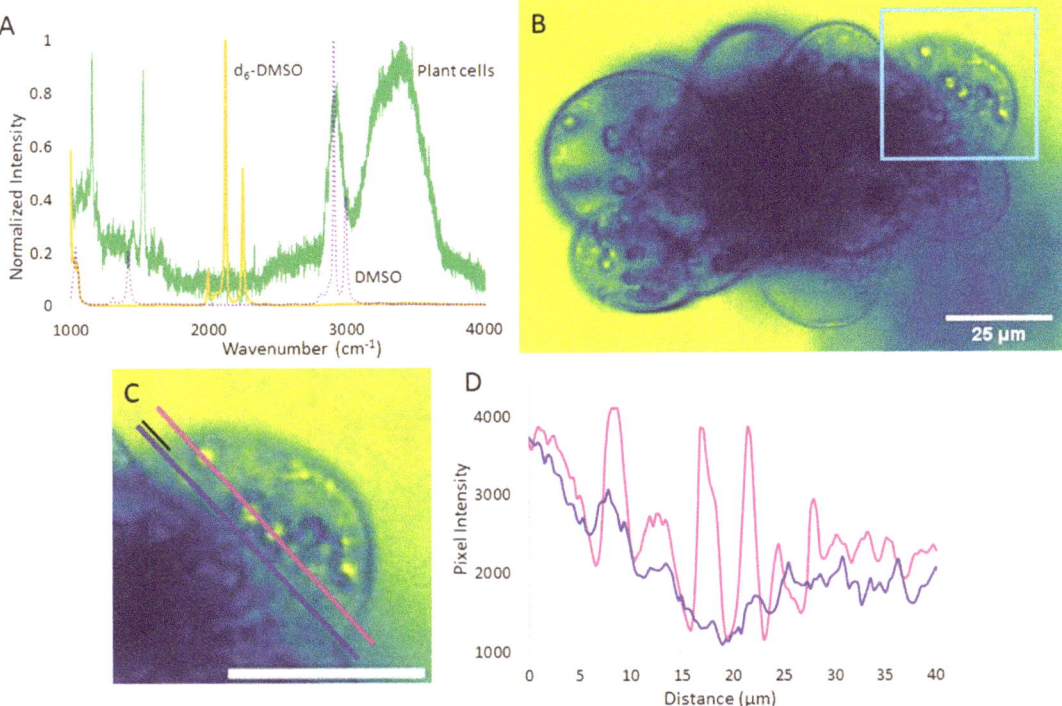

Figure 2. (**A**): Spontaneous Raman spectra of d_6-DMSO (yellow), DMSO (purple, dashed), and the rice cells (green), showing the shift in DMSO vibrational frequency upon deuteration away from significant peaks in the cell spectrum. (**B**): Representative CARS microscopy image of rice cells imaged in resonance with the d_6-DMSO stretching vibrational mode (yellow trace in **A**). Signal from d_6-DMSO appears yellow while places blocking the d_6-DMSO signal appear blue. (**C**): Expanded view of cell outlined in light blue in B. Parallel pink and dark blue line profiles bisect three organelles and the relatively uniform space away from organelles, respectively. Both lines start 5 μm outside of the cell, a distance shown with the small black line, and stretch, parallel, across approximately 40 μm. (**D**): Pixel intensity along pink and dark blue line profiles from C as a function of distance along the line. Scale bars in C and D are 25 μm. Images artificially colored with ImageJ LUT, mpl-viridis [29].

The CARS images collected of d_6-DMSO indicate that DMSO is pooling in specific organelles within the live cells. On the basis of bright field microscopy experiments (shown in SI), we suspect that the organelles preferentially taking-up DMSO are amyloplasts and/or starch bodies. Additionally, there is evidence in the literature suggesting that DMSO interacts with glucose and amylose, both components of starch [30,31]. An increase in DMSO concentration in these organelles may increase the amount of protection afforded to the organelles as DMSO is known to disrupt the hydrogen bond network of water and support vitrification over ice crystallization. This result may also indicate that these organelles are at a higher risk of damage from DMSO toxicity, as DMSO has been shown to cause cell death [32] and, at high concentrations, is presumed to disrupt cell membranes [33,34]. Furthermore, the apparent sequestration of DMSO inside these organelles challenges the assumption of equal DMSO distribution throughout the cell. The preferential localization of the DMSO cryoprotectant has ramifications in all disciplines that use cryopreservation—an unequal distribution of cryoprotectants in cells and tissues have different implications depending on the specific system. In cells that do not contain these organelles, DMSO uptake may be more uniform or different organelles may be preferentially sequestering the cryoprotectant. Previous research on cellular responses to CPAs have made it obvious that assuming CPAs to work equally and effectively in all cell types is flawed [7,8], and this result may partially explain the differences in cellular response seen in brightfield studies like those in Figure 1.

3. Materials and Methods

3.1. Growth and Maintenance of Oryza sativa (Asian Rice) Cells

3.1.1. Rewarming Cells and Initial Plating

Rice callus cells were acquired from the United States Department of Agriculture Agricultural Research Service (USDA-ARS) National Laboratory for Genetic Resources Preservation in Fort Collins, CO. The cell line was originally produced by G. Schaeffer, U.S. Dept. of Agriculture, Beltsville, MD in 1981 [35]. The rice cells used in this work were originally cryopreserved by Finkle and Ulrich in 1981 using PGD (10% w/v polyethylene glycol, 8% w/v glucose, and 10% w/v DMSO, [36]) and a slow-cool procedure. Rice callus cells (A7 line) were removed from liquid nitrogen and immediately warmed in a 40 °C water bath for 2 min until the solid cryoprotectant solution inside the vial was liquid. The cryoprotectant solution and cells were then diluted with 0.5 mL of wash solution made with 30 g L^{-1} sucrose (Alfa Aesar, Ward Hill, MA, USA) and Murashige and Skoog basal plant medium with vitamins (MS Media, M519; PhytoTechnology Laboratories, Lenexa, KS, USA) in distilled water at 22 °C and incubated for 10 min before 1 mL of wash solution was added at 22 °C. The solution was then allowed to sit for 10 min. The vial was then centrifuged for 1 min at 1000 rpm. The supernatant was removed with a pipette and 1 mL of wash solution was added to the remaining cells and they were incubated at 22 °C for 10 min. The liquid was removed from the vial with a pipette and the cells were scooped from the vial onto a sterile filter paper and blotted to remove excess liquid. The semi-dry cells were then plated onto solid modified MS Media (PT046; HiMedia Laboratories, Lincoln University, PA, USA) supplemented with 1 mg L^{-1} each of 2,4-dichlorophenoxyacetic acid (2,4-D, Sigma-Aldrich, St. Louis, MO, USA), indoleacetic acid (IAA, TCI America, Portland, OR, USA), and kinetin (TCI America), 146 mg L^{-1} glutamine (Acros Organics, Geel, Belgium), 30 g L^{-1} sucrose, and 8 g L^{-1} agar (BD Diagnostics, Franklin Lakes, NJ, USA) at pH 5.7. The plated cells were placed in a light-free container and were confirmed to be alive after growth was visible (approximately 3 weeks after plating).

3.1.2. Cell Culture

The cells were grown as suspension cultures or as callus. For both suspension and plated cultures, modified MS medium (PT046) was supplemented with 1 mg L^{-1} each of 2,4-D, IAA and kinetin, 146 mg L^{-1} glutamine, 30 g L^{-1} sucrose, and 8 g L^{-1} agar (removed for suspension cells) at pH 5.7. After autoclaving, 10 mL of medium was placed in 6 cm diameter Petri dishes and allowed to set for 30 min before being stored in the refrigerator. Rice callus cells were transferred to new solid media every 4–6 weeks depending on growth of the callus. When they were transferred, growth appearing the lightest in color was selected from the callus with sterile tweezers and placed onto the new medium. The cells were grown at room temperature in a closed drawer, maintaining constant darkness, and used in experiments as necessary. Suspension cells were grown using the same formulation of modified Murashige and Skoog medium, without the addition of agar. To grow cells in suspension, approximately a gram of the lightest colored rice callus cells was removed from plates and placed in Erlenmeyer flasks with 50 mL liquid media. Cells were culture grown on a shaker continuously rotating at 140 rpm. The Erlenmeyer flasks with suspension cells were wrapped in aluminum foil to ensure the cells would be grown in the dark. Suspension medium was replaced every 1–2 weeks, and the cells were moved to new suspension cultures at that time, depending on growth. When creating a new culture, flasks were removed from the shaker and allowed to settle for about 10 min before excess media was removed with a pipette. Cells were removed from the flask, blotted on sterile filter paper to remove excess medium, and about 1 cm^3 of cells was placed in a new flask.

3.2. Cryoprotectant Solutions

For bright field images, 15% (w/v) DMSO in distilled water was used. For all CPA exposures imaged with the CARS microscope, 15% (w/v) d_6-DMSO (MilliporeSigma, St. Louis, MO, USA) in distilled water was used. This is the concentration of DMSO found in PVS2 (5) and was chosen for the broad applicability of DMSO to both animal and plant cryopreservation.

3.3. Cell Imaging

Bright field images were acquired using an Olympus IX73 fluorescence microscope (Olympus Corporation, Tokyo, Japan) in the Chemistry Department Cell Culture Facility at Colorado State University. A rudimentary perfusion chamber was developed to image cells as exposure to CPA solutions occurred. The rudimentary perfusion chambers were made using a microscope slide, cover slip, and silicone grease, as shown in Figure 3. The microscope slide was coated in 1% poly-L-lysine (Electron Microscopy Sciences, Hatfield, PA) to immobilize the rice cells while solutions flowed through the chamber. The slide was cleaned with methanol, then a large drop, enough to cover the entire area under the coverslip, of poly-L-lysine was placed on the slide and allowed to set for 24 h. After setting, the poly-L-lysine was rinsed away with distilled water and the slide was air dried and immediately used for imaging the cells. Both cells grown on plates and suspension cultures were used. Callus cells were suspended in a small amount of MS Media by placing a small amount of friable callus in a vial with 1 mL of media and vigorously shaking. Suspension cells were placed on microscope slides directly from the cellular suspension. After placing the cells on the microscope slide, a 25 × 50 mm cover slip was placed onto the slide, creating a wide channel through which CPA mixtures could be wicked. Approximately 0.5 mL of the 15% DMSO solution was placed on the edge of the coverslip while a Kimwipe was held to the other end. This created flow and images were acquired as the solution travelled across the perfusion chamber and the CPA solution was allowed to sit in the chamber. This was repeated five times with 15% DMSO in water and a total of 35 cells were clearly visible. Of these cells, 8 appeared dead, 22 cells (61%) had a visible response to the solution, and 3 cells (8%) had a full plasmolysis/deplasmolysis cycle. A visible response was considered a rapid shrinking and expansion where there was no visible plasmolysis. For more on the number of cells responding to CPA mixtures, see the SI.

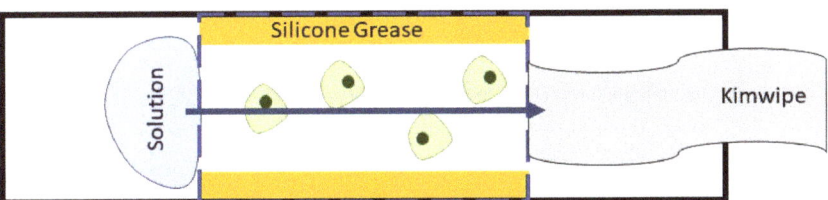

Figure 3. Rudimentary perfusion chamber showing plant cells in green, silicone grease in yellow, wicked solution in blue, and Kimwipe in grey on a white microscope slide (outlined in solid black). 25 × 50 mm cover slip placed on top of grease shown in blue dashed lines.

See Supporting Information for a CARS microscopy instrument description.

3.4. IKI Staining

Cells were stained with an iodine solution prepared with 100 mL distilled water, 1 g of iodine chips (Fisher Scientific, Fair Lawn, NJ, USA), and 2 g KI (Fisher Scientific) for 1.5 to 5 minutes. After staining, the cells were rinsed with water before mounting and imaging.

4. Conclusions

Using CARS microscopy, we visualized sequestration of d_6-DMSO in organelles within living rice suspension cells that were originally cryopreserved in 1981. After 37 years of cryostorage, the cells were thawed in 2018, and grew vigorous, friable calli. Both macroscopic signs of growth and generated autofluorescence within cells demonstrated the viability after rewarming and cryoprotectant exposure experiments. These cells serve as a model system for our first experiments demonstrating the value of direct visualization of cryoprotectants in living systems. Localization of cryoprotectants like DMSO in cell organelles may explain how these substances induce changes in cellular processes [14,32]. The nonuniform distribution of cryoprotectant in the cells implies that any assumption of an equal CPA dispersion within a cell, and consequently equal protection afforded by the CPA throughout the cell, is flawed. Continued investigation into the exact location of various CPAs within a broad range of living cells and tissues will likely illuminate why cellular response varies with CPA exposure, something applicable to both animal and plant cells. The establishment of cellular responses to CPA exposure has the potential to streamline development of cryopreservation protocols for newly endangered plant and animal species, allowing goals such as those set by The Convention on Biological Diversity to be more readily achievable, and enabling more species to be conserved as we face our current climate crisis.

Supplementary Materials: The following are available online at https://www.mdpi.com/2223-7747/10/3/589/s1, Figures S1–S11.

Author Contributions: F.M.D.S.: Conceptualization, Methodology, Validation, Formal Analysis, Investigation, Resources, Writing—Original Draft, Writing—Review and Editing; D.G.S.: Methodology, Resources, Software, Writing—Review and Editing; R.B.: Methodology, Resources, Writing – Review and Editing; G.M.V.: Conceptualization, Writing—Review and Editing, Funding Acquisition; N.E.L.: Conceptualization, Writing—Review and Editing, Funding Acquisition, Supervision, Project Administration. All authors have read and agreed to the published version of the manuscript.

Funding: We gratefully acknowledge financial support from USDA NIFA Grant Number 12835363, NSF Graduate Research Fellowship Program fellow number 2019273956 (FMS), and the Colorado State University Office of the Vice President for Research. Imaging experiments were performed in the Advanced Light Microscopy Core, part of NeuroTechnology Center at University of Colorado Anschutz Medical Campus, supported in part by Rocky Mountain Neurological Disorders Core Grant Number P30 NS048154 and by Diabetes Research Center Grant Number P30 DK116073.

Institutional Review Board Statement: Not applicable.

Informed Consent Statement: Not applicable.

Data Availability Statement: All data are available in the main text or the Supplementary Materials.

Acknowledgments: NEL thanks E. O. Potma and R. C. Prince for their help in launching this project. Additionally, we are grateful for help from Greg Glazner and Radu Moldovan with the CARS microscope at the Advanced Light Microscopy Core, and for assistance with the growth and maintenance of the rice suspension cells from Jean Carlos Bettoni during his time at the USDA ARS facility. Additionally, we acknowledge the primary author's home institution, Colorado State University, as a land-grant institution that would not exist without the sale and exploitation of Indigenous territories in Colorado.

Conflicts of Interest: The authors declare no conflict of interest.

References

1. Coriell, L.L.; Greene, A.E.; Silver, R.K. Historical development of cell and tissue culture freezing. *Cryobiology* **1964**, *1*, 72–79. [CrossRef]
2. Fuller, B.J. Cryoprotectants: The essential antifreezes to protect life in the frozen state. *Cryo Lett.* **2004**, *25*, 375–388.
3. Holt, W.V. Cryobiology, wildlife conservation and reality. *CryoLetters* **2008**, *29*, 43–52. [PubMed]
4. Global Strategy for Plant Conservation. The Targets 2011–2020. Conversion on Biological Diversity. 2011. Available online: https://www.cbd.int/gspc/targets.shtml (accessed on 15 October 2020).

5. Sharrock, S. *Plant Conservation Report 2020: A Review of Progress in Implementation of the Global Strategy for Plant Conservation 2011–2020*; Secretariate of the Convention on Biological Diversity: Montreal, QB, Canada, 2020.
6. Day, J.G.; Stacey, G. (Eds.) *Cryopreservation and Freeze-Drying Protocols*, 3rd ed.; Humana Press: New York, NY, USA, 2015.
7. Sakai, A.; Kobayashi, S.; Oiyama, I. Cryopreservation of nucellar cells of navel orange (*Citrus sinensis* Osb. var. *brasiliensis* Tanaka) by vitrification. *Plant Cell Rep.* **1990**, *9*, 30–33. [CrossRef] [PubMed]
8. Nishizawa, S.; Sakai, A.; Amano, Y.; Matsuzawa, T. Cryopreservation of asparagus (*Asparagus officinalis* L.) embryogenic suspension cells and subsequent plant regeneration by vitrification. *Plant Sci.* **1993**, *91*, 67–73. [CrossRef]
9. Volk, G.M.; Caspersen, A.M. Cryoprotectants and components induce plasmolytic responses in sweet potato (*Ipomoea batatas* (L.) Lam.) suspension cells. *Vitr. Cell. Dev. Biol. Anim.* **2017**, *53*, 363–371. [CrossRef]
10. Volk, G.M.; Caspersen, A.M. Plasmolysis and recovery of different cell types in cryoprotected shoot tips of *Mentha* × *piperita*. *Protoplasma* **2007**, *231*, 215–226. [CrossRef]
11. Salma, M.; Engelmann-Sylvestre, I.; Collin, M.; Escoute, J.; Lartaud, M.; Yi, J.-Y.; Kim, H.-H.; Verdeil, J.-L.; Engelmann, F. Effect of the successive steps of a cryopreservation protocol on the structural integrity of *Rubia akane* Nakai hairy roots. *Protoplasma* **2013**, *251*, 649–659. [CrossRef]
12. Kratochvílová, I.; Kopečná, O.; Bačíková, A.; Pagáčová, E.; Falková, I.; Follett, S.E.; Elliott, K.; Varga, K.; Golan, M.; Falk, M. Changes in Cryopreserved Cell Nuclei Serve as Indicators of Processes during Freezing and Thawing. *Langmuir* **2018**, *35*, 7496–7508. [CrossRef]
13. Zilli, L.; Schiavone, R.; Zonno, V.; Storelli, C.; Vilella, S. Evaluation of DNA damage in *Dicentrarchus labrax* sperm following cryopreservation. *Cryobiology* **2003**, *47*, 227–235. [CrossRef] [PubMed]
14. Verheijen, M.; Lienhard, M.; Schrooders, Y.; Clayton, O.; Nudischer, R.; Boerno, S.; Timmermann, B.; Selevsek, N.; Schlapbach, R.; Gmuender, H.; et al. DMSO induces drastic changes in human cellular processes and epigenetic landscape in vitro. *Sci. Rep.* **2019**, *9*, 1–12. [CrossRef]
15. Kaczmarczyk, A.; Rutten, T.; Melzer, M.; Keller, E.R.J. Ultrastructural changes associated with cryopreservation of potato (*Solanum tuberosum* L.) shoot tips. *CryoLetters* **2008**, *29*, 145–156. [PubMed]
16. Halmagyi, A.; Coste, A.; Tripon, S.; Crăciun, C. Low temperature induced ultrastructural alterations in tomato (*Lycopersicon esculentum* Mill.) shoot apex cells. *Sci. Hortic.* **2017**, *222*, 22–31. [CrossRef]
17. Shi, J.; Wong, T.T.W.; He, Y.; Li, L.; Zhang, R.; Yung, C.S.; Hwang, J.; Maslov, K.; Wang, L.V. High-resolution, high-contrast mid-infrared imaging of fresh biological samples with ultraviolet-localized photoacoustic microscopy. *Nat. Photon.* **2019**, *13*, 609–615. [CrossRef] [PubMed]
18. Cheng, J.-X.; Xie, X.S. Vibrational spectroscopic imaging of living systems: An emerging platform for biology and medicine. *Science* **2015**, *350*, aaa8870. [CrossRef]
19. Zhang, D.; Wang, P.; Slipchenko, M.N.; Cheng, J.-X. Fast Vibrational Imaging of Single Cells and Tissues by Stimulated Raman Scattering Microscopy. *Accounts Chem. Res.* **2014**, *47*, 2282–2290. [CrossRef] [PubMed]
20. Evans, C.L.; Xie, X.S. Coherent Anti-Stokes Raman Scattering Microscopy: Chemical Imaging for Biology and Medicine. *Annu. Rev. Anal. Chem.* **2008**, *1*, 883–909. [CrossRef]
21. Lee, H.J.; Cheng, J.-X. Imaging chemistry inside living cells by stimulated Raman scattering microscopy. *Methods* **2017**, *128*, 119–128. [CrossRef]
22. Zhang, C.; Cheng, J.-X. Perspective: Coherent Raman scattering microscopy, the future is bright. *APL Photon.* **2018**, *3*, 090901. [CrossRef]
23. Potma, E.O.; De Boeij, W.P.; Van Haastert, P.J.; Wiersma, D.A. Real-time visualization of intracellular hydrodynamics in single living cells. *Proc. Natl. Acad. Sci. USA* **2001**, *98*, 1577–1582. [CrossRef] [PubMed]
24. Mokrousova, V.; Okotrub, K.; Amstislavsky, S.; Surovtsev, N. Raman spectroscopy evidence of lipid separation in domestic cat oocytes during freezing. *Cryobiology* **2020**, *95*, 177–182. [CrossRef]
25. Dong, J.; Malsam, J.; Bischof, J.C.; Hubel, A.; Aksan, A. Spatial Distribution of the State of Water in Frozen Mammalian Cells. *Biophys. J.* **2010**, *99*, 2453–2459. [CrossRef] [PubMed]
26. Yu, G.; Li, R.; Hubel, A. Interfacial Interactions of Sucrose during Cryopreservation Detected by Raman Spectroscopy. *Langmuir* **2019**, *35*, 7388–7395. [CrossRef] [PubMed]
27. Yu, G.; Yap, Y.R.; Pollock, K.; Hubel, A. Characterizing Intracellular Ice Formation of Lymphoblasts Using Low-Temperature Raman Spectroscopy. *Biophys. J.* **2017**, *112*, 2653–2663. [CrossRef] [PubMed]
28. Karpegina, Y.; Okotrub, K.; Brusentsev, E.; Amstislavsky, S.; Surovtsev, N. Cryoprotectant redistribution along the frozen straw probed by Raman spectroscopy. *Cryobiology* **2016**, *72*, 148–153. [CrossRef] [PubMed]
29. Schindelin, J.; Arganda-Carreras, I.; Frise, E.; Kaynig, V.; Longair, M.; Pietzsch, T.; Preibisch, S.; Rueden, C.; Saalfeld, S.; Schmid, B.; et al. Fiji: An open-source platform for biological-image analysis. *Nat. Methods* **2012**, *9*, 676–682. [CrossRef]
30. Vasudevan, V.; Mushrif, S.H. Insights into the solvation of glucose in water, dimethyl sulfoxide (DMSO), tetrahydrofuran (THF) and N,N-dimethylformamide (DMF) and its possible implications on the conversion of glucose to platform chemicals. *RSC Adv.* **2015**, *5*, 20756–20763. [CrossRef]
31. Tusch, M.; Krüger, J.; Fels, G. Structural Stability of V-Amylose Helices in Water-DMSO Mixtures Analyzed by Molecular Dynamics. *J. Chem. Theory Comput.* **2011**, *7*, 2919–2928. [CrossRef]

32. Galvao, J.; Davis, B.; Tilley, M.; Normando, E.; Duchen, M.R.; Cordeiro, M.F. Unexpected low-dose toxicity of the universal solvent DMSO. *FASEB J.* **2014**, *28*, 1317–1330. [CrossRef]
33. Cheng, C.-Y.; Song, J.; Pas, J.; Meijer, L.H.; Han, S. DMSO Induces Dehydration near Lipid Membrane Surfaces. *Biophys. J.* **2015**, *109*, 330–339. [CrossRef] [PubMed]
34. Gurtovenko, A.A.; Anwar, J. Modulating the Structure and Properties of Cell Membranes: The Molecular Mechanism of Action of Dimethyl Sulfoxide. *J. Phys. Chem. B* **2007**, *111*, 10453–10460. [CrossRef] [PubMed]
35. Schaeffer, G.W.; Sharpe, F.T., Jr. Lysine in seed protein from S-aminoethyl-L-cysteine resistant anther-derived tissue cultures of rice. *In Vitro* **1981**, *17*, 345–352. [CrossRef]
36. Finkle, B.J.; Ulrich, J.M. Cryoprotectant removal temperature as a factor in the survival of frozen rice and sugarcane cells. *Cryobiology* **1982**, *19*, 329–335. [CrossRef]

Article
Two Advanced Cryogenic Procedures for Improving *Stevia rebaudiana* (Bertoni) Cryopreservation

Carla Benelli [1,*], Lara S. O. Carvalho [2], Soumaya EL merzougui [3] and Raffaella Petruccelli [1]

[1] Institute of BioEconomy, National Research Council (CNR/IBE), 50019 Sesto Fiorentino, Florence, Italy; raffaella.petruccelli@ibe.cnr.it
[2] Department of Biology, Federal University of Lavras, Lavras 3037, Brazil; lcarvalho470@gmail.com
[3] Laboratory of Biotechnology and Valorization of Natural Resources (LBVRN), Faculty of Sciences, Ibn Zohr University, 8106 Agadir, Morocco; soumaya.merzougui@gmail.com
* Correspondence: carla.benelli@ibe.cnr.it

Abstract: Cryopreservation is a useful tool for the long-term storage of plant genetic resources, and different cryogenic procedures have recently been developed. The present study focused on the use of the Droplet-vitrification (DV) and V cryo-plate protocol for the cryopreservation of *Stevia rebaudiana* in vitro-derived apical shoot tips and axillary shoot tips. A preliminary test showed that 90 and 120 min PVS2 (Plant Vitrification Solution 2) treatment significantly reduced the regrowth of the explants before immersion in liquid nitrogen (LN). For both procedures tested, the best osmoprotective condition for obtaining a higher regrowth of cryopreserved explants occurred when explants were PVS2 treated for 60 min. After direct immersion in LN, thawing and plating, the highest regrowth recorded was 80% with DV and 93% with V cryo-plate. Moreover, shoot tips proved to be a more suitable material for Stevia cryopreservation. A satisfactory vegetative regrowth was observed in the subcultures following cryopreservation by DV and V cryo-plate cryogenic procedures.

Keywords: apical shoot tips; axillary shoot tips; droplet-vitrification; long-term conservation; PVS2 vitrification; V cryo-plate

1. Introduction

Stevia rebaudiana (Bertoni) is an herbaceous perennial plant of the Asteraceae family. Its leaves produce diterpene glycosides (stevioside and rebaudiosides), and as stevioside is 300-hundred-fold sweeter than sucrose, it is deemed to be a good natural sugar substitute [1,2]. In addition to its sweetening properties, it has various medicinal properties and actions. For this reason, the Stevia plant is an extremely interesting crop for breeders who select varieties with high diterpene glycosides content as well as for propagators and the target market.

S. rebaudiana is a self-incompatible plant and one of its limiting factors for large-scale cultivation is its poor seed germination [3]. Moreover, plants from seed propagation have a great variability in growth, maturity, and non-uniform plants, with considerable variations in the sweetening level and composition [4,5]. The recent results suggest that seed germination and stem cutting are not cost effective methods for higher biomass production, while the micropropagation can be a promising technique [3].

In vitro conservation and cryopreservation are unconventional biotechnological tools to preserve selected and valuable lines of *S. rebaudiana*, also taking into account the problems relating to its propagation by seed [2,3].

A protocol for in vitro conservation of *S. rebaudiana* under slow growth conditions and mass micropropagation after the storage period was developed by Zayova et al. [6], while the long-term storage, cryopreservation of shoot tips, was carried out using the vitrification method by Shatnawi et al. [7]. Cryopreservation allows the storage of plant material (i.e., seeds, shoot tips, dormant buds, zygotic, and somatic embryos and pollen) at ultra-low temperatures in liquid nitrogen (LN; −196 °C) or in the vapour phase of LN

(−165 °C to −170 °C) [8] and it is becoming a widely practised method for the long-term storage of plant genetic resources [9–12]. An advantage of the cryopreservation is that plant germplasm can theoretically be kept indefinitely in very little space and at low cost, excluding the initial investment. Over the last 30 years, various cryopreservation techniques have been developed using conventional slow freezing methods [13–15] as well as several vitrification-based cryopreservation procedures (encapsulation-dehydration; vitrification; encapsulation-vitrification [16–18], and more recently, droplet-vitrification (DV) [19]. Two recent novel cryopreservation techniques have been identified and have resulted in V cryo-plate [20] and D cryo-plate [21]. The DV uses aluminum foil strips, while the most recent cryogenic procedure (D or V cryo-plate) uses aluminum cryo-plates.

The regrowth rate obtained in Stevia with the vitrification protocol by Shatnawi et al. [7] was 68%. Continuous research and technological evolution have markedly improved the cryogenic methodologies, allowing to enhance the recovery percentage of the species, as has occurred over the years, for example, in Vitis spp. [22,23] and potato [24,25]. The aim of this study was to assess the efficiency of the novel procedures, Droplet-vitrification and V cryo-plate, in order to improve the S. rebaudiana cryopreservation protocol.

2. Results

2.1. Evaluation Plant Vitrification Solution 2 (PVS2) Tolerance

In this study, a preliminary experiment on apical shoot tips (AST) and axillary shoot tips (AxST) to evaluate the effect of exposure duration and optimize temperature of PVS2 treatment showed that exposure to PVS2 induced time-dependent regrowth in both of the explants assessed and that the temperature of the treatment can influence regrowth rates (Figure 1). After 28 days of culture, the best regrowth rates were obtained for the explants treated with PVS2 with exposure times ranging from 20 min to 60 min at 0 °C, and similar trends were observed for AST and AxST. AST treated with PVS2 solution for 60 min at both temperatures had 90% regrowth (Figure 1a), while in the AxST, the regrowth was 90% at 0 °C and 85% at 25 °C (Figure 1b). PVS2 treatments markedly affected regrowth potential after 60 min, when a significant drop was observed, suggesting a toxic response to long-term exposure to PVS2, furthermore regrowth was also affected by treatment temperature. As regards the longer exposure time (120 min), the regrowth AST decreased significantly up to 35% and 20% at 0° C and 25 °C, respectively, while for AxST the regrowth was significantly lower, dropping to 5% at 25 °C.

In the two cryopreservation procedures described below, the PVS2 exposure times of 20, 30, and 60 min at 0 °C will be applied, given that longer incubation times and 25 °C resulted in a low regrowth percentage even in explants without immersion into LN.

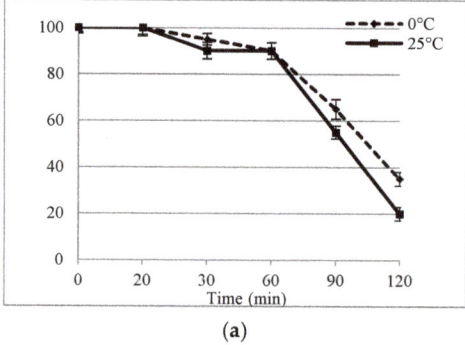

(a)

Figure 1. Cont.

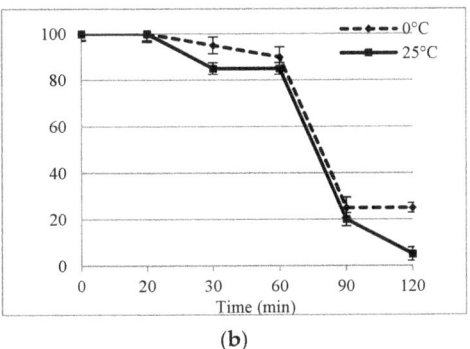

(b)

Figure 1. Effect of PVS2 treatment times on regrowth percentage (± SD) of *S. rebaudiana* apical shoot tips (**a**) and axillary shoot tips (**b**) at 0 °C and 25 °C, after 28 days of culture.

2.2. Droplet-Vitrification Procedure

In the DV procedure, the effect of PVS2 treatment on the explants was more evident after immersion in LN and the survival and regrowth percentage of *S. rebaudiana* depended on PVS2 exposure time (Table 1). The survival rate was lower for all assessed exposure times to PVS2 compared to controls. The post-thaw survival percentages of the cryopreserved explants observed after seven days ranged from 6 to 87% for AST and 0 to 50% for AxST, while the best survival rate was observed for apical shoot tips treated with PVS2 for 60 min (87%). As regards the axillary shoot tips, 30-min and 60-min PVS2 incubation times showed the same osmoprotection effect with 50% survival rate. From the data collected, it is evident that 20 min PVS2 duration time was not osmoprotective for the explants, as none of the axillary shoot tips survived and only 6% of apical shoot tips survived, while no regrowth was observed after 28 days. Due to the lack of shoot development, this PVS2 time will not be considered for evaluating biometric parameters in the subculture. At the end of the experiment (28 days) the regrowth rate of the explants treated with PVS for 60 min remained high, confirming a better apical shoot regeneration response (80%) with respect to axillary shoot tips (50%). The explants that grew into normal shoots were used for the subculture evaluation described below.

Table 1. Droplet-vitrification procedure: effect of different PVS2 exposure times on survival and regrowth of apical shoot tips or axillary shoot tips after cryopreservation. Control (without any treatment).

Treatment	[1] Survival Percentage (7 days)		Regrowth Percentage (28 days)	
	Apical Shoot Tips	Axillary Shoot Tips	Apical Shoot Tips	Axillary Shoot Tips
Control	100.0 a	93.0 a	93.0 a	93.0 a
PVS2 20 min + LN	6.0 c	0.0 c	0.0 d	0.0 c
PVS2 30 min + LN	46.6 bc	50.0 b	43.0 c	46.6 b
PVS2 60 min + LN	87.0 ab	50.0 b	80.0 ab	50.0 b

[1] Statistical analysis in each column was performed by ANOVA. Data followed by different letters are significantly different at $p \leq 0.05$ by Bonferroni's test.

Shoots derived from DV cryopreserved explants after the first subculture showed positive regrowth activity (Figure 2h), especially those obtained from initial explants treated with PVS2 + LN for 60 min, which showed similar or slightly higher values to the control (Table 2). As regards the cryopreserved shoots treated with PVS2 30 + LN, several morphological parameters considered were significantly lower than the control group and the PVS2 60 + LN treatment.

Table 2. First subculture after droplet-vitrification procedure: morphological parameters of *S. rebaudiana* shoots, from control and from cryopreserved apical shoot tips and axillary shoot tips. (Means ± SD).

Treatment	[1] Shoot Length (cm)		Shoots/Explants (n°)		Explants with Shoot (%)	
	Apical Shoot Tips	Axillary Shoot Tips	Apical Shoot Tips	Axillary Shoot Tips	Apical Shoot Tips	Axillary Shoot Tips
Control	4.30 ± 0.48 ab	4.21 ± 0.64 a	0.96 ± 0.63 a	0.89 ± 0.73 a	78.0 a	68.0 a
PVS2 30 min + LN	4.00 ± 0.39 b	3.64 ± 0.44 b	0.84 ± 0.55 a	0.21 ± 0.42 b	77.0 a	21.0 b
PVS2 60 min + LN	4.90 ± 0.44 a	4.03 ± 0.39 ab	0.91 ± 0.58 a	0.80 ± 0.56 a	79.0 a	73.0 a

[1] Statistical analysis in each column was performed by ANOVA. Data followed by different letters are significantly different at $p \leq 0.05$ by Bonferroni's test.

Figure 2. Some steps of droplet-vitrification and V cryo-plate cryopreservation procedures of *Stevia rebaudiana*. Droplet-vitrification: (**a**) PVS2 drops in aluminum foil; (**b**) explants placed into drops of PVS2; (**c**) aluminum foil strip with explants, plunged in cryovial filled with LN; (**d**) thawing of aluminum foil with explants in washing solution. V cryo-plate: (**e**) drops of sodium alginate in wells of cryo-plate; (**f**) explants into wells covered with calcium chloride solution; (**g**) cryo-plate, after PVS2 treatment, placed in cryovial filled with LN; (**h**) cryopreserved shoots with DV procedure after first subculture; (**i**) cryopreserved shoots with V cryo-plate procedure after second subculture.

2.3. V Cryo-Plate Procedure

The V cryo-plate procedure proved to be effective for apical shoot tips and axillary shoot tips of *S. rebaudiana*. High viability was obtained except for the explants processed with PVS2 for 20 min followed by LN exposure. In Table 3 we reported the survival and regrowth percentages after 7 and 28 days of culture post thawing. AST gave the best response to the PVS2 treatments for cryopreservation while the AxST were more sensitive, showing lower survival and regrowth ability. In particular, treatment with PVS2 for 60 min provided better cryo-tolerance capability for the explants immersed in LN and showed the highest regrowth rate at 28 days of culture (93% AST and 67% AxST). Since minimal or no regrowth was observed from explants treated with PVS2 for 20 min + LN, this PVS2 time was not considered for evaluating biometric parameters in the subcultures.

In the first subculture, the shoots from cryopreserved AST showed a more active growth than the shoots from cryopreserved AxST (Table 4). Overall, no significant differences were observed among treatments and with the control, except for the shoot length in AST derived from PVS2 30 min + LN, which had lower values. After 28 days all of the shoots were subcultured into fresh proliferation medium.

Table 3. V cryo-plate procedure: effect of different PVS2 exposure times on survival and regrowth of apical shoot tips or axillary shoot tips after cryopreservation. Control (without any treatment).

Treatment	[1] Survival Percentage (7 days)		Regrowth Percentage (28 days)	
	Apical Shoot Tips	Axillary Shoot Tips	Apical Shoot Tips	Axillary Shoot Tips
Control	97.0 a	95.0 a	93.0 a	93.0 a
PVS2 20 min + LN	10.0 b	0.0 c	3.0 c	0.0 c
PVS2 30 min + LN	87.0 a	64.0 b	70.0 b	46.0 b
PVS2 60 min + LN	93.0 a	69.0 b	93.0 a	67.0 ab

[1] Statistical analysis in each column was performed by ANOVA, Data followed by different letters are significantly different at $p \leq 0.05$ by Bonferroni's test.

Table 4. First subculture after V cryo-plate procedure: morphological parameters of *S. rebaudiana* shoots, from control and from cryopreserved apical shoot tips and axillary shoot tips. (Means ± SD).

Treatment	[1] Shoot Length (cm)		Shoots/Explants (n°)		Explants with Shoot (%)	
	Apical Shoot Tips	Axillary Shoot Tips	Apical Shoot Tips	Axillary Shoot Tips	Apical Shoot Tips	Axillary Shoot Tips
Control	4.52 ± 0.70 a	4.16 ± 0.37 a	1.00 ± 0.74 a	0.97 ± 0.75 a	78.0 a	71.4 a
PVS2 30 min + LN	4.00 ± 0.51 b	3.95 ± 0.42 a	0.90 ± 0.57 a	0.81 ± 0.70 a	77.0 a	66.6 a
PVS2 60 min + LN	4.37 ± 0.78 a	4.20 ± 0.48 a	0.93 ± 0.60 a	0.73 ± 0.52 a	78.6 a	70.0 a

[1] Statistical analysis in each column was performed by ANOVA. Data followed by different letters are significantly different at $p \leq 0.05$ by Bonferroni's test.

During the second subculture, the differences between the V cryo-plate cryopreserved shoots derived from AST and AxST explants incubated in both PVS2 treatments disappeared, for all of the observed parameters (Table 5). However, it is evident that the shoots from AST treated for 30 min and 60 min + LN, even if not significantly different, were longer and produced a greater number of shoots per explant than those of the control, showing good multiplication capacity (Figure 2i).

Table 5. Second subculture after V cryo-plate procedure: morphological parameters of *S. rebaudiana* shoots, from control and from cryopreserved apical shoot tips and axillary shoot tips. (Means ± SD).

Treatment	[1] Shoot Length (cm)		Shoots/Explants (n°)		Explants with Shoot (%)	
	Apical Shoot Tips	Axillary Shoot Tips	Apical Shoot Tips	Axillary Shoot Tips	Apical Shoot Tips	Axillary Shoot Tips
Control	6.69 ± 0.98 a	6.52 ± 0.88 a	1.50 ± 0.68 a	1.43 ± 0.81 a	100.0 a	93.3 a
PVS2 30 min + LN	7.19 ± 1.00 a	6.43 ± 1.10 a	1.53 ± 0.81 a	1.40 ± 0.81 a	96.6 a	83.0 a
PVS2 60 min + LN	7.21 ± 1.40 a	6.44 ± 1.50 a	1.56 ± 0.96 a	1.40 ± 0.96 a	93.3 a	83.0 a

[1] Statistical analysis in each column was performed by ANOVA, Data followed by same letters indicate no significant difference at $p \leq 0.05$ by Bonferroni's test.

In this study, phenotype biometric examination during in vitro subcultures, in both DV and V cryo-plate, did not reveal any morphological abnormalities compared with the control plants.

3. Discussion

In recent years, various biotechnological approaches have been used for conserving endangered and medicinal species, providing conservation of pathogen free plant and biodiversity. Their preservation is essential for plant breeding programs, for maintaining biodiversity, and for utilization as resource of compounds to the medicinal, food, and crop protection industries [26].

A protocol for in vitro conservation of *S. rebaudiana* under slow growth conditions and mass micropropagation after six months of storage without subculturing, was developed by Zayova et al. [6], while cryopreservation to enable the long-term storage of shoot tips was carried out using a vitrification method [7]. Several innovative procedures have recently been implemented for cryopreservation in order to improve the explant physiological state, pre-treatment conditions, time and conditions of the cryoprotectant treatments, increase the cooling and warming rates, and the recovery medium to achieve successful regrowth [12,27–30]. Droplet-vitrification and V cryo-plate are the recent cryogenic procedures available and, in this study, both methods have been applied and evaluated on *S. rebaudiana* apical shoot tips and axillary shoot tips.

In order to implement an efficient cryopreservation protocol, it is essential that plant cells can be cooled in LN and recovered without causing cell damage to maintain the cell viability. The main cause of cell injury is the transition of water into ice crystals during the cooling process [31]. To avoid the formation of ice crystals, cells and tissues have to be adequately dehydrated and/or exposed to cryoprotectant solutions, before immersing them in LN. PVS2 is the most widely used cryoprotectant solution with successful on vitrification-based cryopreservation. It is also vital to carefully control dehydration step and prevent injury by chemical toxicity or excessive osmotic stress during PVS treatment. Possible toxic effects of the PVS2 and potential damage that may occur during vitrification were studied in various species [32–34]. Our findings suggest that 90 min and 120 min of PVS2 treatment, both at 0 °C and 25 °C, adversely affected the *S. rebaudiana* explants, which led to a significant decrease in the apical shoot tips regrowth percentage and axillary shoot tips, thus demonstrating that explants do not tolerate prolonged exposure to this solution. While short exposure duration to PVS2 (20 min) was ineffective in achieving osmoprotection on the tissues; 30 min and 60 min at 0 °C proved to be the best treatments for achieving satisfactory regrowth of shoots tips and nodal segments following both DV and V cryo-plate procedures. This is in line with the results reported for shoot tips of *S. rebaudiana*, which showed 68% of regrowth when subjected to the PVS2 solution for 60 min at 0 °C using a vitrification technique [7] and is also in accordance with the results obtained for other species. For example, the optimal PVS2 treatment time for *Clinopodium odorum* shoot tip cryopreservation was 60 min at 0 °C using the V cryo-plate procedure [35]. However, the exposure time of the samples must be optimized for each species to assure tissue protection, e.g., the optimal exposure time to PVS2 at 0 °C was 30 min for blackberry apices and between 10 min and 30 min for cherry plums shoot tips [36], while in *Limonium serotinum* shoot tip regeneration was highest after PVS2 treatment for 30 min at 0 °C [37]. For two potato cultivars cryopreserved by the droplet-vitrification procedure, exposure to PVS2 for 40 min and 50 min at 0 °C gave the highest recovery rates, moreover apical buds responded better than axillary buds [38]. Irrespective of the vitrification solution used, the exposure time and temperature condition are fundamental and must be accurately determined depending on the plant material since some solutions can be toxic and even slight changes in treatment duration can have a dramatic impact on recovery [19].

The type of explant selected is also an important parameter for successful plant cryopreservation, in addition to size, cellular composition, physiological state and growth phase, which increase the probability of a positive response to various treatments before immersion in LN [9]. According to our cryopreservation data analysis, it seems that cryopreservation conditions proved to be more suitable for Stevia apical shoot tips than for axillary shoot tips. AST response is better for regrowth and in vitro subcultures, especially when a PVS2 exposure period of 60 min + LN was applied. The PVS2-treated AST had a better osmoprotective effect respect to AxST, both in DV (80% and 50%, respectively) and in V cryo-plate, (93% and 67%). In additional, it also showed the higher sensitivity of AxST compared to AST.

The DV procedure [19] is a method obtained by combining the vitrification procedure with the droplet-freezing technique developed by Kartha et al. [39] in which explants are placed in minuscule droplets of vitrification solution on aluminium foil strips, whereas the V cryo-plate procedure [20,40] combines the encapsulation and droplet-vitrification techniques, in which explants are placed on an aluminium cryo-plate and embedded in a thin layer of calcium alginate gel.

The two cryopreservation techniques (DV and V cryo-plate), applied on *S. rebaudiana* explants share the common trait of achieving higher cooling and warming rates compared to other vitrification-based procedures, because the explants placed on aluminium foils or cryo-plates, which are materials with high thermal conductivity, come into direct contact with liquid nitrogen during cooling and with the washing solution during warming [41]. Ultra-rapid cooling is more easily achieved using aluminium foil strips or aluminium cryo-plates rather than cryovials for immersing the explants directly into liquid nitrogen. These upgrades, combined with the optimal PVS2 exposure time, increased the chances of achieving a vitrified state during freezing in Stevia, resulting in superior regrowth after cryopreservation using the DV and V cryo-plate methods compared to results obtained by Shatnawi et coll. [7].

With respect to previous cryogenic methodologies, the simplicity is among the main advantages of using DV, moreover the technique can be successfully replicated by the technical staff [23]. The V cryo-plate method appears to be very systematic and time saving [42]. Both procedures appear highly promising to use large scale.

The DV method was initially used for potato shoot tip cryopreservation [43], but only a few years later its potential usefulness compared to other cryopreservation techniques was reported for several species such as in *Musa* [44], *Rosa* spp. [45–47], potato [25,48], taro [49], *Lilium* [50], *Saccharum* spp. [51], *Thymus* spp. [34], *Mentha* spp. [52], and *Vitis* spp. [29,53,54]. According to our results, applying this method to Stevia, shoot regrowth percentages of 80% and 50% were observed for cryopreserved AST and AxST, respectively. Panis et al. [44] reported that in eight different genomic groups of *Musa* spp., droplet-vitrification procedure increased regrowth by 23-46% compared to standard vitrification. In our study this procedure resulted in an improvement of 23% compared to the traditional vitrification method used by Shatnawi et al. [7] considering apical shoot tips as explants. By histological observations conducted on cryopreserved potato and pineapple shoot tips, some authors [55,56], concluded that droplet-vitrification caused the least injury to the tissues, thus it led to a benefit, increasing the recovery of cryopreserved shoot tips.

The application of V cryo-plate method has been reported for a wide range of species: strawberry [40], Dalmatian chrysanthemum [20], mint [42], mulberry [57], carnation [58], blueberry [59], mat rush [60], and sugarcane [61]. The satisfactory regrowth rates of cryopreserved potato shoot tips (96.7%) in V cryo-plate proved that it is a useful strategy for preserving valuable potato germplasm [62]. Recently, the V cryo-plate method has been successfully applied to grapevine germplasm [63].

Some authors have reported that the cryo-plate method is a user-friendly procedure that minimizes the risk of shoot injury, and permits high and rapid cooling and warming rates of treated materials, which improves recovery [21,59,64,65]. One of the various advantages of using the V cryo-plate procedure is that the explants are attached to the cryo-plates in alginate gel during the whole procedure, including thawing. Handling the explants throughout the procedure is very quick and easy because only the cryo-plates are manipulated, therefore all laboratory personnel can carry out this procedure once they are practiced at mounting the explants on the cryo-plates [20].

The V cryo-plate procedure increased the effectiveness of cryopreservation for *S. rebaudiana*, thus resulting in superior regrowth, 93% in AST and 67% in AxST, compared to the vitrification procedure used by Shatnawi et al. [7], who obtained a 68% shoot regrowth rate. Moreover, the growth of the shoots recovered from V cryo-plate procedure, during the subcultures was very dynamic and some evaluated parameters were slightly higher than the control group, especially for shoots from AST treated with PVS2 + LN for 60 min. However, after the second subculture, no significant differences were observed between the explant types tested and the control shoots, thus proving the effectiveness of the procedure.

In conclusion, Stevia responded well to conservation in LN and applying innovative approaches enabled us to increase the regrowth rate following cryopreservation, using both DV and V cryo-plate. It demonstrated that these techniques are efficient for the long-term storage of *S. rebaudiana*.

4. Materials and Methods

4.1. Plant Material and Culture Conditions

S. rebaudiana shoots (4–5 cm in length) were cultured on 100 mL of Murashige and Skoog (MS) [66] medium with 1 mg L^{-1} indole-3-butyric acid (IBA), 0.1 M sucrose and 7 g L^{-1} plant agar, at pH 5.8 in glass culture vessels (500 mL), with 20 shoots per vessel. The shoot cultures were maintained at 23 ± 1 °C under a 16 h photoperiod (40 μmol m^{-2} s^{-1}). The shoots were subcultured monthly on fresh medium of the same composition (standard culture conditions).

4.2. Cold Hardening of In Vitro Shoot Cultures and Explant Preculture

S. rebaudiana shoot cultures (~4 cm in length with leaves) were transferred to 4 °C in low intensity light (25 μmol m^{-2} s^{-1}) 7 days after the last subculture and maintained in hardening for 2 weeks before conducting cryopreservation experiments. Following preconditioning, apical shoots tips and axillary shoot tips, (explants with size of 1 mm to 1.5 mm for DV and 1 mm to 1.5 mm × 1 mm for V cryo-plate), were excised aseptically and placed in Petri dishes (9 cm Ø) containing hormone-free MS medium supplemented with 0.5 M sucrose at pH 5.8 and precultured for 48 h at 4 °C in darkness. Then, they were placed in loading solution (LS: MS liquid medium containing 2.0 M glycerol and 0.4 M sucrose at pH 5.8) [67] for 20 min at room temperature.

4.3. Evaluation of the Explant Tolerance to PVS2

In order to evaluate the effect of exposure duration to PVS2 solution (PVS2 contains 0.4 M sucrose, 30% (w/v) glycerol, 15% (w/v) ethylene glycol and 15% (w/v) dimethylsulfoxide in liquid MS medium, at pH 5.8) [17], the explants were incubated on sterilized PVS2 for 20, 30, 60, 90, or 120 min at 0 °C and 25 °C. The explants were then rinsed with Washing Solution (WS: liquid MS medium, containing 1.2 M sucrose [17] at 25 °C for 20 min and plated on proliferation medium (MS medium with 1 mg IBA, 0.1 M sucrose, 7 g L^{-1} plant agar, pH 5.8) for recovery under standard conditions.

4.4. Cryopreservation of S. rebaudiana Explants

The DV and V cryo-plate procedures were applied to the apical shoot tips and axillary shoot tips of *S. rebaudiana* (Figure 2).

4.4.1. Droplet-Vitrification (DV) Procedure

The apical shoot tips (AST) and axillary shoot tips (AxST) were kept in LS at room temperature for 20 min and then placed into droplets of PVS2 (4 µL to 5 µL; each drop containing one explant) on sterilised aluminium foil strips (~6 mm × 25 mm, Figure 2a,b) in an open Petri dish at 0 °C and maintained under these conditions for 20, 30, or 60 min. Following PVS2 exposure, the aluminium foil strips with the explants were immersed into 2 mL Nalgene® cryovials filled with liquid nitrogen (Figure 2c) and then plunged into LN for at least 1 h. For thawing, the frozen aluminium foils were removed from the cryovial and immersed immediately into Washing Solution at room temperature for 20 min (Figure 2d). Explants were placed on hormone-free MS and kept in the dark for at 22 ± 1 °C for 24 h and then transferred to standard culture conditions for recovery. The control group was composed of explants without any treatment.

4.4.2. V Cryo-Plate Procedure

The aluminium cryo-plates (7 mm × 37 mm × 0.5 mm) with 12 wells (Ø 1.5 mm, depth 0.75 mm) were used. Droplets of approximately 2 µL of sodium alginate solution (the solution contains 2% (w/v) sodium alginate (Carlo Erba, Cornaredo, Milan, Italy; medium viscosity) in calcium-free MS basal medium with 0.4 M sucrose, at pH 5.8) were poured into the cryo-plate wells (Figure 2e). The explants were placed individually into each well with a scalpel blade and fully immersed in one drop of sodium alginate solution (Figure 2f). The calcium chloride solution (0.1 M calcium chloride in MS medium with 0.4 M sucrose, at pH 5.8) was poured dropwise onto the section of the cryo-plates until the explants were covered and then left for 15 min to achieve complete polymerization. The excess calcium solution was removed by sucking it with a micropipette. The cryo-plates with explants were placed in LS for 20 min at room temperature, the LS was then removed and the cryo-plates were filled with PVS 2 for 20, 30, 60, min at 0 °C. After each PVS2 treatment, the cryo-plates were transferred into 2 mL Nalgene® cryovials placed in LN (Figure 2g), and then plunged directly into LN for at least 1 h. The cryo-plates with cryopreserved explants were thawed by immersing them in 2 mL of WS solution for 15 min. For recovery, the explants were placed on MS proliferation medium, in the dark for 24 h and then transferred to the light conditions described above. The control group was composed of explants without any treatment.

4.5. Data Collection and Statistical Analysis

In order to evaluate PVS2 incubation times, the explant regrowth percentage was determined 4 weeks after the treatments, while for cryopreservation experiments, post-thaw survival rate data (i.e., percentage of explants that maintained their green colour and vigour) were recorded 7 days after transfer to proliferation medium, and the regrowth rate (the percentage of explants forming shoots ≥ 1 cm) was recorded after 28 days. Each experimental treatment was replicated three times with each replication containing 10 shoot tips and nodal segments in the DV procedure and with 10 shoot tips and 15 nodal segments in the V cryo-plate procedure. For both procedures the following biometric parameters were recorded from regenerated cryopreserved shoots after the first subculture (28 days): (i) mean number of shoots, (ii) mean shoot length, and (iii) percentage of explants with shoots. In V cryo-plate procedure the biometric observations also involved the second subculture (28 days) with 3 replicates, 10 explants per each replicate. The data collected were subjected to one-way analysis of variance (ANOVA) followed by the Bonferroni post hoc test to determine the significance level between means at 95% confidence level ($p \leq 0.05$). The percentage data were transformed in Arcsine before applying one-way ANOVA. The data were statistically analyzed using Statgraphics Centurion XVI (Stat Point, Inc., Herndon, VA, USA).

Author Contributions: Conceptualization, C.B.; methodology, C.B., L.S.O.C.; investigation, L.S.O.C., S.E.m.; resources, C.B.; data curation, R.P.; writing—original draft preparation, C.B.; writing—review and editing, R.P., C.B.; project administration, C.B.; All authors have read and agreed to the published version of the manuscript.

Funding: This research received no external funding.

Institutional Review Board Statement: Not applicable.

Informed Consent Statement: Not applicable.

Data Availability Statement: The data presented in this study are available on request from the corresponding author.

Conflicts of Interest: The authors declare no conflict of interest.

References

1. Lemus-Mondaca, R.; Vega-Gálvez, A.; Zura-Bravo, L.; Ah-Hen, K. *Stevia rebaudiana* Bertoni, source of a high-potency natural sweetener: A comprehensive review on the biochemical, nutritional and functional aspects. *Food Chem.* **2012**, *132*, 1121–1132. [CrossRef] [PubMed]
2. Yadav, A.K.; Singh, S.; Dhyani, D.; Ahuja, P.S. A review on the improvement of stevia [*Stevia rebaudiana* (Bertoni)]. *Can. J. Plant Sci.* **2011**, *91*, 1–27. [CrossRef]
3. Khalil, S.A.; Zamir, R.; Ahmad, N. Selection of suitable propagation method for consistent plantlets production in *Stevia rebau-diana* (Bertoni). *Saudi J. Biol. Sci.* **2014**, *21*, 566–573. [CrossRef] [PubMed]
4. Nakamura, S.; Tamura, Y. Variation in the main glycosides of Stevia (*Stevia rebaudiana* Bertoni). *Jpn. J. Trop. Agric.* **1985**, *29*, 109–116. [CrossRef]
5. Sivaram, L.; Mukundan, U. *In vitro* culture studies on *Stevia rebaudiana*. *Vitr. Cell. Dev. Biol. Anim.* **2003**, *39*, 520–523. [CrossRef]
6. Zayova, E.; Nedev, T.; Dimitrova, L. *In vitro* storage of *Stevia rebaudiana* Bertoni under slow growth conditions and mass multiplication after storage. *Bio Bull.* **2017**, *3*, 30–38.
7. Shatnawi, M.A.; Shibli, R.A.; Abu-Romman, S.M.; Al-Mazra, M.S.; Al Ajlouni, Z.I.; Shatanawi, A.; Odeh, W.H. Clonal prop-agation and cryogenic storage of the medicinal plant *Stevia rebaudiana*. *Span. J. Agricul. Res.* **2011**, *1*, 213–220. [CrossRef]
8. Reed, B.M. Cryopreservation—Practical Considerations. In *Plant Cryopreservation: A Practical Guide*; Reed, B.M., Ed.; Springer: New York, NY, USA, 2008; pp. 3–13.
9. Niino, T. Developments in plant genetic resources cryopreservation technologies. In *Effective Genebank Management in APEC Member Economies*; Jung-Hoon, K., Ed.; NIAB: Suwon, Korea, 2006; pp. 197–217.
10. Wang, M.-R.; Lambardi, M.; Engelmann, F.; Pathirana, R.; Panis, B.; Volk, G.M.; Wang, Q.-C. Advances in cryopreservation of *in vitro*-derived propagules: Technologies and explant sources. *Plant Cell Tissue Organ Cult. (PCTOC)* **2021**, *144*, 7–20. [CrossRef]
11. Panis, B. Sixty years of plant cryopreservation: From freezing hardy mulberry twigs to establishing reference crop collections for future generations. *Acta Hortic.* **2019**, *1234*, 1–8. [CrossRef]
12. Benelli, C.; De Carlo, A.; Engelmann, F. Recent advances in the cryopreservation of shoot-derived germplasm of economically important fruit trees of *Actinidia, Diospyros, Malus, Olea, Prunus, Pyrus* and *Vitis*. *Biotechnol. Adv.* **2013**, *31*, 175–185. [CrossRef]
13. Withers, L.A.; Engelmann, F. In vitro Conservatipon of Plant Genetic Resources. In *Biotechnology in Agriculture*; Altman, A., Ed.; Marcel Dekker Inc.: New York, NY, USA, 1998; pp. 57–88.
14. Reed, B.M.; Uchendu, E. Controlled Rate Cooling. In *Plant Cryopreservation: A Practical Guide*; Reed, B.M., Ed.; Springer: New York, NY, USA, 2008; pp. 77–92.
15. Engelmann, F. Importance of Cryopreservation for the Conservation of Plant Genetic Resources. In *Cryopreservation of Tropical Plant Germplasm: Current Research Progress and Application*; Engelmann, F., Takagi, H., Eds.; Japan International Research Center for Agricultura: Tsukuba, Japan; International Plant Genetic Resources Institute: Rome, Italy, 2000; pp. 8–20.
16. Ono, M.; Baak, S.J. Revisiting the J-Curve for Japan. *Mod. Econ.* **2014**, *5*, 32–47. [CrossRef]
17. Sakai, A.; Kobayashi, S.; Oiyama, I. Cryopreservation of nucellar cells of navel orange (*Citrus sinensis* Osb. var. brasiliensis Tanaka) by vitrification. *Plant Cell Rep.* **1990**, *9*, 30–33. [CrossRef] [PubMed]
18. Kartha, K.K.; Engelmann, F. Cryopreservation and Germplasm Storage. In *Plant Cell and Tissue Culture*; Vasil, I.K., Thorpe, T.A., Eds.; Kluwer Press: Dordrecht, Germany, 1994; pp. 195–230.
19. Sakai, A.; Engelmann, F. Vitrification, encapsulation-vitrification and droplet-vitrification: A review. *CryoLetters* **2007**, *28*, 151–172. [PubMed]
20. Yamamoto, S.; Rafique, T.; Priyantha, W.S.; Fukui, K.; Matsumoto, T.; Niino, T. Development of a cryopreservation procedure using aluminium cryo-plates. *CryoLetters* **2011**, *3*, 256–265.
21. Niino, T.; Yamamoto, S.; Fukui, K.; Castillo Martínez, C.R.; Arizaga, M.V.; Matsumoto, T.; Engelmann, F. Dehydration im-proves cryopreservation of mat rush (*Juncus decipiens* Nakai) basal stem buds on cryo-plates. *CryoLetters* **2013**, *34*, 549–560. [PubMed]
22. Matsumoto, T. Cryopreservation of axillary shoot tips of *in vitro*-grown grape (*Vitis*) by a two-step vitrification protocol. *Euphytica* **2003**, *131*, 299–304. [CrossRef]
23. Bettoni, J.C.; Kretzschmar, A.A.; Bonnart, R.; Shepherd, A.; Volk, G.M. Cryopreservation of 12 *Vitis* Species Using Apical Shoot Tips Derived from Plants Grown *In vitro*. *HortScience* **2019**, *54*, 976–981. [CrossRef]
24. Gonzalez-Arnao, M.T.; Panta, A.; Roca, W.M.; Escobar, R.H.; Engelmann, F. Development and large scale application of cryopreservation techniques for shoot and somatic embryo cultures of tropical crops. *Plant Cell Tissue Organ Cult. (PCTOC)* **2007**, *92*, 1–13. [CrossRef]
25. Niino, T.; Arizaga, M.V. Cryopreservation for preservation of potato genetic resources. *Breed. Sci.* **2015**, *65*, 41–52. [CrossRef]
26. Panis, B.; Lambardi, M. Status of cryopreservation technologies in plants (crops and forest trees). In *The Role of Biotechnology in Exploring and Protecting Agricultural Genetic Resources*; Ruane, J., Sonnino, A., Eds.; FAO: Rome, Italy, 2006; pp. 61–78.

27. Mathew, L.; McLachlan, A.; Jibran, R.; Burritt, D.J.; Pathirana, R.N. Cold, antioxidant and osmotic pre-treatments maintain the structural integrity of meristematic cells and improve plant regeneration in cryopreserved kiwifruit shoot tips. *Protoplasma* **2018**, *255*, 1065–1077. [CrossRef]
28. Marković, Z.; Chatelet, P.; Preiner, D.; Sylvestre, I.; Kontić, J.K.; Engelmann, F. Effect of shooting medium and source of material on grapevine (*Vitis vinifera* L.) shoot tip recovery after cryopreservation. *CryoLetters* **2014**, *35*, 40–47. [PubMed]
29. Bettoni, J.C.; Bonnart, R.; Volk, G.M. Challenges in implementing plant shoot tip cryopreservation technologies. *Plant Cell Tissue Organ Cult. (PCTOC)* **2021**, *144*, 21–34. [CrossRef]
30. Volk, G.M.; Harris, J.L.; Rotindo, K.E. Survival of mint shoot tips after exposure to cryoprotectant solution components. *Cryobiology* **2006**, *52*, 305–308. [CrossRef] [PubMed]
31. Fuller, B.J. Cryoprotectants: The essential antifreezes to protect life in the frozen state. *CryoLetters* **2004**, *25*, 375–388. [PubMed]
32. Towill, L.E.; Bonnart, R. Cracking in a vitrification solution during cooling or warming does not effect growth of cryo-preserved mint shoot tips. *CryoLetters* **2003**, *24*, 341–346. [PubMed]
33. Volk, G.M.; Walters, C. Plant vitrification solution 2 lowers water content and alters freezing behavior in shoot tips during cryoprotection. *Cryobiology* **2006**, *52*, 48–61. [CrossRef]
34. Ozudogru, E.A.; Kaya, E. Cryopreservation of *Thymus cariensis* and *T. vulgaris* shoot tips: Comparison of three vitrification-based methods. *CryoLetters* **2012**, *33*, 363–375. [PubMed]
35. Engelmann-Sylvestre, I.; Engelmann, F. Cryopreservation of *in vitro*-grown shoot tips of *Clinopodium odorum* using aluminium cryo-plates. *Vitr. Cell. Dev. Biol. Anim.* **2015**, *51*, 185–191. [CrossRef]
36. Vujović, T.; Sylvestre, I.; Ružić, D.; Engelmann, F. Droplet-vitrification of apical shoot tips of *Rubus fruticosus* L. and *Prunus cerasifera* Ehrh. *Sci. Hortic.* **2011**, *130*, 222–228. [CrossRef]
37. Barraco, G.; Sylvestre, I.; Iapichino, G.; Engelmann, F. Cryopreservation of *Limonium serotinum* apical meristems from *in vitro* plantlets using droplet-vitrification. *Sci. Hortic.* **2011**, *130*, 309–313. [CrossRef]
38. Panta, A.; Panis, B.; Ynouye, C.; Swennew, R.; Roca, W. Development of a PVS2 droplet vitrification method for potato cryopreservation. *CryoLetters* **2014**, *35*, 255–266. [PubMed]
39. Kartha, K.K.; Leung, N.L.; Mroginski, L.A. *In vitro* growth responses and plant regeneration from cryopreserved meristems of cassava (Manihot esculenta Crantz). *Z. Pflanzenphysiol.* **1982**, *107*, 133–140. [CrossRef]
40. Yamamoto, S.; Fukui, K.; Rafique, T.; Khan, N.I.; Castillo Martinez, C.R.; Sekizawa, K.; Matsumoto, T.; Niino, T. Cryopreservation of *in vitro*-grown shoot tips of strawberry by the vitrification method using aluminium cryo-plates. *Plant Genet. Resour.* **2012**, *10*, 14–19. [CrossRef]
41. Engelmann, F. Cryopreservation of Clonal Crops: A Review of Key Parameters. *Acta Hortic.* **2014**, *1039*, 31–39. [CrossRef]
42. Yamamoto, S.; Rafique, T.; Fukui, K.; Sekizawa, K.; Niino, T. V-cryo-plate procedure as an effective protocol for cryobanks: Case study of mint cryopreservation. *CryoLetters* **2012**, *33*, 12–23. [PubMed]
43. Schäfer-Menuhr, A.; Schumacher, H.M.; Mix-Wagner, G. Cryopreservation of Potato Cultivars—Design of a Method for Routine Application in Genebanks. *Acta Hortic.* **1997**, *447*, 477–482. [CrossRef]
44. Panis, B.; Piette, B.M.A.G.; Swennen, R. Droplet vitrification of apical meristems: A cryopreservation protocol applicable to all Musaceae. *Plant Sci.* **2005**, *168*, 45–55. [CrossRef]
45. Halmagyi, A.; Pinker, I. Plant regeneration from *Rosa* shoot tips cryopreserved by a combined droplet vitrification method. *Plant Cell Tissue Organ Cult. (PCTOC)* **2006**, *84*, 145–153. [CrossRef]
46. Pawłowska, B.; Szewczyk-Taranek, B. Droplet vitrification cryopreservation of *Rosa canina* and *Rosa rubiginosa* using shoot tips from in situ plants. *Sci. Hortic.* **2014**, *168*, 151–156. [CrossRef]
47. Le Bras, C.; Le Besnerais, P.-H.; Hamama, L.; Grapin, A. Cryopreservation of ex-vitro-grown *Rosa chinensis* 'Old Blush' buds using droplet-vitrification and encapsulation-dehydration. *Plant Cell Tissue Organ Cult. (PCTOC)* **2013**, *116*, 235–242. [CrossRef]
48. Kim, H.H.; Yoon, J.W.; Park, Y.E.; Cho, E.G.; Sohn, J.K.; Kim, T.S.; Engelmann, F. Cryopreservation of potato cultivated va-rieties and wild species: Critical factors in droplet vitrification. *CryoLetters* **2006**, *27*, 223–234. [PubMed]
49. Sant, R.; Panis, B.; Taylor, M.; Tyagi, A. Cryopreservation of shoot-tips by droplet vitrification applicable to all taro (*Colocasia esculenta* var. esculenta) accessions. *Plant Cell Tissue Organ Cult. (PCTOC)* **2007**, *92*, 107–111. [CrossRef]
50. Chen, X.-L.; Li, J.-H.; Xin, X.; Zhang, Z.-E.; Xin, P.-P.; Lu, X.-X. Cryopreservation of *in vitro*-grown apical meristems of *Lilium* by droplet-vitrification. *S. Afr. J. Bot.* **2011**, *77*, 397–403. [CrossRef]
51. Kaya, E.; Souza, F.V.D. Comparison of two PVS2-based procedures for cryopreservation of commercial sugarcane (*Saccharum* spp.) germplasm and confirmation of genetic stability after cryopreservation using ISSR markers. *Vitr. Cell. Dev. Biol.-Anim.* **2017**, *53*, 410–417. [CrossRef]
52. Senula, A.; Keller, E.R.J.; Sanduijav, T.; Yohannes, T. Cryopreservation of cold-acclimated mint (*Mentha* spp.) shoot tips using a simple vitrification protocol. *CryoLetters* **2007**, *28*, 1–12.
53. Bi, W.-L.; Hao, X.-Y.; Cui, Z.-H.; Volk, G.M.; Wang, Q. Droplet-vitrification cryopreservation of *in vitro*-grown shoot tips of grapevine (*Vitis* spp.). *Vitr. Cell. Dev. Biol.-Anim.* **2018**, *54*, 590–599. [CrossRef]
54. Volk, G.M.; Shepherd, A.N.; Bonnart, R. Successful Cryopreservation of Vitis Shoot Tips: Novel Pre-treatment Combinations Applied to Nine Species. *CryoLetters* **2019**, *39*, 322–330.

55. Wang, B.; Li, J.-W.; Zhang, Z.; Wang, R.-R.; Ma, Y.-L.; Blystad, D.-R.; Keller, E.J.; Wang, Q.-C. Three vitrification-based cryopreservation procedures cause different cryo-injuries to potato shoot tips while all maintain genetic integrity in regenerants. *J. Biotechnol.* **2014**, *184*, 47–55. [CrossRef]
56. Souza, F.V.D.; Kaya, E.; Vieira, L.D.J.; De Souza, E.H.; Amorim, V.B.D.O.; Skogerboe, D.; Matsumoto, T.; Alves, A.A.C.; Ledo, C.A.D.S.; Jenderek, M.M. Droplet-vitrification and morphohistological studies of cryopreserved shoot tips of cultivated and wild pineapple genotypes. *Plant Cell Tissue Organ Cult. (PCTOC)* **2015**, *124*, 351–360. [CrossRef]
57. Yamamoto, S.; Rafique, T.; Sekizawa, K.; Koyama, A.; Ichihashi, T.; Niino, T. Development of an effective cryopreservation protocol using aluminum cryo-plates for *in vitro*-grown shoot tips of mulberries (*Morus* spp.) originated from the tropics and subtropics. *Sanshi Konchu Biotec (J. Insect Biotech. Sericology)* **2012**, *81*, 57–62.
58. Sekizawa, K.; Yamamoto, S.-I.; Rafique, T.; Fukui, K.; Niino, T. Cryopreservation of *in vitro*-grown shoot tips of carnation (*Dianthus caryophyllus* L.) by vitrification method using aluminium cryo-plates. *Plant Biotechnol.* **2011**, *28*, 401–405. [CrossRef]
59. Matsumoto, T.; Yamamoto, S.; Fukui, K.; Niino, T. Cryopreservation of blueberry dormant shoot tips using V cryo-plate method. *HortScience* **2014**, *49*, S337–S338.
60. Niino, T.; Watanabe, K.; Nohara, N.; Rafique, T.; Yamamoto, S.-I.; Fukui, K.; Arizaga, M.V.; Martinez, C.R.C.; Matsumoto, T.; Engelmann, F.; et al. Cryopreservation of mat rush lateral buds by air dehydration using aluminum cryo-plate. *Plant Biotechnol.* **2014**, *31*, 281–287. [CrossRef]
61. Rafique, T.; Yamamoto, S.I.; Fukui, K.; Mahmood, Z.; Niino, T. Cryopreservation of sugarcane using the V cryo-plate technique. *CryoLetters* **2015**, *36*, 51–59. [PubMed]
62. Yamamoto, S.-I.; Wunna; Rafique, T.; Arizaga, M.V.; Fukui, K.; Gutierrez, E.J.C.; Martinez, C.R.C.; Watanabe, K.; Niino, T. The Aluminum Cryo-plate Increases Efficiency of Cryopreservation Protocols for Potato Shoot Tips. *Am. J. Potato Res.* **2015**, *92*, 250–257. [CrossRef]
63. Bettoni, J.C.; Bonnart, R.; Shepherd, A.N.; Kretzschmar, A.A.; Volk, G.M. Modifications to a *Vitis* Shoot Tip Cryopreservation Procedure: Effect of Shoot Tip Size and Use of Cryoplates. *CryoLetters* **2019**, *40*, 103–112.
64. Salma, M.; Fki, L.; Engelmann-Sylvestre, I.; Niino, T.; Engelmann, F. Comparison of droplet-vitrification and D-cryoplate for cryopreservation of date palm (*Phoenix dactylifera* L.) polyembryonic masses. *Sci. Hortic.* **2014**, *179*, 91–97. [CrossRef]
65. Matsumoto, T.; Yamamoto, S.-I.; Fukui, K.; Rafique, T.; Engelmann, F.; Niino, T. Cryopreservation of Persimmon Shoot Tips from Dormant Buds Using the D Cryo-plate Technique. *Hortic. J.* **2015**, *84*, 106–110. [CrossRef]
66. Murashige, T.; Skoog, F. A Revised Medium for Rapid Growth and Bio Assays with Tobacco Tissue Cultures. *Physiol. Plant.* **1962**, *15*, 473–497. [CrossRef]
67. Matsumoto, T.; Sakai, A.; Yamada, K. Cryopreservation of *in vitro*-grown apical meristems of wasabi (*Wasabia japonica*) by vitrification and subsequent high plant regeneration. *Plant Cell Rep.* **1994**, *13*, 442–446. [CrossRef]

Article

Efficient Protocol for Improving the Development of Cryopreserved Embryonic Axes of Chestnut (*Castanea sativa* Mill.) by Encapsulation–Vitrification

Mariam Gaidamashvili [1,*], Eka Khurtsidze [1], Tamari Kutchava [1], Maurizio Lambardi [2] and Carla Benelli [2]

1. Department of Biology, Faculty of Exact and Natural Sciences, Iv. Javakhishvili Tbilisi State University, 1, Chavchavadze Ave., 0179 Tbilisi, Georgia; eka.khurtsidze@tsu.ge (E.K.); tamari.kutchava2013@ens.tsu.edu.ge (T.K.)
2. Institute of BioEconomy, National Research Council (CNR/IBE), Sesto Fiorentino, 50019 Florence, Italy; maurizio.lambardi@ibe.cnr.it (M.L.); carla.benelli@ibe.cnr.it (C.B.)
* Correspondence: mariam.gaidamashvili@tsu.ge

Abstract: An optimized cryopreservation protocol for embryonic axes (EAs) of chestnut (*Castanea sativa* Mill.) has been developed based on the encapsulation–vitrification procedure. EAs of mature seeds were aseptically dissected and encapsulated in alginate beads with or without 0.3% (*w/v*) activated charcoal (AC). Embedded EAs were dehydrated with Plant Vitrification Solution 2 for different treatment times up to 120 min, followed by direct immersion in liquid nitrogen. Cryopreserved embryonic axes encapsulated with AC showed higher survival (70%) compared to those encapsulated without AC (50%). Sixty-four percent of embryonic axes, from synthetic seeds with AC, subsequently developed as whole plants. Plantlet regrowth was faster in AC-encapsulated EAs and showed enhanced postcryopreservation shoot and root regrowth over 2 cm after five weeks from rewarming. Results indicate that encapsulation–vitrification with activated charcoal added to the beads is an effective method for the long-term preservation of *Castanea sativa* embryonic axes.

Keywords: activated charcoal; alginate; cryopreservation; European chestnut; zygotic embryo

1. Introduction

European chestnut or sweet chestnut (*Castanea sativa* Mill.), belonging to the genus *Castanea*, is dominant in the mountainous forests of Western Georgia (150–1800 m), occupying the highest percentage of areas covered with forests (approx. 75%). Chestnut forests are developed in both West and East Georgia, but to the West of the country, they occupy larger areas. Chestnut trees generally extend from 100 m (Western Georgia) up to 900–1000 m a.s.l., reaching the absolute upper limit at 1400 m in sporadic locations of West and East Georgia [1,2]. According to the official International Union for Conservation of Nature (IUCN) list, *Castanea sativa* has been assessed as Least Concern [3,4]. Because of low self-renewal and pathogenic diseases, the large massifs of chestnut forests in Georgia are on the verge of destruction [5]. Therefore, sweet chestnut has been included in the Red List of Georgia under state Vulnerable (VU), according to the IUCN Red List Categories and Criteria [6,7]. The reason for including *Castanea sativa* in the Red List is the fragmentation and decreased distribution range. Hereafter, the development of efficient conservation measures is essential for both economic and wildlife protection commitments.

The ex situ conservation of chestnut in seed banks is limited due to nonresistance to storage at low-temperature conditions of partially dehydrated recalcitrant seeds [8,9]. Medium- (by in vitro slow-growth storage) and long-term preservation techniques (by cryopreservation in liquid nitrogen (LN)) are currently widely used for selected germplasm collections of various woody perennials [10–12]. In chestnut, Janeiro et al. reported the successful medium-term preservation of chestnut hybrid clones since the mid-1990s [13]. *Castanea* shoot cultures remained viable between 5 and 18 months of slow-growth storage at 4–8 °C [14–16]. Depending on storage conditions, up to 82% of explants survived and resumed normal growth [17]. As for cryopreservation, this method has been broadly used for the preservation of different biological materials of chestnut species and hybrid clones, such as shoot tips [18,19], zygotic embryonic axes [8,20,21] and embryogenic cultures [22–24], where desiccation and Plant Vitrification Solution 2 (PVS2)-based vitrification techniques [25] have been practiced on naked explants. On the other hand, since its proposal in the early 1990s [26,27], the use of dehydrated encapsulated explants (generally named synthetic seeds) has become a valid alternative for the cryopreservation of many plant species [28,29]. Following the "encapsulation–dehydration" method, a new variant, termed "encapsulation–vitrification," was proposed [27,30]. This method combines the advantages of the vitrification and encapsulation of explants, greatly reducing the time required to apply a protocol in comparison to the "encapsulation–dehydration" method. So far, the "encapsulation–vitrification" technique has successfully been applied to the cryopreservation of the shoot tips of several fruit crops [31–33] and embryogenic cell suspensions of grapevine (*Vitis* spp.) [34].

The present study describes, for the first time, a protocol for the cryopreservation of the excised embryonic axes (EAs) of *Castanea sativa* L. by using the "encapsulation–vitrification" approach to evaluate its effectiveness in long-term conservation. Furthermore, the addition of activated charcoal (AC) as a component of the artificial matrix of synthetic seeds to diminish the polyphenol toxicity was tested with the aim to facilitate the optimization of the tested new cryoprocedure.

2. Results

Effect of LS and PVS2 Treatments on Survival and Regrowth of Encapsulated EAs

Treatment of alginate-coated EAs with only a loading solution (LS, containing 2 M glycerol and 0.4 M sucrose) for 60 min at 25 °C induced a small reduction of survival from 100% (control, nontreated and noncryopreserved) to 88.9% (time 0); however, treatment with LS positively influenced the survival rate of noncryostored (LN−) encapsulated EAs after treatment with PVS2. EA survival remained between 85.7% (30 and 60 min of treatment) and 81.3% (90 min), with no significant differences in percentage values. Only the survival of EAs treated for 120 min was reduced to 53.8% (Figure 1A). Similar findings were observed with the regrowth rates of noncryopreserved encapsulated EAs. EA regrowth remained in the range of 72.2% to 68.8% (30 and 90 min of treatment), and the regrowth of EAs treated for 120 min was reduced to 46.2% (120 min) (Figure 1B). However, only the loading treatment induced tolerance to ultrarapid cooling in LN, as survival and plantlet regrowth of encapsulated EAs passed from nil to almost 16.7% and 8.3%, respectively (Figure 1A,B, 0 treatment).

The PVS2 treatment duration significantly affected the survival of cryopreserved (LN+) encapsulated EAs. The 30 min treatment with PVS2 induced the highest (50%) survival of EAs. A further increase in PVS2 treatment time to 120 min resulted in a significant decline of cryopreserved EA survival up to a minimum of 13.3% (Figure 1A). Referring to the regrowth of plantlets, derived from "germinated" synthetic seeds, noncryopreserved encapsulated EAs exhibited a decrease in regrowth, ranging from 91.6% in control plants to 46.2% after treatment with PVS2 at 120 min (Figure 1B). Plantlet regrowth rates were significantly reduced after cryogenic storage at −196 °C. The 30 min PVS2 treatment was the most effective in inducing tolerance to ultrarapid cooling in LN, resulting in 50% plantlet regrowth in postcryopreservation (Figure 1B). A further increase in PVS2 treatment time yielded a significant reduction of cryopreserved EA regrowth up to a minimum of 13.3% (Figure 1B).

Survival values were significantly different when EAs were encapsulated in alginate beads containing 0.3% activated charcoal (AC) in the artificial matrix. The best results were achieved after the 30 min treatment with PVS2 with a survival rate of 70% in cryopreserved EAs, significantly higher than in non-AC beads (50%; Figure 2A).

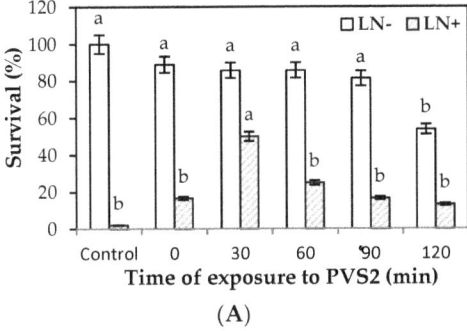
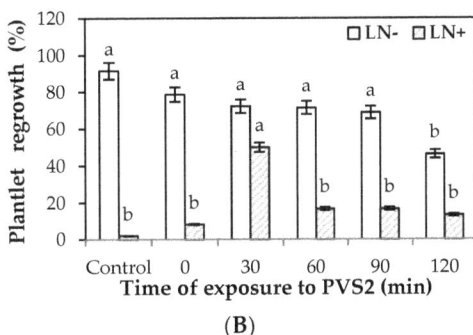

Figure 1. Percentages of survival (**A**) and plantlet regrowth (**B**) of encapsulated *Castanea sativa* embryonic axes (EAs) after exposure to Plant Vitrification Solution 2 (PVS2) for increasing times, with (LN+) or without (LN−) subsequent immersion in liquid nitrogen. EAs were encapsulated in WPM medium containing 2.5% (w/v) sodium alginate. Encapsulated EAs were treated for 60 min with a loading solution (2.0 M glycerol, 0.4 M sucrose), followed by treatment with PVS2 at 0 °C for 30–120 min, prior to direct immersion in LN for 1 h. Control EAs received no LS and PVS2 treatments. Within each line (LN− and LN+), data followed by different letters are significantly different at $p \leq 0.05$ by LSD test (bars, SE of means).

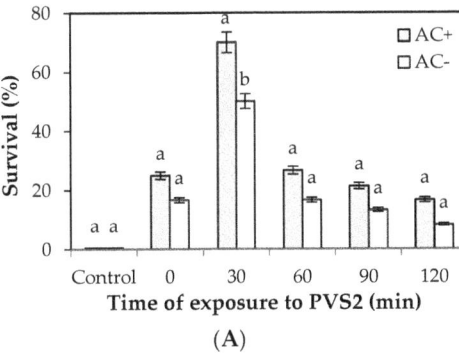
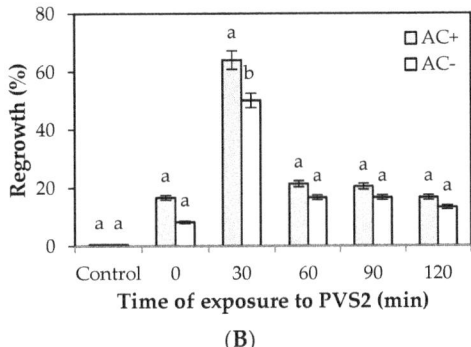

Figure 2. Effect of activated charcoal on the survival (**A**) and regrowth (**B**) of cryopreserved (LN+) chestnut embryonic axes by encapsulation–vitrification. EAs were encapsulated in WPM solution containing 2.5% (w/v) sodium alginate with (AC+) or without (AC−) 0.3% (w/v) activated charcoal in the artificial matrix. Encapsulated EAs were treated for 60 min with LS (2.0 M glycerol, 0.4 M sucrose), followed by treatment with PVS2 at 0 °C for 30–120 min, prior to direct immersion in liquid nitrogen for 1 h. Control EAs received no LS and PVS2 treatments. Within each exposure time, different letters indicate significant differences between AC+ and AC− at $p \leq 0.05$ by chi-squared test (bars, SE of means).

The presence of AC significantly influenced regrowth rates in EAs. The regrowth of AC-encapsulated EAs after treatment with PVS2 for increasing times was in the range of 21.4% and 20.5% for 60 and 90 min of treatment, respectively, whereas it was 16.7% for 120 min of treatment, with no significant differences in percentage values (Figure 2B). In comparison, treatment with PVS2 for 30 min yielded 64% regrowth, showing to be the best recovery rate of AC-added cryopreserved encapsulated EAs, significantly higher than EAs encapsulated without AC (50%, Figure 2B).

The shoot and root length data summarized in Table 1 also show the effect of AC on the plantlet regrowth of encapsulated EAs subjected to various vitrification times with PVS2, five weeks after cryostorage, rewarming and plating.

All surviving cryostored EAs produced roots and shoots, and their development was clearly pronounced in both (AC−) and (AC+) synthetic seeds with the 30 min PVS2 treatment time (Table 1). Moreover, it is noteworthy that with AC added to the synthetic seed, an appreciable shoot and root length was highlighted during postcryopreservation

(Figure 3D), with 14.5 and 22.8 cm, respectively, after five weeks, whereas without AC, the cryopreserved plantlets had 9.2 mm and 10.2 mm of the shoot and root length at the same period. The germination of noncryopreserved synthetic seeds started after one week of culture, and it was delayed up to four weeks after cryogenic storage in LN. However, the regrowth initiation time was shorter in encapsulated explants containing AC, i.e., 20 days in postcryopreservation (Figure 3C). After eight weeks of culture, all (AC+)-derived plantlets showed well-developed roots and shoots that allowed their transfer in greenhouse conditions (Figure 3D–F).

Table 1. Effect of activated charcoal (AC) on the plantlet regrowth of encapsulated *Castanea sativa* EAs subjected to various dehydration times with PVS2 following immersion in LN evaluated 5 weeks after cryostorage, rewarming and plating.

PVS2 (min)	Regrowth [a] (AC−)		Regrowth [a] (AC+)	
	Shoot Length (mm)	Root Length (mm)	Shoot Length (mm)	Root Length (mm)
0	4.0 b	5.2 b	5.0 c	6.3 c
30	9.2 a	10.2 a	14.5 a	22.8 a
60	6.5 b	8.8 a	8.7 b	9.8 b
90	5.9 b	8.9 a	8.3 b	10.4 b
120	4.9 b	7.8 a	7.9 b	9.5 b

[a] Mean of 90 plantlets tested. Data were recorded after 5 weeks of culture following cryopreservation. Statistical analysis in each column was performed by ANOVA. Data followed by different letters are significantly different at $p \leq 0.05$ by LSD test.

Figure 3. Plant regeneration from cryopreserved embryonic axes of *Castanea sativa* by encapsulation–vitrification. (**A**) Excised embryonic axes (EAs) used for cryopreservation. (**B**) Encapsulated EAs in 2.5% sodium alginate with (AC+) or without (AC−) 0.3% (w/v) activated charcoal in the artificial matrix. (**C**) Survived encapsulated (AC+) EAs after cryopreservation and 20 days of postculture. (**D**) Primary plantlet development 2 weeks after survival assessment cryopreserved by encapsulation–vitrification procedure. (**E**) Elongated root and shoot 8 weeks of postculture after cryopreservation. (**F**) Plantlets established under greenhouse conditions 4 weeks after transfer to soil.

3. Discussion

Synthetic seed technology, in addition to fulfilling needs related to micropropagation [35–39], can prove to be an efficient tool for the storage of rare and commercially important species at low temperatures. It has the potential for the medium-term and long-term preservation of plant explants encapsulated in synthetic seeds, without losing viability after immersion in LN when cryopreservation is applied [26–28,40,41].

The "encapsulation–vitrification" cryoprocedure [27] has been used for the cryopreservation of the shoot tips of several woody fruit crops [28,31–33] and embryogenic cell suspensions [34]. Although it requires a long treatment time compared to the vitrification of naked explants, the encapsulation of plant germplasm makes for less damage to samples during the vitrification procedures [42,43]. In our encapsulation–vitrification experiment with chestnut EAs, after treatment with LS for 60 min, the 30 min exposure time of PVS2 showed the best regrowth rate (50%). Optimizing the time of exposure to PVS2 is most important for producing a satisfactory level of regrowth after cryopreservation, and the PVS2 osmoprotection effect can change among different species [28,44]. For example, the duration of PVS2 treatment was up to 200 min in the encapsulated shoot tips of *Dianthus caryophyllus* L [45].

In the following experiment, the addition of AC into synthetic seeds treated with the same conditions positively affected plantlet initiation and regrowth from chestnut EAs, with the concentration amended with 0.3% (w/v) AC. In a previous research, AC was added in a culture medium to overcome the onset of browning, shortly after the excision of EAs, and promoted their germination [21]. In another study, AC added in the artificial endosperm of synthetic seeds containing somatic embryos of hybrid rice improved their germination and conversion to plantlets [46]. Furthermore, the germination and root development of encapsulated somatic embryos of *Picea glauca* and *Picea mariana* enhanced with the addition of 0.05 gL^{-1} AC to the beads [47]. Therefore, as also shown in this study, AC represents a component that can improve the development of explants even after their ultrarapid cooling in LN. Indeed, the survival and regrowth rates of cryopreserved encapsulated EAs were markedly increased when AC was included in the bead composition.

It is also noteworthy that the results obtained here showed an improvement in the survival and regrowth of cryostored chestnut EAs by 15% and 10%, respectively, in comparison with a previous study concerning the vitrification procedure of naked EAs [21]. Although the survival and regrowth rates of encapsulated (AC−) AEs were lower than the same parameters obtained in a previous study by the "desiccation–one-step cooling" protocol (70% and 64%, respectively) [21], it should be noted that the overall ratio between embryo survival and plantlet regrowth appreciably improved with the presence of AC in synthetic seeds. Thus, the survival/regrowth ratio of AC-encapsulated AEs after cryopreservation was 91.4% versus 83% obtained by the "desiccation–one-step freezing procedure," at the best treatment times. Corredoira et al., [20] reported a 63.3% recovery applied to *Castanea sativa* zygotic embryos by the desiccation procedure, which was still lower than the regrowth percentage obtained in our experiment with (AC+) encapsulation–vitrification.

The conversion of synthetic seeds into plants after germination is a fundamental aspect of the success of the encapsulation–vitrification technique. In this study, the development of cryopreserved encapsulated EAs with (AC+), after treatment with 30 min PVS2, was faster by 6–7 days with respect to EAs encapsulated without AC. Furthermore, the root and shoot length of (AC+) EAs achieved 22.8 mm and 14.5 mm, respectively, five weeks after rewarming and plating, whereas the (AC−) EAs showed less development at the same period (Table 1). Evident differences were also found in the postcryopreservation initiation times of plantlet formation with respect to previous cryopreservation procedures applied on EAs. Indeed, the plantlet development of cryopreserved EAs synthetic seeds with (AC+) started two weeks earlier than naked vitrified EAs, where the full germination of EAs (expressed as plantlet regrowth) required eight weeks and eight days earlier than desiccated by dehydration–"one-step freezing" EAs [21]. Notably, even root and shoot elongation from encapsulated (AC+) EAs was considerably faster, exceeding 2 cm root and 1.5 cm shoot length in five weeks.

AC seems to play a role to keep nutrients within the artificial matrix, releasing them slowly during the development of embryos. The absorption of detrimental polyphenolic exudates released by encapsulated explants is also facilitated by AC [48].

4. Materials and Methods

4.1. Plant Material

Chestnut fruits were collected in Western Georgia at the beginning of October 2019 from the open-pollinated trees of *Castanea sativa*. Mature fruits were stored at 4 °C for a maximum of 1 month until use in the cryopreservation trials. For the cryopreservation experiments, the fruits were washed in 2% (v/v) household detergent and rinsed three times under tap water. Then, the pericarp, seed coat and part of the kernel were removed. The remaining embryo axes along with the part of the kernel were surface-sterilized by successive immersion in 70% (v/v) ethanol with a few drops of Tween 20 for 2 min, followed by decontamination with a 10% (v/w) solution of sodium hypochlorite (Sigma-Aldrich®, Saint Louis, MO, USA) for 20 min. After being rinsed in sterile distilled water three times, EAs, composed of the zygotic embryos along with 2–3 mm long cotyledon residuals, were dissected from the seeds (Figure 3A).

4.2. Encapsulation

Dissected EAs were immersed in a calcium-free liquid woody plant medium (WPM) [49] without plant growth regulators, supplemented with 2.5% (w/v) sodium alginate (Bioworld®, Dublin, OH, USA) and 0.3 M sucrose. The mixture (including dissected EAs) was dropped with a sterile pipette into WPM liquid medium containing 100 mM calcium chloride, forming beads about 4–5 mm in diameter (Figure 3B). The drops with EAs were maintained in the solution for 20 min to achieve polymerization.

In one specific experiment, 0.3% (w/v) activated charcoal (AC; Sigma-Aldrich, DARCO®, Saint Louis, MO, USA) was added to the sodium alginate solution to assess its influence on the survival and regrowth rate of encapsulated EAs after PVS2 treatment and subsequent cooling. All operations were performed under sterile conditions. After the incubation period in the complexion agent, the encapsulated explants were rinsed three times in sterile distilled water.

4.3. Encapsulation–Vitrification Technique for EA Cryopreservation

Encapsulated EAs were transferred to LS containing 2.0 M glycerol and 0.4 M sucrose for 60 min at 25 °C, followed by treatment with PVS2 [25] (30% w/v glycerol, 15%, w/v DMSO, 15% w/v ethylene glycol in WPM medium containing 0.4 M sucrose) for 0, 30, 60, 90, 120 min treatment times at 0 °C. Then, synthetic seeds were placed in 2 mL cryovials (5 in each) and immersed in LN for 24 h (LN+). For rewarming, the cryovials were rapidly immersed in a water bath at 40 °C for 2 min. Encapsulated EAs were rinsed in a washing solution containing the WPM liquid medium and 1.2 M sucrose (two times of 10 min each, at 25 °C), and then LN+ and LN− (synthetic seeds without cooling) samples were cultured in test tubes (20 mm × 150 mm) in WPM supplemented with 30 g L^{-1} sucrose and 0.4 µM 6-benzylaminopurine (BAP). The medium was solidified with 6 g L^{-1} agar (PlantMedia™, Dublin, OH, USA) and adjusted to pH 5.7 before autoclaving. Cultured tubes were maintained in a growth chamber at 24 ± 0.5 °C under a 16/8 h light/dark regime with an irradiance of 40 µmol m^{-2} s^{-1} in cool-white fluorescent light. After 2 and 4 weeks, survival and regrowth were recorded, respectively. After 8 weeks of in vitro culture, plantlets (i.e., "germinated" EAs) were washed thoroughly in running tap water; the root length was measured and transferred to plastic cups filled with a mixture of 100% sphagnum peat/perlite at a ratio of 2:1. The plantlets were relocated for acclimatization in controlled chambers at 23 ± 1 °C under 60 ± 5% moisture content and 16 h photoperiod with an irradiance of 40 µmol m^{-2} s^{-1} in cool-white fluorescent light over the following three weeks. After the emergence of new leaves, the plants were transplanted in bigger pots containing peat, soil and perlite at a ratio of 1:2:1 and transferred to natural greenhouse conditions.

4.4. Data Collection and Statistical Analysis

The total number of embryos used for the experiment was 240. Each treatment, with (AC+) or without (AC−)-activated charcoal, included 50 noncryopreserved (LN−) encapsulated EAs (10 EAs for each condition from 0 to 120 min treatment time) and 50 cryopreserved (LN+) encapsulated EAs (10 EAs for each condition). Control EAs (20 for each, AC− and AC+ treatments), receiving no LS, PVS2 or LN treatment, were also included. For a 0 h-min PVS2 treatment time, encapsulated EAs were only loaded in LS solution and cryopreserved without PVS2 or directly cultured in test tubes for synthetic seed "germination." Each treatment consisted of 3 replicates, and all experiments were repeated 3 times.

Survival was recorded after two weeks of culture and defined as the percentage of the total number of encapsulated EAs, which showed initial normal germination and development (i.e., root and shoot emission) or only root development. The regrowth of encapsulated EAs was assessed after four weeks of culture. Plant regrowth rate was estimated as a percentage of whole plantlets (retaining normal shoots and roots ≥ 5 mm in length) developing from encapsulated EAs relative to the total number of synthetic seeds cultured after cryopreservation. Root and shoot length were recorded weekly. Statistical analysis of percentages was performed by ANOVA (when comparing multiple treatments), followed by the LSD test at $p \leq 0.05$ for mean separation or chi-squared test (when comparing pairs of treatments). Percentage data used in ANOVA were subjected to arcsine transformation prior to analysis. The bars in the figures represent standard errors (SE) of means.

5. Conclusions

The present study has clearly demonstrated the feasibility of the long-term preservation of *Castanea sativa* germplasm by the encapsulation–vitrification of EAs. The acquisition of suitable dehydration tolerance with PVS2 to survive after the cryopreservation of EA synthetic seeds and their germination and regrowth under optimized conditions (AC+) promoted growth by shortening the development times and limiting the loss of explants; therefore, the overall performance of the cryopreserved EAs appears to be improved in comparison with previous studies. The protocol described in this study will now be tested on a wide range of chestnut cultivars and hybrid clones to achieve the practical long-term cryopreservation of *Castanea* genus germplasm.

Author Contributions: Conceptualization, M.G.; methodology, M.G., C.B.; software, E.K.; validation, M.G.; formal analysis, M.G., E.K. and T.K.; investigation, E.K. and T.K.; resources, M.G.; data curation, M.G. and E.K.; writing—original draft preparation, M.G.; writing—review and editing, M.G., C.B., M.L.; visualization, M.G. and E.K., supervision, M.L.; project administration, M.G.; funding acquisition, M.G. All authors have read and agreed to the published version of the manuscript.

Funding: This work was supported by Shota Rustaveli National Science Foundation of Georgia (SRNSFG), grant number FR17-444. The APC was funded by Shota Rustaveli National Science Foundation of Georgia (SRNSFG).

Institutional Review Board Statement: Not applicable.

Informed Consent Statement: Informed consent was obtained from all subjects involved in the study.

Acknowledgments: Shota Rustaveli National Science Foundation of Georgia (SRNSFG) and National Council Research of Italy (Bilateral Project: Developing efficient cryopreservation procedures for the long-term storage of endangered plant genetic resources of Georgia) are acknowledged for financial support.

Conflicts of Interest: The authors declare no conflict of interest. The funders had no role in the design of the study; in the collection, analyses, or interpretation of data; in the writing of the manuscript, or in the decision to publish the results.

References

1. Dolukhanov, A. *Rastitel'nost'Gruzii (Vegetation of Georgia)*; Metsniereba: Tbilisi, Georgia, 1989; Volume 1. (In Russian)
2. Nakhutsrishvili, G. Forest vegetation of Georgia. In *the Vegetation of Georgia (South Caucasus)*; Nakhutsrishvili, G., Ed.; Springer: Berlin/Heidelberg, Germany, 2013; pp. 35–87.
3. IUCN Red List of Threatened Species. Version 2020-3. Available online: https://www.iucnredlist.org (accessed on 18 December 2020).
4. Barstow, M.; Khela, S. Castanea Sativa. The IUCN Red List of Threatened Species. 2018. Available online: https://dx.doi.org/10.2305/IUCN.UK.2018-1.RLTS.T202948A67740523.en (accessed on 18 December 2020).
5. Tavadze, B.; Supatashvili, A.; Kapanadze, G.; Mamukashvili, T. Pathological status of chestnut stands in Tkibuli region (Georgia). *Ann. For.* **2012**, *5*, 21–32.
6. Red List of Georgia. Edict of the President of Georgia #303 on Approval of the Red List of Georgia. Tbilisi. 2006. Available online: https://www.matsne.gov.ge/ka/document/view/97288?publication=0 (accessed on 18 December 2020).
7. IUCN Red List Categories and Criteria 2012 Version 3.1, 2nd ed. Available online: https://www.iucn.org/content/iucn-red-list-categories-and-criteria-version-31-second-edition (accessed on 18 December 2020).
8. Pence, V.C. Desiccation and survival of *Aesculus, Castanea and Quercus* embryo axis through cryopreservation. *Cryobiology* **1992**, *29*, 391–399. [CrossRef]
9. Westengen, O.T.; Jeppson, S.; Guarino, L. Global Ex-Situ Crop Diversity Conservation and the Svalbard Global Seed Vault: Assessing the Current Status. *PLoS ONE* **2013**, *8*, e64146. [CrossRef] [PubMed]
10. Lambardi, M.; De Carlo, A. Application of tissue culture to germplasm conservation of temperate broad-leaf trees. In *Micropropagation of Woody Trees and Fruits*; Jain, S.M., Ishii, K., Eds.; KluwerAcademic: Dordrecht, The Netherlands, 2003; pp. 815–840.
11. Panis, B.; Lambardi, M. Status of cryopreservation technologies in plants (crops and forest trees). In *The Role of Biotechnology in Exploring and Protecting Agricultural Genetic Resources*; Ruane, J., Sonnino, A., Eds.; FAO: Rome, Italy, 2006; pp. 61–78.
12. Panis, B. Sixty years of plant cryopreservation: From freezing hardy mulberry twigs to establishing reference crop collections for future generations. *Acta Hortic.* **2019**, *1234*, 1–8. [CrossRef]
13. Janeiro, L.V.; Vieitez, A.M.; Ballester, A. Cold storage of the in vitro cultures of wild cherry, chestnut and oak. *Ann. Sci. For.* **1995**, *52*, 287–293. [CrossRef]
14. Lambardi, M.; Benelli, C.; De Paoli, G.; Battistini, A. Biotechnologie per la conservazione del Castagno. In Proceedings of the Convegno Nazionale Castagno, Firenze, Italy, 25–27 October 2001; Bellini, E., Ed.; Università di Firenze: Marradi (Firenze), Italy, 2001; pp. 86–91.
15. Corredoira, E.; Valladares, S.; Martinez, M.T.; Couselo, J.L.; San Jose, M.C.; Ballester, A.; Vieitez, A.M. Conservación de germoplasma en especies leñosas con técnicas de cultivo in vitro y almacenamiento en frío (Span). *J. Rural Dev.* **2011**, *2*, 15–24.
16. Corredoira, E.; Martinez, M.T.; Cernadas, M.J.; San-Jose, M.C. Application of biotechnology in the conservation of the genus *Castanea*. *Forests* **2017**, *8*, 394. [CrossRef]
17. Capuana, M.; Di Lonardo, S. In vitro conservation of chestnut (*Castanea sativa*) by slow growth. *In Vitro Cell. Dev. Biol. Plant* **2013**, *49*, 605–610. [CrossRef]
18. Vidal, N.; Sanchez, C.; Jorquera, L.; Ballester, A.; Vieitez, A.M. Cryopreservation of chestnut by vitrification of in vitro-grown shoot tips. *In Vitro Cell. Dev. Biol.-Plant* **2005**, *41*, 63–68. [CrossRef]
19. Jorquera, L.; Vidal, N.; Sánchez, C.; Vieitez, A.M. Optimizing conditions for successful plant regeneration from cryopreserved *Castanea sativa* shoot tips. *Acta Hortic.* **2005**, *693*, 511–518. [CrossRef]
20. Corredoira, E.; San-Jose, M.C.; Ballester, A.; Vieitez, A.M. Cryopreservation of zygotic embryo axes and somatic embryos of European chestnut. *CryoLetters* **2004**, *25*, 33–42. [PubMed]
21. Gaidamashvili, M.; Khurtsidze, E.; Benelli, K.; Lambardi, M. Development of an Efficient 'One-Step Freezing' Cryopreservation Protocol for a Georgian Provenance of Chestnut (*Castanea sativa* Mill.) Zygotic Embryos. *Not. Bot. Hortic. Agrobo.* **2019**, *47*, 1047–1054. [CrossRef]
22. Holliday, C.; Merkle, S.A. Preservation of American chestnut germplasm by cryostorage of embryogenic cultures. *J. Am. Chestnut Found.* **2000**, *14*, 46–52.
23. San Jose, M.C.; Jorquera, L.; Vidal, N.; Corredoira, E.; Sanchez, C. Cryopreservation of European chestnut germplasm. *Acta Hortic.* **2005**, *693*, 225–232. [CrossRef]
24. Vieitez, A.M.; San Jose, M.C.; Corredoira, E. Cryopreservation of zygotic embryonic axes and somatic embryos of European chestnut. In *Plant Embryo Cultures: Methods and Protocols, Methods in Molecular Biology*; Thorpe, T.A., Yeung, E.C., Eds.; Springer: New York, NY, USA, 2011; Volume 710, pp. 201–213.
25. Sakai, A.; Kobayashi, S.; Oiyama, I. Cryopreservation of nucellar cells of navel orange (*Citrus sinensis* Osb. var. *brasiliensis* Tanaka) by vitrification. *Plant Cell Rep.* **1990**, *9*, 30–33. [CrossRef] [PubMed]
26. Fabre, J.; Dereuddre, J. Encapsulation-dehydration: A new approach to cryopreservation of *Solanum* shoot tips. *CryoLetters* **1990**, *11*, 413–426.
27. Matsumoto, T.; Sakai, A.; Takahashi, C.; Yamada, K. Cryopreservation of in-vitro grown apical meristems of wasabi (*Wasabia japonica*) by encapsulation-vitrification method. *CryoLetters* **1995**, *16*, 189–196.
28. Sakai, A.; Engelmann, F. Vitrification, encapsulation-vitrification and droplet-vitrification: A review. *CryoLetters* **2007**, *28*, 151–172.

29. Kulus, D. Application of synthetic seeds in propagation, storage, and preservation of *Asteraceae* plant species. In *Synthetic Seeds Germplasm Regeneration, Preservation and Prospects*; Faisal, A., Alatar, A., Eds.; Springer: Berlin, Germany, 2019; pp. 155–179.
30. Sakai, A.; Hirai, D.; Niino, T. Development of PVS-Based Vitrification and Encapsulation–Vitrification Protocols. In *Plant Cryopreservation: A Practical Guide*; Reed, B.M., Ed.; Springer: New York, NY, USA, 2008; pp. 33–57.
31. Niino, T.; Sakai, A. Cryopreservation of alginate-coated in vitro grown shoot tips of apple, pear and mulberry. *Plant Sci.* **1992**, *87*, 199–206. [CrossRef]
32. Paul, H.; Daigny, G.; Sangwan-Norreel, B.S. Cryopreservation of apple (*Malus domestica* Borkh.) shoot tips following encapsulation-dehydration or encapsulation-vitrification. *Plant Cell Rep.* **2000**, *19*, 768–774. [CrossRef]
33. Wang, Q.; Laamanen, J.; Uosukainen, M.; Valkonen, J.P.T. Cryopreservation of in vitro-grown shoot tips of raspberry (*Rubus idaeus* L.) by encapsulation–vitrification and encapsulation–dehydration. *Plant Cell Rep.* **2005**, *24*, 280–288. [CrossRef] [PubMed]
34. Wang, Q.C.; Mawassi, M.; Sahar, N.; Li, P.; Violeta, C.-T.; Gafny, R.; Sela, I.; Tanne, E.; Perl, A. Cryopreservation of grapevine (*Vitis* spp.) embryogenic cell suspensions by encapsulation–vitrification. *Plant Cell Tissue Organ Cult.* **2004**, *77*, 267–275. [CrossRef]
35. Lambardi, M.; Halmagyi, A.; Benelli, C.; De Carlo, A. Seed cryopreservation for conservation of ancient Citrus germplasm. *Adv Hortic Sci.* **2007**, *21*, 198–202.
36. Rai, M.K.; Asthana, P.; Singh, S.K.; Jaiswal, V.S.; Jaiswal, U. The encapsulation technology in fruit plants—A review. *Biotechnol. Adv.* **2009**, *27*, 671–679. [CrossRef] [PubMed]
37. Sharma, S.; Shahzad, A.; da Silva, J.A.T. Synseed technology—a complete synthesis. *Biotechnol. Adv.* **2013**, *31*, 186–207. [CrossRef] [PubMed]
38. Benelli, C.; Micheli, M.; De Carlo, A. An improved encapsulation protocol for regrowth and conservation of four ornamental species. *Acta Soc. Bot. Pol.* **2017**, *86*, 3559. [CrossRef]
39. Micheli, M.; Standardi, A.; da Silva, D.F. Encapsulation and Synthetic Seeds of Olive (*Olea europaea* L.): Experiences and Overview. In *Synthetic Seeds—Germplasm Regeneration, Preservation and Prospects*; Faisal, M., Alatar, A., Eds.; Springer: Cham, Switzerland, 2019; pp. 347–361.
40. Carlo, A.D.; Benelli, C.; Lambardi, M. Development of a shoot-tip vitrification protocol and comparison with encapsulation-based procedures for plum (*Prunus domestica* L.) cryopreservation. *CryoLetters* **2000**, *21*, 215–222.
41. Kulus, D. Effect of bead composition, PVS type, and recovery medium in cryopreservation of bleeding heart 'Valentine'—Preliminary Study. *Agronomy* **2020**, *10*, 891. [CrossRef]
42. Kulus, D. Shoot Tip Cryopreservation of *Lamprocapnos spectabilis* (L.) fukuhara using different approaches and evaluation of stability on the molecular, biochemical, and plant architecture levels. *Int. J. Mol. Sci.* **2020**, *21*, 3901. [CrossRef]
43. Kami, D. Cryopreservation of plant genetic resources. In *Current Frontiers in Cryobiology*; Katkov, I., Ed.; IntechOpen: London, UK, 2012; pp. 439–456.
44. Kulus, D. Application of cryogenic technologies and somatic embryogenesis in the storage and protection of valuable genetic resources of ornamental plants. In *Somatic Embryogenesis in Ornamentals and Its Applications*; Mujib, A., Ed.; Springer: New Delhi, India, 2016.
45. Halmagyi, A.; Deliu, C. Cryopreservation of carnation (*Dianthus caryophyllus* L.) shoot tips by encapsulation-vitrification. *Sci. Hortic.* **2007**, *113*, 300–306. [CrossRef]
46. Kumar, M.B.A.; Vakeswaran, V.; Krishnasamy, V. Enhancement of synthetic seed conversion to seedlings in hybrid rice. *Plant Cell Tissue Organ Cult.* **2005**, *81*, 97–100. [CrossRef]
47. Lulsdorf, M.M.; Tautorus, T.E.; Kikcio, S.I.; Bethune, T.D.; Dunstan, D.I. Germination of encapsulated embryos of interior spruce (*Picea glauca engelmannii* complex) and black spruce (*Picea mariana* Mill.). *Plant Cell Rep.* **1993**, *12*, 385–389. [CrossRef] [PubMed]
48. George, E.F.; Sherrington, P.D. *Plant Propagation by Tissue Culture—Handbook and Directory of Commercial Laboratories*; Exegetics Ltd.: Eversley, UK, 1984.
49. Lloyd, G.; McCown, B.H. Woody Plant Medium (WPM)—A mineral nutrient formulation for microculture of woody plant species. *HortScience* **1981**, *16*, 453.

Review

Vitrification Solutions for Plant Cryopreservation: Modification and Properties

Jiri Zamecnik *, Milos Faltus and Alois Bilavcik

Crop Research Institute, Drnovska 507, 16106 Prague, Czech Republic; faltus@vurv.cz (M.F.); bilavcik@vurv.cz (A.B.)
* Correspondence: zamecnik@vurv.cz

Abstract: Many plants cannot vitrify themselves because they lack glassy state-inducing substances and/or have high water content. Therefore, cryoprotectants are used to induce vitrification. A cryoprotectant must have at least the following primary abilities: high glass-forming property, dehydration strength on a colligative basis to dehydrate plant cells to induce the vitrification state, and must not be toxic for plants. This review introduces the compounds used for vitrification solutions (VSs), their properties indicating a modification of different plant vitrification solutions, their modifications in the compounds, and/or their concentration. An experimental comparison is listed based on the survival or regeneration rate of one particular species after using more than three different VSs or their modifications. A brief overview of various cryopreservation methods using the Plant Vitrification Solution (PVS) is also included. This review can help in alert researchers to newly introduced PVSs for plant vitrification cryoprotocols, their properties, and the choice of their modifications in the compounds and/or their concentration.

Keywords: cryoprotectant; ultra-low temperature; glassy state; toxicity

1. Introduction

The cryopreservation of plant genetic resources aims to ensure the long-term storage of viable and genetically stable plant material at an ultra-low temperature using liquid nitrogen (LN, −196 °C) or liquid nitrogen vapour (LNV, −165 to −190 °C). At these temperatures, plant tissues are preserved in a state where cellular divisions and metabolic activities are minimized [1,2], thus preserving the genetic integrity for a longer duration [3,4]. The process of cryopreservation ensures the viability of plant tissues for a theoretically unlimited period [5].

Vitrification—glass formation without crystallization [6,7]—is one of the basic principles used in plant cryopreservation methods. Only a few plants can form a vitreous state naturally [8]. Most plants cannot vitrify themselves because they lack glassy state inducing substances and/or have high water content. The cryoprotectant used for vitrification should have at least three primary abilities: a high glass-forming ability, dehydration strength on a colligative basis to dehydrate plant cells to induce the vitrification state, and the cryoprotectant concentration used must not result in excessive toxicity to the plants. Despite the toxicity of Plant Vitrification Solution 2 (PVS2), it remains a highly effective vitrification solution for plant shoot tip systems [9].

Vitrification cryoprotective solutions reduce the risk of damage of the organelle structures by avoid forming ice crystals [10]; this is achieved by increasing the cell viscosity to the point at which ice formation is inhibited both inside and outside the cell [11]. Thus, cryoprotective solutions protect cell membranes with a gelatinous fluid and can form a glassy state in the cells, which helps plants survive at ultra-low temperatures. In addition, they can prevent further lethal water loss and maintain the percentage of regeneration after cryopreservation [12].

Cryopreservation through the application of vitrification solutions was first reported in plant cells [13–15]. Currently, the vitrification-based methods that use vitrification solutions are considered the most widely applied for plant cryo-biologists [16,17]. There are more than 800 papers on shoot tip cryopreservation using vitrification solutions in the literature [18,19]. In addition, several vitrification solutions with different compositions have been reported, but most of the vitrification-based protocols use only a few key solutions [19–22].

Cryopreservation procedures are currently available for many essential plant species [23–25]. For successful vitrification protocols using Cryo Protective Agent (CPA), all the freezable water must be removed from the cells through the use of Plant Vitrification Solution (PVS) before LN exposure. Furthermore, other steps are essential to the success of cryopreservation protocols such as preconditioning, hardening, pre-loading, loading, osmoprotection with various substances before cryopreservation [26,27] and unloading after cryopreservation [28,29]. PVS treatment time, concentration, temperature, as well as shoot tip size, age, its physiology are also very important [30]. All these points are beyond the scope of this review.

The issues addressed by this review are the comparison of different PVSs, their differences in the concentration of the substances used, or the modification of their composition graphically in tables for a quick orientation. Modification is a way to improve previously used PVSs for new genotypes or genotypes with low regeneration rates. A case study on 13 different plant species using at minimum four different PVSs and their modifications is listed. In addition, a brief overview of cryopreservation methods using the PVSs is also included. We assume that this review will also help better select PVSs and their modifications for plant cryopreservation progress.

2. Cryoprotective Substances

The successful cryoprotection by vitrification is based on eliminating the formation of ice crystals and on reducing the toxicity of cryoprotective substances [31]. Cryopreservation is a reversible process, as long as the optimal combination and concentration of cryoprotective solution effective enough to form vitrified plant tissues are used. Vitrification refers to the physical process of supercooling a liquid to low temperatures and finally solidifying into a metastable glass without undergoing crystallization at a applied cooling rate [32]. The basic characteristics of substances used in cryopreservation are summarized in Table 1. The glass transition temperature tends to increase with increasing the relative molecular mass as opposed to the melting point [33]. In this order, the most commonly used cryoprotective substances are glycerol (Gly), dimethyl sulfoxide (DMSO), ethylene glycol (EG), propylene glycol (PG), sucrose (Suc), and sorbitol (Sor), at different concentrations in combination with MS medium [34] to allow sufficient dehydration of plant material and also induce cryoprotective processes to the cells [35]. DMSO was later discovered to be an active cryoprotective substance, the alkaline in the main, a good ligand, easily alkalized by a strong base, and subject to deprotonation. Some key features of cryoprotective substances are mentioned below and in the Table 1.

The cryoprotective substances are classified into several groups according to the way of penetrating the cells: (a) small substances penetrating the cells through the cell wall and plasma membrane, such as EG, PG, Gly and DMSO; (b) substances penetrating only through the cell wall, e.g., oligosaccharides such as sucrose (Suc), sorbitol (Sor), mannitol (Man), or amino acids such as proline (Pro), and relatively low molecular mass polymers such as polyethylene glycol (PEG1000); (c) substances that do not penetrate through the cell walls or the plasma membrane, such as relatively high molecular mass polymers (soluble proteins, polysaccharides and, polyethylene glycol PEG 6000) [36–38].

2.1. Substances That Can Penetrate through the Cell Wall and into the Protoplast

2.1.1. Glycerol

Gly was used as the first for mouse embryo cryopreservation [39]. Gly was also the first cryoprotectant to see widespread use in human cryobiology [40]. Due to its higher relative molecular mass and viscosity, it penetrates membranes [35,41,42], but slower than DMSO and EG [42]. Gly occurs in plants in relevant quantities. Fifteen crop plants grown under field conditions had leaf concentrations of Gly between 10 and 39 $\mu g\ g^{-1}$ wet weight of tissue [43]. Gly is a polar molecule, freely miscible with water and simple alcohols [44]. Gly shows lower toxicity because it has a low ability to penetrate the membrane, which at high concentrations can lead to osmotic shock [45]. Gly concentration used for cryopreservation is dependent on the specific vitrification solution and ranges from 20% (weight in volume, w/v) [46] to 50% (w/v) [47].

Table 1. Characteristics of commonly used vitrification solutions and their chemical and physical properties in PVS.

Substances	Abr.	M_r g mol^{-1}	T_m °C	T_g °C	Density g cm^{-3}	LD$_{50}$
	1	2	3	4	5	6	7
Sulfoxides						
Dimethyl sulfoxide [48]	DMSO	78.13	18.45	−132.15	1.10	**
Diols						
Ethylene glycol [49]	EG	62.07	−13	−113.15	1.11	*
Propylene glycol [50]	PG	76.06	−59	−100.65	1.4	**
PEG 8000 [51]	PEG	8000	63	54.82	1.21	***
Triols						
Glycerol [52]	Gly	92.09	19	−83.15	1.26	*
Polyalcohols						
Sorbitol [53]	Sor	182.17	111.5	−6	1.49	**
Monosaccharides						
Glucose [54]	Glu	164.16	147	22.85	1.4	*
Disaccharides						
Sucrose [54]	Suc	342.3	186	59.85	1.587	***
Proteins						
Bovine serum albumin [55]	BSA	66.5 kDa	69.8	§	-	-
Amide						
Formamide [56]		45.04	2.55	-	1.13	*
Plant Vitrification Solutions						
Plant vitrification solution 1 [46]	PVS1	42.48	−41	−122	1.15	-
Plant vitrification solution 1 [57]	PVS1			−155		-
Plant vitrification solution 2 [46]	PVS2	37.51	−44	−119	1.14	-
Plant vitrification solution 2 [58]	PVS2			−115		-
Plant vitrification solution 3 [47]	PVS3	56.29	−35.4	−93.9	1.29	-
Plant vitrification solution 4 [59]	PVS4	73.53	−33.9	−112	1.31	-
Plant vitrification solution N [58]	PVSN	-	−50	−110	-	-
Vitrification solution L [46]	VSL	32.97	−41	−125	1.9	-
Vitrification solution L [46]	VSL+	-	−47	−121	-	-

§—the glass transition is depending on the rate of cooling [60];. Abr.—abbreviation; Mr—relative molecular mass; Tg—glass transition temperature; Tm—temperature of melting point equilibrium; LD50—median dose (*dosis letalis* media) * <5000, ** 5001–20,000, *** >20,001 mg kg^{-1}, mL kg^{-1} (mouse, rat, rabbit). The data in columns 3–5 are from the citations mentioned in column No. 1 as upper index; data in columns 6–7 are from the safety sheets.

2.1.2. Dimethyl Sulfoxide

Another common cryoprotectant is DMSO [61], which is used, like Gly, not only for freezing both animal and plant tissues but also in the cryopreservation of microorganisms [41]. DMSO enhances the passage of water molecules across the cell wall and cytoplasmic membrane [62]. The uptake dynamics of DMSO, Gly, and sucrose during dehydration of garlic shoot tips displayed a biphasic nature, with an initial rapid influx followed by a slower, gradual increase in DMSO [63], Gly, and sucrose (Suc) [64]. Room temperature increased the membrane permeability in contrast to a temperature close to zero. The reverse efflux pattern during unloading was similarly temperature dependent. DMSO is commonly used in combination with glycol-type compounds as this combination interacts with water and biological materials slightly different way than these components alone. DMSO is hydrophobic at higher temperatures, meaning it is less toxic at lower temperatures, which leads to lower, slower, and easily controllable tissue dehydration and oxidation of sulphide groups [65]. In addition, DMSO is an effective solvent and has a high osmolality. In contrast, it is harmful and may cause somaclonal variability or mutagenesis. Due to the persisting uncertainties regarding mutagenesis [66,67] or high acute and chronic toxicity, as shown in the case of rhesus monkeys [68] and low order phytotoxicity, some current cryobanks do not use vitrification solutions containing DMSO, as a precaution, not denying the fact that DMSO can be an excellent cryoprotective substance.

2.1.3. Ethylene Glycol

EG acts as a dehydration substance before cryopreservation. Due to its low freezing point, EG is used for its rapid penetration into the cells and its ability to block the ice crystal formation. EG in cryoprotective solutions is usually used at a half concentration in comparison with Gly [69,70]. EG is metabolized to oxalic acid, which increases the acidity of the organism and it is harmful. The oxalic acid reacts with the calcium contained mainly in the cell wall and forms insoluble calcium oxalate crystals stored in vacuoles [71]. High concentrations of the calcium oxalate, crystallized in various crystals, form raphides, druses, or others [72]. EG is commonly used in combination with glycol-type compounds as its combination interacts with water and biological materials [73].

2.1.4. Propylene Glycol

PG is also a commonly used liquid in the cryopreservation process. PG is metabolized to oxalic acid in the metabolic process and acts like EG. However, as a cryoprotectant in warm conditions, PG is non-toxic while EG is metabolized to toxic elements [74].

2.1.5. Polyethylene Glycol

PEG is a liquid or solid, depending on the molecular mass (which typically ranges from 300 g mol^{-1} to 10,000,000 g mol^{-1}). The chemical properties are almost the same, but forms with different molecular mass (from 600 to 8000 g mol^{-1}), and different physical properties are usually used for cryopreservation [75]. PEG used for plant cryopreservation, e.g., PEG 6000 normally has a concentration of 15% (w/v) [76].

2.2. Substances That Can Penetrate through the Cell Wall

High levels of carbohydrates and sugar alcohols occur in plants as natural, non-toxic cryoprotective substances [77]. Monosaccharides are readily dissolved in cryoprotective solutions and can vitrify plant tissues at a lower concentration level than disaccharides. Therefore, the disaccharide sucrose is often used as an antifreeze agent. Compared to cryoprotection using monosaccharide glucose, sucrose has a higher efficiency [78]. In addition, carbohydrates contribute to the dehydration of samples and are added to the cryoprotective mixture to increase the protection of the membrane integrity in a dehydrated state [70,79,80].

2.2.1. Glucose

Glucose belongs to a group of monosaccharides that reducing sugar due to the presence of an aldehyde group, which is oxidized to a carboxylic acid group to form D-gluconic acid [81]. On the contrary, reducing the aldehyde group of glucose to the primary alcoholic group forms D-Glucitol, called sorbitol. D-Glucose monohydrate is produced in green plants during the photosynthesis process, which is a fast and basic energy supply. Due to its molecular size, it penetrates the cell membrane faster than sucrose, but in an experimental comparison of plant regeneration after cryopreservation, sucrose, as well as the sugar alcohol mannitol, proved to be more useful [69].

2.2.2. Sorbitol

Sorbitol is used in cryopreservation because of its lower melting point [82]. It provides less energy than sucrose (1 g of sorbitol gives up to 10,886 kJ of energy). Göldner et al. [83] introduced a range of carbohydrates to increase the frost resistance of plants (*Digitalis lanata*) used for plant pre-cultivation. They showed that the most damaged plant cells were cultured on sorbitol and proline medium. In contrast, the smallest cell damage occurred when sucrose was used. The concentration of 0.4 M and 0.8 M sorbitol in the pre-cultivation embryogenic tissues of hybrid firs (*Abies alba* × *A. cephalonica*, *Abies alba* × *A. numidica*) has been found acceptable for subsequent survival and regeneration of the plants [84]. High levels of sugars and sugar alcohols are found in many polar plants, insects, fungi, etc., as non-toxic cryoprotectants [85].

2.2.3. Sucrose

Sucrose, composed of fructose and glucose molecules, is the most widespread disaccharide. It is easily hydrolyzable, dissociated by glycosidase invertase to the laevorotatory glucose and dextrorotatory fructose. These translocated sugars are photosynthetically metabolized in the Calvin cycle. Due to their molecular size, these carbohydrates are preferable for the transport assimilated over long distances. Sucrose is energetically abundant (1 g of sucrose provides 16,747 kJ of energy), and it acts as an energy source in heterotrophic nutrition after plant rewarming during its regeneration. The disaccharide sucrose is more effective than the monosaccharide glucose for vitrification [47,86–88]. Sucrose is used to promote dehydration before and/or during cryopreservation. Sucrose is normally membrane-impermeable and has low toxicity. The concentration of sucrose used in cryopreservation processes varies from 5% (w/v) [46] to 50% (w/v) [47], but most often 40% sucrose (w/v) [89,90] is used. For mint shoot tips, sucrose reduces the toxicity of ethylene glycol and DMSO at 22 °C and Gly at 0 °C [65].

2.2.4. Amides

Amides are weak cryoprotectants compared to polyols (formamide is too weak to vitrify itself, but can assist vitrification by other cryoprotectants). Adding methyl groups increases the effectiveness of cryoprotectants [91]. Amides, compared to polyols, generally have weak cryoprotective effects [45].

2.2.5. Bovine Serum Albumin (BSA)

BSA decreases the kinetic constant value determined for concentrated EG solutions. However, BSA's effect was small compared to that which could be produced by a slight increase in EG concentration [92]. On the contrary, Rall [93] suggest that the inclusion of BSA in vitrification solutions may be an effective means of increasing the stability of the amorphous state of vitrification solutions.

3. Substances That Do Not Penetrate through the Cell Wall

Substances that do not penetrate through the cell wall are polymers with high molecular weight such as soluble proteins, polysaccharides, mucilage, PEG1000.

Turner et al. [94] proposed that the mode of action of polyalcohols (in our enumeration Gly, mannitol, sorbitol) is not based on molarity, but rather on the total number of hydroxyl (OH) groups present in the medium. Furthermore, based on their results, they propose that the orientation of OH groups is a determining factor in effective cryopreservation [94].

The development of cryogenic technologies is facilitated by biophysical studies capable of monitoring glass stability during cryopreservation [33]. The glass transition temperature of substances depends on the concentration of an aqueous solution, cooling/warming rates, annealing temperature, and type of mixture. For example, three different glass transitions were found in the subzero temperature range of −163, −138, and −93 °C at 20% BSA (w/w) [60]. Sucrose has also glass transitions at the three different temperatures ranges; T_{g1} (−50 to −45 °C), T_g' (−36 °C), and T_g (−83 to −57 °C); all sucrose glass transitions are concentration-dependent and the first two are cooling rate-independent [88]. The thermal analysis of plant vitrification solution: PVS1, PVS2, Towill's, Fahy's, or Steponkus' vitrification solutions reveals only a small water peak detected in shoot tips after 120 minutes dehydration duration. Still, recovery of cryopreserved garlic shoot tips exposure to these vitrification solutions was low (from 0 to 25%), in comparison to 80% regeneration after PVS3 [95].

4. Vitrification Solutions and Modifications

Many PVS solution variations have been reported to be suitable for cryopreservation of several plants. During their testing, a number of their successful modifications were published. In this section, an attempt is made to give an overview of the most important of them. The original Plant Vitrification Solution 1 (PVS1) was firstly used by Uragami [14] for cultured cells and somatic embryos derived from the mesophyll tissue of asparagus (*Asparagus officinalis* L.). The original composition of PVS1 is in Table 2.

Table 2. The concentration of substances of the original Plant Vitrification Solution numbered one (PVS1) uses Uragami [14].

PVS1	DMSO (%)	Suc (%)	Gly (%)	EG (%)	PG (%)	PEG (%)	Sor (%)	Total (%)	Plant
Uragami	7		22	15	15		9.1	68.1	*Asparagus officinalis* L. [14]
PVS1-M1	6		22	13	13			54	*Malus* sp. [96]
PVS1-M2	6		19	13	13		9.1	60.1	*Porphyra yezoensis* [94]
PVS1-M3	6	13.7	22	13	13			67.7	*Allium sativum* L. [97]
PVS1-M4	10		22	13		13		58	*Dioscorea opposite* [98]
PVS1-M5	7		22	15		15		59	*Citrus madurensis* [99]
PVS1-M6		31.1	18.4	15				64.5	*Allium sativum* L. [100]
PVS1-M7	7	15	22	30				74	*Bletila strata* [101]
PVS1-M8	5	13.7	13.7					32.4	*Citrus madurensis* [99]

DMSO—dimethyl sulfoxide, Suc—sucrose, Gly—glycerol, EG—ethylene glycol, PG—propylene glycol, PEG—polyethylene glycol 8000 m.w., Sor—sorbitol, Total—total concentration of all substances. Significant composition changes added or omitted substances and/or modification in the concentrations of original PVS1 in % (w/v) used in plant cryopreservation. The shaded area expresses no changes concerning the original PVS1.

The PVS1 was used in modifications PVS1-M1 to PVS1-M3 and PVS1-M8 with a lower concentration of DMSO at a concentration up to 6% and in modifications PVS1-M4 with a higher concentration of DMSO (10%). Sucrose was not used in the original PVS1, but it was used at a concentration of 13.7% (w/v) in the modifications (PVS1-M3 and PVS1-M8). The Gly was in the modification PVS1-M2, PVS-M6, and PVS1-M8 in lower concentrations and higher concentrations in the PVS1-M6 than in the original PVS1. EG was used less concentrated (13% w/v) in (PVS1-M1 to PVS1-M4) and more concentrated (PVS1-M7) in comparison with the original. PG was used less concentrated in PVS1-M1 to PVS1-M3 and without change in modifications PVS1-M4 to PVS1-M8. Sorbitol was omitted in PVS1-M1 and PVS1-M3 to PVS1-M8. These modifications were also used for shoot tip cryopreservation e.g., *Rauvolfia serpentine* [102] and *Cocos nucifera* L. [103].

Among several PVSs, PVS2 (Table 3) and PVS3 (Table 4) are the most frequently used vitrification solutions. The PVS2 was firstly used at a concentration of 60% [22]. The PVS2 in full-strength [13] (first row in Table 3) is also widely used. Several modifications of PVS2 with different substances and their concentration used have been published: DMSO was used in a lower concentration, from 7.5 to 13% (w/v), in modifications PVS2-M2 to PVS2-M6 (Table 3). The exact concentration of sucrose (0.4 M) used in the original PVS2 was also in modifications from PVS2-M1 to PVS2-M3 and in PVS2-M10. Sucrose was used in a higher concentration from 15 to 34.2% (w/v) in the modifications from PVS2-M6 to PVS2-M9 and no sugar was used in PVS2-M4 and PVS2-M5. Gly was used in the modification PVS2-M3, PVS2-M6 in a lower concentration, and PVS2-M8 higher than in the original PVS2. The concentration of EG was unchanged. PG was added in PVS2-M1 and PVS2-M2 at 15 and 7.5% (w/v), respectively. PEG 8000 was added in PVS2-M3 and PVS2-M10 as 3% (w/v) solution and PVS2-M6 as 2% (w/v) solution. Instead of sucrose, sorbitol in PVS2-M5 was added at 15% (w/v) (Table 3).

PVS2-M1 is a widely used plant vitrification solution [104] in several cryoprotocols for various plant species e.g., *Photinia × fraseri* Dress. [105], *Allium sativum* L. [95,106], *Rauvolfia serpentine* [102], *Cocos nucifera* L. [103], *Porphyra yezoensis* [94], *Mentha piperita* L. [65], *Dioscorea* spp. [15]. PVS2-M3 was used for cryopreservation of e.g. *Prunus avium* L. [107], and PVS2-M6 was used for cryopreservation of *Bromus inermis* Leyss [46]. PVS2-M8 was used for cryopreservation of *Clinopodium odorum* [108].

Incubation time in PVS2 varies according to species, temperature conditions, shoot tip size, pretreatment, preculture and, cryopreservation protocol. Therefore, there is no generic time for PVS2. For example, in apple the droplet-vitrification method had the highest regrowth percentage after 30–50 min PVS2 exposure at room temperature [109]; in potato droplet-vitrification had the highest regrowth percentage after 50 min PVS2 exposure at 0 °C [110]; in shallot droplet-vitrification had the highest regrowth percentage after 40–60 min PVS2 exposure at 0 °C [111]; in grapevine droplet-vitrification had the highest regrowth percentage after 90 min PVS2 exposure at 0 °C [112], and in yacon droplet-vitrification had the highest regrowth percentage after 60 min PVS2 exposure at 0 °C [22,113]. DMSO and Gly penetrate the cell wall membrane and increase cellular osmolality avoiding ice formation [7,38,114].

Volk and Walters [9] proposed according to their differential scanning calorimeter measure that the PVS2 operates through two cryoprotective mechanisms: (a) it replaces cellular water, and (b) it changes the freezing behaviour of any water remaining in the cells. They expressed the theory that the penetration of some of the components (e.g., DMSO) of PVS2 into the cell is essential to its cryoprotective efficacy. Significantly, the assumption that the mode of action of PVS2 is primarily caused by osmotic dehydration cannot explain its high effectiveness. Cell-penetrating constituents of PVS2 replace water as the cells become dehydrated and prevent injurious cell shrinkage caused by dehydration [9].

Table 3. Composition and modification in concentration of substances of Plant Vitrification Solution 2 (PVS2) Sakai [13].

PVS2	DMSO (%)	Suc (%)	Gly (%)	EG (%)	PG (%)	PEG (%)	Sor (%)	Total (%)	Plant
Sakai [§]	15	13.7	30	15				73.7	*Citrus sinensis* [13]
PVS2-M1		13.7	30	15	15			73.7	*Porphyra yezoensis* [94]
PVS2-M2	7.5	13.7	30	15	7.5			73.7	*Porphyra yezoensis* [94]
PVS2-M3	12.5	13.7	25	15		3 *		69.2	*Prunus salicina* Lindley cv. Methley x *Prunus spinosa* L. [115]
PVS2-M4	15		30	15				60	*Malus* [96]
PVS2-M5	15		30	15			15	75	*Tetraclinis articulata* (Vahl.) [116]
PVS2-M6	13	15	25	15		2		70	*Guazuma crinita* Mart. [117]
PVS2-M7	15	34.2	30	15				94.2	*Poncirus trifoliata* (L.) Raf. × *Citrus sinensis* (L.) Osbeck. [118]
PVS2-M8 [§§]	15	22.5	37.5	15				52.5	*Allium sativum* L. and *Dendranthema grandiflora* [61], *Rubus fruticosus* L. and *Prunus cerasifera* Ehrh [119]
PVS2-M9 [§§§]	15	15	30	15				75	*Guazuma crinita* Mart. [13]
PVS2-M10	15	13.7	30	15		3 **		76.7	*Populus alba* L. [120]

Important composition changes, added or omitted substances, and/or modification in the concentrations of original PVS2 in % (w/v) used in the plant cryopreservation. All substances were dissolved in MS medium with 0.4 M of sucrose. The sucrose concentration in PVS2 was approximately 0.15 M. The shaded area expresses no changes concerning the original PVS2. DMSO—dimethyl sulfoxide, Suc—sucrose, Gly—glycerol, EG—ethylene glycol, PG—propylene glycol, PEG—polyethylene glycol, Sor—sorbitol, Total—total concentration of all substances. § termed '100%' of PVS2; §§ termed PVS2-A3 [61]; §§§—'60%' of PVS2; *—PEG 8000 m.w., **—PEG 4000 m.w.

The first use of PVS3 was reported on *Asparagus officinalis* L. by Nishizawa et al. [47]. The original PVS3 plant vitrification solution contained 50% to 50% (w/v) sucrose and Gly (Table 4).

Table 4. The concentration of substances of the original Plant Vitrification Solution 3 (PVS3) [47]. Important composition changes added or omitted substances, and/or modification in the concentrations of original PVS3 in % (w/v) used in plant cryopreservation.

PVS3	DMSO (%)	Suc (%)	Gly (%)	EG (%)	Total (%)	Plant
Nishizawa		50	50		100	*Asparagus officinalis* L. [47]
PVS3-M1	5	50	50		100	*Porphyra yezoensis* [94]

Table 4. Cont.

PVS3	DMSO (%)	Suc (%)	Gly (%)	EG (%)	Total (%)	Plant
PVS3-M2		50	30		80	*Malus domestica* Borkh. [69]
PVS3-M3		45	45		90	*Kalopanax septemlobus* [121]
PVS3-M4		40	40		80	Gentian, Wasabi, *Malus* [22,122] *Fragaria ananassa* [123]
PVS3-M5		60	35	20	105	*Asparagus officinalis* L. [47]

DMSO—dimethyl sulfoxide, Suc—sucrose, Gly—glycerol, EG—ethylene glycol, Total—total concentration of all substances. The shaded area expresses no changes concerning the original PVS3.

Sakai [119] and Benson [11,22] indicated 40% to 40% ratio like an original PVS3. We label this ratio as modification 4 to the original PVS3; this modification is listed as PVS3-M4 in Table 4. Since the first description of PVS3, there have been many reports of vitrification methods using PVS3. PVS3 was used for 136 different plant species till 2007 [22]. The published modifications concern is lowering the concentration of sucrose, mostly in the same ratio, e.g., [22,69,121], to Gly, e.g., PVS3-M3 [121] and PVS3-M4 [22]. PVS3 is the plant vitrification solution without DMSO and EG in the original composition compared to PVS2; curiously, the first modification, PVS3-M1, contains 5% (w/v) of DMSO in addition to the basic substances. In PVS3-M2, there is only a change in Gly as a 30% (w/v) concentrated solution [69]. PVS3-M5 contains 20% (w/v) of ethylene glycol in addition together with an increase of the content of sucrose to 60% (w/v) and a decrease of glycerol to 35% (w/v).

The original PVS3 is also widely used in the cryopreservation of *Lithodora rosmarinifolia* (Ten.) [122], *Photinia x fraseri* Dress. [105], *Allium sativum* L. [95], *Rauvolfia serpentine* [102], *Cocos nucifera* L. [103], *Bromus inermis* Leyss [103], *Porphyra yezoensis* [94], *Mentha piperita* L. [65], *Dioscorea* spp. [15], *Malus* [96] and other species, but somewhat less than PVS2, the most frequently used vitrification solution [22].

The original PVS4 is without DMSO (Table 5). The two following modifications (PVS4-M1) and PVS4-M2) are with the addition of DMSO. In PVS4-M1 there are only two compounds, DMSO and Gly, at 5% (w/v); this low concentration of cryoprotectants combined with a slow cooling rate (0.1–0.2 °C min^{-1}) act rather as dehydration solution than vitrification solution.

Table 5. Plant Vitrification Solution 4 (PVS4), Steponkus', Towill's and their modifications in concentration, composition, and some omitted and added substances in % (w/v).

	DMSO (%)	Suc (%)	Gly (%)	EG (%)	PEG * (%)	Sor (%)	BSA (%)	CaCl$_2$ (mM)	Total (%)	Plant
PVS4		20.5	35	20					75.5	Various plants [124]
PVS4-M1	5		5						10	*Malus* [96]
PVS4-M2 **	10	15	20	30				10	75	*Citrus madurensis* [99] *Bromus inermis* Leyss [46]
PVS4-M3 ***	10	5	20	30				10	65	*Bromus inermis* Leyss [46]
Steponkus				43.5		16	6		65.5	*Secale cereale* L. [125]
Steponkus-M1		13.7		50		15	6		84.7	*Allium sativum* L. [126]
Towill	7.8			35	10				52.8	*Mentha aquatica* × *M. spicata* [127]
Towill-M1	10			35	5				50	*Guazuma crinita* Mart. [117]
Towill-M2	6.8	13.7		35	10				65.5	*Allium sativum* L. [126]

DMSO—dimethyl sulfoxide, Suc—sucrose, Gly—glycerol, EG—ethylene glycol, PG—propylene glycol, PEG—polyethyleneglycol, Sor—sorbitol, BSA—bovine serum albumin, Total—total concentration of all substances. * PEG 8000 m.w., ** known also as VSL, *** known also as VSL+ [46]. The shaded area expresses no changes concerning the original PVS.

The PVS proposed by Steponkus' vitrification solution is without sucrose. Sucrose is used in the first modification of Steponkus' vitrification solution (Steponkus-M1) in the concentration of 13.7% (w/v). The PVS proposed by Towill's is slightly modified by increasing the DMSO to 10% and decreasing PEG 8000 to a half. Sucrose is used in the second modification of Towill's vitrification solution (Towill-M2) in the concentration of 13.7% (w/v) together with 6.8 % (w/v) of DMSO influencing the regeneration up to 39% of *Allium sativum* shoot tips [126].

Thermal analysis of PVSs revealed that increasing Gly concentration reduced endothermic peaks, indicating the ice-blocking property of Gly [61]. Increasing sucrose concentration in PVSs also decreased endothermic enthalpies by decreasing explant moisture content and increasing the influx of cryoprotectants [64]. Therefore, balancing the Gly and sucrose concentration in the design of PVSs is also crucial to increase recovery. A limitation of the use of PVS3 is the high osmotic stress increasing during the action; therefore, induction of desiccation tolerance during preconditioning of samples is essential if they are not inherently tolerant [61,128].

5. Comparison of Vitrification Solutions on Regeneration

Evaluation of different cryoprotectant solutions and their modification are ordered in Table 6. There is a comparison of three or more different cryoprotective vitrification mixtures.

Table 6. Regeneration rate (%) after application of Plant Vitrification Solutions and their modifications (M1_M4 for details see Tables 2–5). Three or more Plant Vitrification Solutions or their modifications at one particular species.

PVS1	Mod.	PVS2	Mod.	PVS3	PVS4	PVS5	VSL	Steponkus	Mod.	Towill	Mod.	Fahy	R/S **	Plant
0	M3	0		80	0			0	M1	0	M2	0	R	*Cocos nucifera* L. [103]
11	M1	27		80	25			23		39	M2	11	R	*Allium sativum* L. [126]
0		20		0			0	0 §		0		0	R	*Elaeis guineensis* [129]
0	M1	0	M4	70	0	0							S	*Malus* [96]
36		30		20	28								S	*Centaurea ultreia* [130]
65		75		65	65								S	*Citrus madurensis* [99]
55		32					80						S	*Gentian* [46]
80		20		0			14						R	*Fraser Photinia* [105]
18	M2	24	M1,M2	15									R	*Porphyra yezoensis* [131]
34		49		25									R	*Solanum tuberosum* L. [132]
0	M1	87		0									R	*Rauvolfia serpentina* L. [102]
		59			38								R	*Discorea* [15]
92	M1	82		52 §§§§									83 *	S	*Ipomoea batatas* [133]

§ according to Watanabe and Steponkus [134]; §§§§ 88% of PVS3; * PVS N (1 M sucrose + 15% glycerol + 14% ethylene glycol [133]; ** R—stands for regeneration, regrowth, S—stands for survival after cryopreservation. The PVS3, PVS4, PVS5 [96], VSL [46], and Fahy's vitrification solution [31] are unmodified compared to other PVSs.

PVS3, PVS4, PVS5, and Fahy are without modification in Table 6, even though they have several modifications (see Tables 3 and 4). Among the PVSs, PVS2 and PVS3 are the most frequently used [104]. In some cases, PVS3 may be less toxic to plant species sensitive to PVS2, such as *Allium* sp. [111]. Based on the results presented over the years and also from Tables 2–6, it is evident that the optimum cryoprotectant solution treatment is species or cultivar-specific. Furthermore, the PVSs exposure duration and temperature conditions during incubation are related to the size of the shoot tip, as well as to the preculture and pretreatment conditions [19,135].

The original composition is listed in the first row in each table (Tables 2–5) and its modifications in concentration and/or in compounds used are followed. Sakai et al. [82] were the first to report a PVS2-vitrification cryopreservation protocol for nucellar cells of *Citrus sinensis*. The modifications of PVS2 were applied on other plants as presented in the paper by Uragami et al. [14] and Maruyama et al. [117].

Modifications of PVS and its influence on the viability of explants have been reported. Suzuki [46] in addition to the three original vitrification solutions (PVS1, PVS2, VSL) (Table 6) presented the effect of 12 other combinations of cryoprotective substances on gentian axillary buds. The best regeneration of 79.7% after liquid nitrogen treatment was achieved with the original VSL. Cho et al. [99] used four original PVSs (PVS1, PVS2, VSL, and VSL+) (Table 6). The best one for *Citrus madurensis* embryonic axes survival after liquid nitrogen treatment was PVS2. Kim [61] modified the PVS2 in nine modifications and PVS3 in four modifications in concentrations of substances in the droplet-vitrification procedure. The best one was the PVS3 without any modifications for shoot tips harvested from in vitro conditions of *Dendranthema grandiflora* T. and garlic clove shoot apices of *Allium sativum* L.

6. Vitrification Solution and Cryopreservation Methods

Increased vitrification method efficiency was achieved by treating plants in a pretreatment and preculture steps before cryopreservation of plant shoot tips [22,89,136–139].

Pretreatment/preculture increases tolerance to PVSs during the dehydration process. Pretreatment conditioning differs by species, and then preculture for some species is crucial [140]. During pretreatments the sucrose intake mostly takes place, and increased content of proline and other protective substances accumulates in the plant shoot tips while growing in the carbohydrate enriched culture medium. The temperature during the incubation of plants in PVSs is important for both toxicity and dehydration. The temperature close to 0 °C for plants treated in PVS2 is crucial and had significantly lower lethality than at 22 °C [65,141–143]. When the temperature is subsequently lowered, the penetrating components of PVS2 cryoprotect the cells by restricting the molecular mobility of water molecules and preventing them from nucleating ice crystals [11].

In PVSs vitrification-based methods, most or all of the freezable water is removed by using highly concentrated and viscous cryoprotectant mixtures which, after rapid cooling in LN, form a glass [11,144]. The amount of water in the cells is decreasing due to an accumulation of these substances, and the central vacuole is divided into several smaller ones.

Cryoprotective substances help ensure the stability of membranes and enzymes in subsequent dehydration by vitrification solutions and avoid the formation of ice crystals [145,146]. In this case, the samples are exposed to minutes–a few hours long treatment by several cryoprotective substances before LN exposure. The effect of cryoprotective solution composition for plant regeneration was studied in different plant species [11,61,126,139,147,148]. The published results indicate the importance of PVS compositions, the vitrification protocol, pre-culture, regrowth media, and the application of an appropriate vitrification technique to achieve optimum post-cryopreservation recovery [105,131].

With the combination of the composition of cryopreservation solution (15% DMSO + 3% sucrose) and subsequent slow cooling, a droplet-freezing method was developed for cassava shoot tips [146].

The droplet-vitrification method is derived from the DMSO droplet methods proposed by Kartha et al. [146], and Kaczmarczyk et al. [149] and Schaefer-Menhur et al. [150]. The procedure is similar to the droplet method but with highly concentrated cryoprotective solution PVS2 [104,147,150–152] or PVS3 either in original or in its modification PVS3-M4 [122] or both PVS2 and PVS3 [97,113,126] before ultra-fast cooling. Rewarming of the samples is usually done in unloading solution tempered in a 40 °C sterile water bath for 1–2 min. When using potentially phytotoxic DMSO, the cryoprotective mixture is washed out in unloading solution tempered in a water bath with solutions of decreasing concentrations of sucrose or sorbitol as unloading solutions. This method was successfully applied many of plant species and is widely used in genebanks for cryopreserving vegetatively propagated crop collections [153,154].

Other cryopreservation methods use the PVS for inducing vitrification, such as encapsulation-dehydration and encapsulation-vitrification method with PVS2, PVS3 [139,155] foil-vitrification, droplet-vitrification, and droplet-freezing methods with PVS1, PVS2, PVS3, and VSL [105]. In addition other methods use the vitrification solutions PVS2 and PVS3 in V cryo-plate and D cryo-plate methods [27], PVS2 in cryo-mesh method [156,157], and PVS2 in vacuum infiltration vitrification method (VIV) [158].

The determination of plant survival and regeneration level is done by a visual evaluation of growing the plants in vitro conditions. The ratio of regenerated to the total number of cryopreserved plants is expressed as a success of the cryoprotocol (see Table 6).

Cryoprotectants can change the biophysical properties of plant parts. Cryoprotectants are selected based on their potential non-toxicity, high osmolality, and ability to penetrate as a particular component of vitrification solution into the cell. A low survival and regeneration of plants can also be caused by insufficient osmotic adjustment of plant material, excessive shrinkage of cells in hypertonic conditions, the toxicity of the vitrification solution, low penetration ability of the cryoprotective solution into the plant tissue, low dehydration of the plant tissue and subsequent formation of intracellular ice crystals during freezing [6]. During slow cooling, the sample may reduce its cell surface due to the loss of cytoplasmic membrane, and the cell lysis can occur upon returning to a normal state [64,79]. The toxicity of cryoprotective substances can be associated with the denaturation of proteins, which are damaged either by low temperature or high cryoprotectant concentration necessary for plant tissue vitrification. The strongest vitrification is achieved with cryoprotective substances, which can bind hydrogen bonds to water molecules. They make the interaction of hydrogen bonds water-water, which is the basis for forming ice crystal structure. These substances can be bound by hydrogen bonds in proteins, causing their denaturation [35]. Reduction of water bound to proteins can damage the cells. Dehydration to the level of bound water is essential for successful cryopreservation without the use of cryoprotective substances.

The following steps are recommended for cryopreservation of a new plant species. First, the new PVS should be chosen from cryoprotective solutions close to the species family with the highest regeneration ability. The second possibility is to use the PVS widely used for most plants, e.g., PVS2, PVS3, etc. After choosing the PVS, it is necessary to test the toxicity level following the growing test and level of dehydration [159] according to their regeneration. If the thermal analysis is available, it will help a lot at this step [160]. The difference between the regeneration rate of control and ultra-low temperature treated plants is the potential to improve regeneration by improved vitrification solution.

7. Conclusions

Cryopreservation methods allow long-term storage of genetically unique plant material in the vitreous state at ultra-low temperatures of LN, which leads to the suppression of all biochemical reactions. Vitrification solutions as a mixture of two to seven substances induce a glassy state in plant tissues and prevent the ice crystal formation during the cooling and warming process. The cryoprotective mixture toxicity can be reduced by an appropriate combination or decrease in the concentration of cryoprotective substances and/or physical condition, mainly low temperature at which those are applied. The best cryoprotective solutions can reduce the toxicity of the vitrification mixture. Easier and faster cryoprotectant penetration into the cells and tissue dehydration to the optimal level for cryopreservation will increase the survival and regeneration of plants and extend cryopreservation methods for other plant species and genotypes. The widely used vitrification solutions meet these demands for high regeneration (over the minimum standard of cryobank) after cryopreservation.

Author Contributions: Writing and review of literature, J.Z.; writing—review and editing, A.B. and M.F.; project and funding acquisition, J.Z. and M.F. All authors have read and agreed to the published version of the manuscript.

Funding: This research was funded by the Ministry of Agriculture of the Czech Republic, projects number MZERO0418, QK1910476, and QK1910277.

Institutional Review Board Statement: Not applicable.

Informed Consent Statement: Not applicable.

Data Availability Statement: The data presented in this study are available on request from the corresponding author.

Acknowledgments: The authors would like to acknowledge Renata Kotkova for first establishing the requirement of this topic during her doctoral studies and to Stacy Hammond Hammond for the English corrections.

Conflicts of Interest: The authors declare no conflict of interest.

References

1. Engelmann, F. Use of biotechnologies for the conservation of plant biodiversity. *Vitr. Cell. Dev. Biol. Anim.* **2011**, *47*, 5–16. [CrossRef]
2. Wang, M.-R.; Bi, W.; Shukla, M.R.; Ren, L.; Hamborg, Z.; Blystad, D.-R.; Saxena, P.K.; Wang, Q.-C. Epigenetic and Genetic Integrity, Metabolic Stability, and Field Performance of Cryopreserved Plants. *Plants* **2021**, *10*, 1889. [CrossRef] [PubMed]

3. Engelmann, F. In vitro conservation methods. In *Biotechnology and Plant Genetic Resources*; Callow, J.A., Ford Lloyd, B.V., Newbury, H.J., Eds.; CAB International: Oxford, UK, 1997; pp. 119–162.
4. Wang, M.-R.; Chen, L.; Da Silva, J.A.T.; Volk, G.M.; Wang, Q.-C. Cryobiotechnology of apple (*Malus* spp.): Development, progress and future prospects. *Plant Cell Rep.* **2018**, *37*, 689–709. [CrossRef]
5. Panis, B. Sixty years of plant cryopreservation: From freezing hardy mulberry twigs to establishing reference crop collections for future generations. *Acta Hortic.* **2019**, 1–8. [CrossRef]
6. Zámečník, J.; Šesták, J. Constrained States Occurring in Plants Cryo-Processing and the Role of Biological Glasses. In *Hot Topics in Thermal Analysis and Calorimetry*; Springer: Singapore, 2010; Volume 8, pp. 291–310.
7. Benson, E.E. Cryopreservation of Phytodiversity: A Critical Appraisal of Theory & Practice. *Crit. Rev. Plant Sci.* **2008**, *27*, 141–219. [CrossRef]
8. Hirsh, A.G. Vitrification in plants as a natural form of cryoprotection. *Cryobiology* **1987**, *24*, 214–228. [CrossRef]
9. Volk, G.M.; Walters, C. Plant vitrification solution 2 lowers water content and alters freezing behavior in shoot tips during cryoprotection. *Cryobiology* **2006**, *52*, 48–61. [CrossRef]
10. Grout, B.W.W. Introduction to the in Vitro Preservation of Plant Cells, Tissues and Organs. In *Genetic Preservation of Plant Cells in Vitro*; Springer: Singapore, 1995; pp. 1–20.
11. Benson, E.E. Cryopreservation theory. In *Plant Cryopreservation: A Practical Guide*; Springer: New York, NY, USA, 2008; pp. 15–32.
12. Benson, E.E. Cryopreservation. In *Plant Conservation Biotechnology*; CRC Press: Boca Raton, FL, USA, 1999; pp. 109–122.
13. Sakai, A.; Kobayashi, S.; Oiyama, I. Cryopreservation of nucellar cells of navel orange (*Citrus sinensis* Osb. var. brasiliensis Tanaka) by vitrification. *Plant Cell Rep.* **1990**, *9*, 30–33. [CrossRef]
14. Uragami, A.; Sakai, A.; Nagai, M.; Takahashi, T. Survival of cultured cells and somatic embryos of *Asparagus officinalis* cryopreserved by vitrification. *Plant Cell Rep.* **1989**, *8*, 418–421. [CrossRef]
15. Leunufna, S.; Keller, E.R.J. Investigating a new cryopreservation protocol for yams (*Dioscorea* spp.). *Plant Cell Rep.* **2003**, *21*, 1159–1166. [CrossRef]
16. Jiroutová, P.; Sedlák, J. Cryobiotechnology of Plants: A Hot Topic not Only for Gene Banks. *Appl. Sci.* **2020**, *10*, 4677. [CrossRef]
17. Roque-Borda, C.; Kulus, D.; de Souza, A.V.; Kaviani, B.; Vicente, E. Cryopreservation of Agronomic Plant Germplasm Using Vitrification-Based Methods: An Overview of Selected Case Studies. *Int. J. Mol. Sci.* **2021**, *22*, 6157. [CrossRef]
18. Agrawal, A.; Singh, S.; Malhotra, E.V.; Meena, D.P.S.; Tyagi, R.K. In Vitro Conservation and Cryopreservation of Clonally Propagated Horticultural Species. In *Conservation and Utilization of Horticultural Genetic Resources*; Rajasekharan, P., Rao, V., Eds.; Springer: New York, NY, USA, 2019; pp. 529–578.
19. Bettoni, J.C.; Bonnart, R.; Volk, G.M. Challenges in implementing plant shoot tip cryopreservation technologies. *Plant Cell Tissue Organ Cult. (PCTOC)* **2021**, *144*, 21–34. [CrossRef]
20. Malik, S.K.; Chaudhury, R. Cryopreservation Techniques for Conservation of Tropical Horticultural Species Using Various Explants. In *Conservation and Utilization of Horticultural Genetic Resources*; Springer: Singapore, 2019; pp. 579–594.
21. Panis, B.; Lambardi, M. Status of cryopreservation technologies in plants (crops and forest trees). *Role Biotechnol.* **2005**, *5*, 43–54.
22. Sakai, P.A.; Hirai, D.; Niino, T. Development of PVS-Based Vitrification and Encapsulation–Vitrification Protocols. In *Plant Cryopreservation: A Practical Guide*; Springer: Singapore, 2008; pp. 33–57.
23. Höfer, M.; Hanke, M.-V. Cryopreservation of fruit germplasm. *Vitr. Cell. Dev. Biol. Anim.* **2017**, *53*, 372–381. [CrossRef]
24. Kulus, D.; Zalewska, M. Cryopreservation as a tool used in long-term storage of ornamental species—A review. *Sci. Hortic.* **2014**, *168*, 88–107. [CrossRef]
25. Bi, W.-L.; Pan, C.; Hao, X.-Y.; Cui, Z.-H.; Kher, M.M.; Marković, Z.; Wang, Q.-C.; da Silva, J.A.T. Cryopreservation of grapevine (*Vitis* spp.)—A review. *In Vitro Cell. Dev. Biol. Plant* **2017**, *53*, 449–460. [CrossRef]
26. Yamamoto, S.; Rafique, T.; Fukui, K.; Sekizawa, K.; Niino, T. V-cryo-plate procedure as an effective protocol for cryobanks: Case study of mint cryopreservation. *Cryo Lett.* **2012**, *33*, 12–23.
27. Yamamoto, S.-I.; Rafique, T.; Priyantha, W.S.; Fukui, K.; Matsumoto, T.; Niino, T. Development of a cryopreservation procedure using aluminium cryo-plates. *Cryo Lett.* **2011**, *32*, 256–265.
28. Kim, H.H.; Yoon, J.W.; Park, Y.E.; Cho, E.G.; Sohn, J.K.; Kim, T.K.; Engelmann, F. Cryopreservation of potato cultivated varieties and wild species: Critical factors in droplet vitrification. *Cryo Lett.* **2006**, *27*, 223–234.
29. Panis, B.; Nguyễn, T.n.T. *Cryopreservation of Musa Germplasm*; Bioversity International: Rome, Italy, 2001; Volume 5.
30. Carra, A.; Carimi, F.; Bettoni, J.C.; Pathirana, R. Progress and Challenges in the Application of Synthetic Seed Technology for Ex Situ Germplasm Conservation in Grapevine (*Vitis* spp.). In *Synthetic Seeds*; Springer: Singapore, 2019; pp. 439–467.
31. Fahy, G.M.; Wowk, B.; Wu, J.; Paynter, S. Improved vitrification solutions based on the predictability of vitrification solution toxicity. *Cryobiology* **2004**, *48*, 22–35. [CrossRef] [PubMed]
32. Teixeira, A.S.; González-Benito, M.E.; Molina-García, A.D. Glassy State and Cryopreservation of Mint Shoot Tips. *Biotechnol. Prog.* **2013**, *29*, 707–717. [CrossRef] [PubMed]
33. Zámečník, J.; Faltus, M.; Bilavčík, A.; Kotková, R. Comparison of cryopreservation methods of vegetatively propagated crops based on thermal analysis. In *Current Frontiers Cryopreservation*; IntechOpen: London, UK, 2012; pp. 333–358.
34. Murashige, T.; Skoog, F. A Revised Medium for Rapid Growth and Bio Assays with Tobacco Tissue Cultures. *Physiol. Plant.* **1962**, *15*, 473–497. [CrossRef]

35. Santarius, K.A. Freezing of Isolated Thylakoid Membranes in Complex Media. VII. The Effect of Bovine Serum Albumin. *Biochem. Physiol. Pflanz.* **1991**, *187*, 149–162. [CrossRef]
36. Elmoazzen, H.; Elliott, J.; McGann, L. Cryoprotectant equilibration in tissues. *Cryobiology* **2005**, *51*, 85–91. [CrossRef] [PubMed]
37. Kanaze, F.I.; Kokkalou, E.; Niopas, I.; Georgarakis, M.; Stergiou, A.; Bikiaris, D. Thermal analysis study of flavonoid solid dispersions having enhanced solubility. *J. Therm. Anal. Calorim.* **2006**, *83*, 283–290. [CrossRef]
38. Tao, D.; Li, P.H. Classification of plant cell cryoprotectants. *J. Theor. Biol.* **1986**, *123*, 305–310. [CrossRef]
39. Rall, W.F.; Fahy, G.M. Ice-free cryopreservation of mouse embryos at −196 °C by vitrification. *Nat. Cell Biol.* **1985**, *313*, 573–575. [CrossRef]
40. Gao, D.; Liu, J.; Liu, C.; McGann, L.; Watson, P.; Kleinhans, F.; Mazur, P.; Critser, E.; Critser, J. Andrology: Prevention of osmotic injury to human spermatozoa during addition and removal of glycerol. *Hum. Reprod.* **1995**, *10*, 1109–1122. [CrossRef]
41. Hubálek, Z. Protectants used in the cryopreservation of microorganisms. *Cryobiology* **2003**, *46*, 205–229. [CrossRef]
42. Golan, M.; Jelinkova, S.; Kratochvilova, I.; Skládal, P.; Pešl, M.; Rotrekl, V.; Pribyl, J. AFM Monitoring the Influence of Selected Cryoprotectants on Regeneration of Cryopreserved Cells Mechanical Properties. *Front. Physiol.* **2018**, *9*, 804. [CrossRef]
43. Gerber, D.W.; Byerrum, R.U.; Gee, R.W.; Tolbert, N. Glycerol concentrations in crop plants. *Plant Sci.* **1988**, *56*, 31–38. [CrossRef]
44. Sillanpää, M.; Ncibi, C. Biochemicals. In *A Sustainable Bioeconomy*; Springer: Berlin/Heidelberg, Germany, 2017; pp. 141–183.
45. Warner, R.M.; Ampo, E.; Nelson, D.; Benson, J.D.; Eroglu, A.; Higgins, A.Z. Rapid quantification of multi-cryoprotectant toxicity using an automated liquid handling method. *Cryobiology* **2021**, *98*, 219–232. [CrossRef]
46. Suzuki, M.; Tandon, P.; Ishikawa, M.; Toyomasu, T. Development of a new vitrification solution, VSL, and its application to the cryopreservation of gentian axillary buds. *Plant Biotechnol. Rep.* **2008**, *2*, 123–131. [CrossRef]
47. Nishizawa, S.; Sakai, A.; Amano, Y.; Matsuzawa, T. Cryopreservation of asparagus (*Asparagus officinalis* L.) embryogenic suspension cells and subsequent plant regeneration by vitrification. *Plant Sci.* **1993**, *91*, 67–73. [CrossRef]
48. Lu, Z.; Liu, C.T. A new approach to understanding and measuring glass formation in bulk amorphous materials. *Intermetallics* **2004**, *12*, 1035–1043. [CrossRef]
49. Kuleshova, L.; Mac Farlaneb, D.R.; Trounson, A.; Shaw, J. Sugars Exert a Major Influence on the Vitrification Properties of Ethylene Glycol-Based Solutions and Have Low Toxicity to Embryos and Oocytes. *Cryobiology* **1999**, *38*, 119–130. [CrossRef] [PubMed]
50. Murthy, S.; Singh, G. Examination of the concentration dependence of Tg of binary aqueous solutions. *Thermochim. Acta* **2008**, *469*, 116–119. [CrossRef]
51. Jonnalagadda, S.; Robinson, D.H. Effect of the inclusion of PEG on the solid-state properties and drug release from polylactic acid films and microcapsules. *J. Appl. Polym. Sci.* **2004**, *93*, 2025–2030. [CrossRef]
52. Zondervan, R.; Kulzer, F.; Berkhout, G.C.G.; Orrit, M. Local viscosity of supercooled glycerol near Tg probed by rotational diffusion of ensembles and single dye molecules. *Proc. Natl. Acad. Sci. USA* **2007**, *104*, 12628–12633. [CrossRef]
53. Talja, R.A.; Roos, Y.H. Phase and state transition effects on dielectric, mechanical, and thermal properties of polyols. *Thermochim. Acta* **2001**, *380*, 109–121. [CrossRef]
54. Simperler, A.; Kornherr, A.; Chopra, R.; Bonnet, P.A.; Jones, W.; Motherwell, A.W.D.S.; Zifferer, G. Glass Transition Temperature of Glucose, Sucrose, and Trehalose: An Experimental and in Silico Study. *J. Phys. Chem. B* **2006**, *110*, 19678–19684. [CrossRef]
55. Roberts, A.; Finnigan, W.; Kelly, P.; Faulkner, M.; Breitling, R.; Takano, E.; Scrutton, N.; Blaker, J.; Hay, S. Non-covalent protein-based adhesives for transparent substrates—Bovine serum albumin vs. recombinant spider silk. *Mater. Today Bio* **2020**, *7*, 100068. [CrossRef]
56. Anonym. Data Safety Sheet—Formamide. 2015. Available online: https://www.carlroth.com/medias/SDB-4095-IE-EN.pdf?context=bWFzdGVyfHNlY3VyaXR5RGF0YXNoZWV0c3wyNjc1NzV8YXBwbGljYXRpb24vcGRmfHNlY3VyaXR5RGF0YXNoZWV0cy9oNzIvaDc2Lzkw NDYwODU3MzAzMzQuc GRmfDUyOGVmNzI4NzM3MmU1NDQwYmIzZDYwODI5 OTYxNDU2NmZhOWJlNTVkMzVlOWRlZTk2NjQyNjNkYzliMzI0OTk (accessed on 27 November 2021).
57. Matsumoto, T. Cryopreservation of Plant Genetic Resources: Conventional and New Methods. *Rev. Agric. Sci.* **2017**, *5*, 13–20. [CrossRef]
58. Vozovyk, K.; Bobrova, O.; Prystalov, A.; Shevchenko, N.; Kuleshova, L. Amorphous state stability of plant vitrification solutions. *Biologija* **2020**, *66*, 66. [CrossRef]
59. Matsumoto, T.; Sakai, A.; Yamada, K. Cryopreservation of in vitro-grown apical meristems of wasabi (*Wasabia japonica*) by vitrification and subsequent high plant regeneration. *Plant Cell Rep.* **1994**, *13*, 442–446. [CrossRef]
60. Kawai, K.; Suzuki, T.; Oguni, M. Low-Temperature Glass Transitions of Quenched and Annealed Bovine Serum Albumin Aqueous Solutions. *Biophys. J.* **2006**, *90*, 3732–3738. [CrossRef]
61. Kim, H.-H.; Lee, Y.-G.; Shin, D.-J.; Ko, H.-C.; Gwag, J.-G.; Cho, E.-G.; Engelmann, F. Development of alternative plant vitrification solutions in droplet-vitrification procedures. *Cryo Lett.* **2009**, *30*, 320–334. [CrossRef]
62. Notman, R.; Noro, M.; O'Malley, B.; Anwar, J. Molecular Basis for Dimethylsulfoxide (DMSO) Action on Lipid Membranes. *J. Am. Chem. Soc.* **2006**, *128*, 13982–13983. [CrossRef]
63. Kim, H.-H.; Kim, J.-B.; Baek, H.-J.; Cho, E.-G.; Chae, Y.-A.; Engelmann, F. Evolution of DMSO concentration in garlic shoot tips during a vitrification procedure. *Cryo Lett.* **2004**, *25*, 91–100.
64. Kim, J.-B.; Kim, H.-H.; Baek, H.-J.; Cho, E.-G.; Kim, Y.-H.; Engelmann, F. Changes in sucrose and glycerol content in garlic shoot tips during freezing using PVS3 solution. *Cryo Lett.* **2005**, *26*, 103–112.

65. Volk, G.M.; Harris, J.L.; Rotindo, K.E. Survival of mint shoot tips after exposure to cryoprotectant solution components. *Cryobiology* **2006**, *52*, 305–308. [CrossRef]
66. Hakura, A.; Mochida, H.; Yamatsu, K. Dimethyl sulfoxide (DMSO) is mutagenic for bacterial mutagenicity tester strains. *Mutat. Res. Lett.* **1993**, *303*, 127–133. [CrossRef]
67. Kapp, R., Jr.; Eventoff, B. Mutagenicity of dimethylsulfoxide (DMSO): In vivo cytogenetics study in the rat. *Teratog. Carcinog. Mutagenesis* **1981**, *1*, 141–145. [CrossRef]
68. Vogin, E.E.; Carson, S.; Cannon, G.; Linegar, C.R.; Rubin, L.F. Chronic toxicity of DMSO in primates. *Toxicol. Appl. Pharmacol.* **1970**, *16*, 606–612. [CrossRef]
69. Halmagyi, A.; Valimareanu, S.; Coste, A.; Deliu, C.; Isac, V. Cryopreservation of Malus shoot tips and subsequent plant regeneration. *Rom. Biotechnol. Lett.* **2010**, *15*, 80.
70. Volk, G. Application of Functional Genomics and Proteomics to Plant Cryopreservation. *Curr. Genom.* **2010**, *11*, 24–29. [CrossRef] [PubMed]
71. Franceschi, V.R.; Horner, H.T. Calcium oxalate crystals in plants. *Bot. Rev.* **1980**, *46*, 361–427. [CrossRef]
72. Prychid, C.J.; Jabaily, R.S.; Rudall, P. Cellular Ultrastructure and Crystal Development in Amorphophallus (Araceae). *Ann. Bot.* **2008**, *101*, 983–995. [CrossRef]
73. Ranjbar, H.; Ahmadi, H.; Sheshdeh, R.K.; Ranjbar, H. Application of relative sensitivity function in parametric optimization of a tri-ethylene glycol dehydration plant. *J. Nat. Gas Sci. Eng.* **2015**, *25*, 39–45. [CrossRef]
74. Bhattacharya, S. Cryoprotectants and their usage in cryopreservation process. In *Cryopreservation Biotechnology in Biomedical and Biological Sciences*; IntechOpen: London, UK, 2018; p. 7.
75. Steuter, A.A.; Mozafar, A.; Goodin, J.R. Water Potential of Aqueous Polyethylene Glycol. *Plant Physiol.* **1981**, *67*, 64–67. [CrossRef]
76. Popova, E.; Bukhov, N.; Popov, A.; Kim, H.-H. Cryopreservation of protocorm-like bodies of the hybrid orchid Bratonia (*Miltonia flavescens* × *Brassia longissima*). *Cryo Lett.* **2010**, *31*, 426–437.
77. Fuller, B.J. Cryoprotectants: The essential antifreezes to protect life in the frozen state. *Cryo Lett.* **2004**, *25*, 375–388.
78. Sipen, P.; Anthony, P.; Davey, M.R. Cryopreservation of scalps of Malaysian bananas using a pre-growth method. *Cryo Lett.* **2011**, *32*, 197–205.
79. Acker, J.P.; McGann, L.E. Protective effect of intracellular ice during freezing? *Cryobiology* **2003**, *46*, 197–202. [CrossRef]
80. Bryant, G.; Koster, K.L.; Wolfe, J. Membrane behaviour in seeds and other systems at low water content: The various effects of solutes. *Seed Sci. Res.* **2001**, *11*, 17–25. [CrossRef]
81. Shendurse, A.; Khedkar, C. Glucose: Properties and analysis. *Encycl. Food Health* **2016**, *3*, 239–247.
82. Bhandari, B.R.; Roos, Y.H. Dissolution of sucrose crystals in the anhydrous sorbitol melt. *Carbohydr. Res.* **2003**, *338*, 361–367. [CrossRef]
83. Göldner, E.M.; Seitz, U.; Reinhard, E. Cryopreservation of *Digitalis lanata* Ehrh. cell cultures: Preculture and freeze tolerance. *Plant Cell Tissue Organ Cult.* **1991**, *24*, 19–24. [CrossRef]
84. Salaj, T.; Matusikova, I.; Panis, B.; Swennen, R.; Salaj, J. Recovery and characterisation of hybrid firs (*Abies alba* × *A. cephalonica*, *Abies alba* × *A. numidica*) embryogenic tissues after cryopreservation. *Cryo Lett.* **2010**, *31*, 206–217.
85. Subramanian, S.; Raj, A.; Kumar, R.; Rana, S.K.; Jha, A.K.; Gautam, S. Isolation, Culturing and cryopreservation of putative granulosa stem cells from buffalo ovaries. *Int. J. Cell Sci. Biotechnol.* **2014**, *4*, 20–25.
86. Carpenter, J.F.; Crowe, J.H. The mechanism of cryoprotection of proteins by solutes. *Cryobiology* **1988**, *25*, 244–255. [CrossRef]
87. Santarius, K.A.; Giersch, C. Cryopreservation of spinach chloroplast membranes by low-molecular-weight carbohydrates: II. Discrimination between colligative and noncolligative protection. *Cryobiology* **1983**, *20*, 90–99. [CrossRef]
88. Sikora, A.; Dupanov, V.O.; Kratochvíl, J.; Zamecnik, J. Transitions in Aqueous Solutions of Sucrose at Subzero Temperatures. *J. Macromol. Sci. Part B* **2007**, *46*, 71–85. [CrossRef]
89. Sakai, A.; Kobayashi, S.; Oiyama, I. Survival by Vitrification of Nucellar Cells of Navel Orange (*Citrus sinensis* var. brasiliensis Tanaka) Cooled to −196 °C. *J. Plant Physiol.* **1991**, *137*, 465–470. [CrossRef]
90. Sopalun, K.; Kanchit, K.; Ishikawa, K. Vitrification-based cryopreservation of *Grammatophyllum speciosum* protocorm. *Cryo Lett.* **2010**, *31*, 347–357.
91. Horvath, A.; Wayman, W.R.; Urbányi, B.; Ware, K.M.; Dean, J.C.; Tiersch, T.R. The relationship of the cryoprotectants methanol and dimethyl sulfoxide and hyperosmotic extenders on sperm cryopreservation of two North-American sturgeon species. *Aquaculture* **2005**, *247*, 243–251. [CrossRef]
92. Bronshteyn, V.L.; Steponkus, P.L. Nucleation and Growth of Ice Crystals in Concentrated Solutions of Ethylene Glycol. *Cryobiology* **1995**, *32*, 1–22. [CrossRef]
93. Rall, W. Factors affecting the survival of mouse embryos cryopreserved by vitrification. *Cryobiology* **1987**, *24*, 387–402. [CrossRef]
94. Turner, S.; Senaratna, T.; Touchell, D.; Bunn, E.; Dixon, K.; Tan, B. Stereochemical arrangement of hydroxyl groups in sugar and polyalcohol molecules as an important factor in effective cryopreservation. *Plant Sci.* **2001**, *160*, 489–497. [CrossRef]
95. Kim, H.-H.; Yoon, J.-W.; Kim, J.-B.; Engelmann, F.; Cho, E.-G. Thermal analysis of garlic shoot tips during a vitrification procedure. *Cryo Lett.* **2005**, *26*, 33–44.
96. Wu, Y.; Zhao, Y.; Zhou, M.; Engelmann, F. Cryopreservation of temperate fruit tree germplasm. In *Plant Genetic Resources Network in East Asia. Proceedings of the Meeting for the Regional Network for Conservation and Use of Plant Genetic Resources in East Asia, Ulaanbaatar, Mongolia, 13–16 August 2001*; International Plant Genetic Resources Institute (IPGRI): Rome, Italy, 2002; pp. 77–88.

97. Kim, H.-H.; Popova, E.V.; Yi, J.-Y.; Cho, G.-T.; Park, S.-U.; Lee, S.-C.; Engelmann, F. Cryopreservation of hairy roots of Rubia akane (Nakai) using a droplet-vitrification procedure. *Cryo Lett.* **2011**, *31*, 473–484.
98. Hong, S.; Yin, M.; Shao, X.; Wang, A.; Xu, W. Cryopreservation of embryogenic callus of *Dioscorea bulbifera* by vitrification. *Cryo Lett.* **2009**, *30*, 64–75.
99. Cho, E.G.; Hor, Y.L.; Kim, H.H.; Rao, V.R.; Engelmann, F. Cryopreservation of Citrus madurensis zygotic embryonic axes by vitrification: Importance of pregrowth and preculture conditions. *Cryo Lett.* **2002**, *22*, 391–396.
100. Ivchenko, T.V.; Vitsenya, T.I.; Shevchenko, N.A.; Bashtan, N.O.; Kornienko, S.I. Hypothermic and Low-Temperature Storage of Garlic (*Allium sativum* L.) for in Vitro Collections. *Probl. Cryobiol. Cryomedicine* **2017**, *27*, 110–120. [CrossRef]
101. Ishikawa, K.; Harata, K.; Mii, M.; Sakai, A.; Yoshimatsu, K.; Shimomura, K. Cryopreservation of zygotic embryos of a Japanese terrestrial orchid (*Bletilla striata*) by vitrification. *Plant Cell Rep.* **1997**, *16*, 754–757. [CrossRef]
102. Ray, A.; Bhattacharya, S. Cryopreservation of in vitro grown nodal segments of *Rauvolfia serpentina* by PVS2 vitrification. *Cryo Lett.* **2009**, *29*, 321–328.
103. Sajini, K.K.; Karun, A.; Amamath, C.H.; Engelmann, F. Cryopreservation of coconut (*Cocos nucifera* L.) zygotic embryos by vitrification. *Cryo Lett.* **2011**, *32*, 317–328.
104. Panis, B.; Swennen, R. Plant cryopreservation: Applications, constraints and prospects. In *Society for Low Temperature Biology. Annual Scientific Meeting, AGM and Symposium. Validation, Safety and Ethical Issues Impacting the Low Temperature Storage of Biological Resources*; Cryo Letters: Lewes, UK, 2007; pp. 1–29.
105. Tokatli, Y.O.; Akdemir, H. Cryopreservation of Fraser photinia (*Photinia* × *fraseri* Dress.) via vitrification-based one-step freezing techniques. *Cryo Lett.* **2010**, *31*, 40–49.
106. Volk, G.M.; Maness, N.; Rotindo, K. Cryopreservation of garlic (*Allium sativum* L.) using plant vitrification solution 2. *Cryo Lett.* **2004**, *25*, 219–226.
107. March, G.G.-D.; De Boucaud, M.-T.; Chmielarz, P. Cryopreservation of *Prunus avium* L. embryogenic tissues. *Cryo Lett.* **2006**, *26*, 341–348.
108. Engelmann-Sylvestre, I.; Engelmann, F. Cryopreservation of in vitro-grown shoot tips of *Clinopodium odorum* using aluminium cryo-plates. *Vitr. Cell. Dev. Biol. Anim.* **2015**, *51*, 185–191. [CrossRef]
109. Li, B.-Q.; Feng, C.-H.; Wang, M.-R.; Hu, L.-Y.; Volk, G.; Wang, Q.-C. Recovery patterns, histological observations and genetic integrity in Malus shoot tips cryopreserved using droplet-vitrification and encapsulation-dehydration procedures. *J. Biotechnol.* **2015**, *214*, 182–191. [CrossRef]
110. Vollmer, R.; Villagaray, R.; Castro, M.; Anglin, N.; Ellis, D. Cryopreserved potato shoot tips showed genotype-specific response to sucrose concentration in rewarming solution (RS). *Plant Cell Tissue Organ Cult. (PCTOC)* **2018**, *136*, 353–363. [CrossRef]
111. Wang, M.-R.; Zhang, Z.; Zámečník, J.; Bilavčík, A.; Blystad, D.-R.; Haugslien, S.; Wang, Q.-C. Droplet-vitrification for shoot tip cryopreservation of shallot (*Allium cepa* var. aggregatum): Effects of PVS3 and PVS2 on shoot regrowth. *Plant Cell Tissue Organ Cult.* **2020**, *140*, 185–195. [CrossRef]
112. Bettoni, J.C.; Kretzschmar, A.A.; Bonnart, R.; Shepherd, A.; Volk, G.M. Cryopreservation of 12 Vitis Species Using Apical Shoot Tips Derived from Plants Grown In Vitro. *HortScience* **2019**, *54*, 976–981. [CrossRef]
113. Hammond, S.D.H.; Viehmannova, I.; Zamecnik, J.; Panis, B.; Faltus, M. Droplet-vitrification methods for apical bud cryopreservation of yacon [*Smallanthus sonchifolius* (Poepp. and Endl.) H. Rob.]. *Plant Cell Tissue Organ Cult. (PCTOC)* **2021**, *147*, 197–208. [CrossRef]
114. Niedermeyer, W.; Parish, G.R.; Moor, H. Reactions of yeast cells to glycerol treatment alterations to membrane structure and glycerol uptake. *Protoplasma* **1977**, *92*, 177–193. [CrossRef]
115. Brison, M.; de Boucaud, M.-T.; Dosba, F. Cryopreservation of in vitro grown shoot tips of two interspecific *Prunus* rootstocks. *Plant Sci.* **1995**, *105*, 235–242. [CrossRef]
116. Serrano-Martinez, F.; Casas, J.L. Cryopreservation of *Tetraclinis articulata* (vahl.) Masters. *Cryo Lett.* **2011**, *32*, 248–255.
117. Maruyama, E.; Kinoshita, I.; Ishii, K.; Ohba, K.; Sakai, A. Germplasm conservation of Guazuma crinita, a useful tree in the Peru-Amazon, by the cryopreservation of in vitro-cultured multiple bud clusters. *Plant Cell Tissue Organ Cult. (PCTOC)* **1997**, *48*, 161–165. [CrossRef]
118. Wang, Q.; Batuman, Ö.; Li, P.; Bar-Joseph, M.; Gafny, R. A simple and efficient cryopreservation of in vitro-grown shoot tips of Troyer'citrange [*Poncirus trifoliata* (L.) Raf. × *Citrus sinensis* (L.) Osbeck.] by encapsulation-vitrification. *Euphytica* **2002**, *128*, 135–142. [CrossRef]
119. Vujović, T.; Jevremović, D.; Marjanović, T.; Ružić, Đ. Cryopreservation of Serbian autochthonous plum 'Crvena Ranka' using aluminium cryo-plates. *Genetika* **2021**, *53*, 283–294. [CrossRef]
120. Lambardi, M.; Fabbri, A.; Caccavale, A. Cryopreservation of white poplar (*Populus alba* L.) by vitrification of in vitro-grown shoot tips. *Plant Cell Rep.* **2000**, *19*, 213–218. [CrossRef]
121. Shin, D.J.; Kong, H.; Popova, E.V.; Moon, H.K.; Park, S.Y.; Park, S.-U.; Lee, S.C.; Kim, H.H. Cryopreservation of *Kalopanax septemlobus* embryogenic callus using vitrification and droplet-vitrification. *Cryo Lett.* **2012**, *33*, 402–410.
122. Barraco, G.; Sylvestre, I.; Iapichino, G.; Engelmann, F. Investigating the cryopreservation of nodal explants of *Lithodora rosmarinifolia* (Ten.) Johnst., a rare, endemic Mediterranean species. *Plant Biotechnol. Rep.* **2012**, *7*, 141–146. [CrossRef]
123. Lee, Y.-Y.; Balaraju, K.; Song, J.-Y.; Yi, J.-Y.; Lee, S.-Y.; Lee, J.-R.; Yoon, M.; Kim, H.-H. Cryopreservation of in vitro grown shoot tips of strawberry (*Fragaria* × *ananassa* Duch.) genetic resources by droplet-vitrification. *Korean J. Plant Resour.* **2019**, *32*, 689–697.

124. Sakai, A. Development of cryopreservation techniques. In *Cryopreservation of Tropical Plant Germplasm: Current Research Progress and Application*; CGIAR: Montpellier, France, 2000; pp. 1–7.
125. Langis, R.; Schnabel, B.; Earle, E.; Steponkus, P. Cryopreservation of *Brassica campestris* L. cell suspensions by vitrification. *Cryo Lett.* **1989**, *10*, 421–428.
126. Kim, H.-H.; Cho, E.-G.; Baek, H.-J.; Kim, C.-Y.; Keller, E.R.J.; Engelmann, F. Cryopreservation of garlic shoot tips by vitrification: Effects of dehydration, rewarming, unloading and regrowth conditions. *Cryo Lett.* **2004**, *25*, 59–70.
127. Towill, L. Cryopreservation of isolated mint shoot tips by vitrification. *Plant Cell Rep.* **1990**, *9*, 178–180. [CrossRef]
128. Grospietsch, M.; Stodulkova, E.; Zamecnik, J. Effect of osmotic stress on the dehydration tolerance and cryopreservation of *Solanum tuberosum* shoot tips. *Cryo Lett.* **1999**, *20*, 339–346.
129. Suranthran, P.; Gantait, S.; Sinniah, U.R.; Subramaniam, S.; Alwee, S.S.R.S.; Roowi, S.H. Effect of loading and vitrification solutions on survival of cryopreserved oil palm polyembryoids. *Plant Growth Regul.* **2012**, *66*, 101–109. [CrossRef]
130. Mallon, R.; Bunn, E.; Turner, S.R.; Gonzalez, M.L. Cryopreservation of *Centaurea ultreiae* (Compositae) a critically endangered species from Galicia (Spain). *Cryo Lett.* **2008**, *29*, 363–370.
131. Turner, S.R.; Senaratna, T.; Bunn, E.; Tan, B.; Dixon, K.; Touchell, D.H. Cryopreservation of Shoot Tips from Six Endangered Australian Species using a Modified Vitrification Protocol. *Ann. Bot.* **2001**, *87*, 371–378. [CrossRef]
132. Halmagyi, A.; Deliu, C.; Coste, A.; Keul, M.; Cheregi, O.; Cristea, V. Vitrification of potato shoot tips for germplasm cryopreservation. *Contrib. Bot.* **2004**, *39*, 187–193.
133. Shevchenko, N.; Mozgovska, A.; Bobrova, O.; Bashtan, N.; Kovalenko, G.; Ivchenko, T. Post-Thaw Survival of Meristems from In Vitro Sweet Potato (*Ipomoea batatas* (L.) Lam.) Plants. *Biol. Life Sci. Forum* **2020**, *4*, 43. [CrossRef]
134. Watanabe, K.; Steponkus, P.L. Vitrification of *Oryza sativa* L. cell suspensions. *Cryo Lett.* **1995**, *16*, 255–262.
135. Folgado, R.; Panis, B.; Sergeant, K.; Renaut, J.; Swennen, R.; Hausman, J.-F. Unravelling the effect of sucrose and cold pretreatment on cryopreservation of potato through sugar analysis and proteomics. *Cryobiology* **2015**, *71*, 432–441. [CrossRef]
136. Dumet, D.; Grapin, A.; Bailly, C.; Dorion, N. Revisiting crucial steps of an encapsulation/desiccation based cryopreservation process: Importance of thawing method in the case of Pelargonium meristems. *Plant Sci.* **2002**, *163*, 1121–1127. [CrossRef]
137. Matsumoto, T. An approach to enhance dehydration tolerance of alginate-coated dried meristems cooled to −196 °C. *Cryo Lett.* **1995**, *16*, 299–306.
138. Reed, B.M. Cryopreservation—Practical considerations. In *Plant Cryopreservation: A Practical Guide*; Springer: Berlin/Heidelberg, Germany, 2008; pp. 3–13.
139. Sakai, A.; Engelmann, F. Vitrification, encapsulation-vitrification and droplet-vitrification: A review. *Cryo Lett.* **2007**, *28*, 151–172.
140. Volk, G.M.; Shepherd, A.N.; Bonnart, R. Successful Cryopreservation of Vitis Shoot Tips: Novel Pre-treatment Combinations Applied to Nine Species. *Cryo Lett.* **2019**, *39*, 322–330.
141. Benelli, C.; Carvalho, L.; EL Merzougui, S.; Petruccelli, R. Two Advanced Cryogenic Procedures for Improving *Stevia rebaudiana* (Bertoni) Cryopreservation. *Plants* **2021**, *10*, 277. [CrossRef]
142. O'Brien, C.; Hiti-Bandaralage, J.C.A.; Folgado, R.; Lahmeyer, S.; Hayward, A.; Folsom, J.; Mitter, N. First report on cryopreservation of mature shoot tips of two avocado (*Persea americana* Mill.) rootstocks. *Plant Cell Tissue Organ Cult. (PCTOC)* **2021**, *144*, 103–113. [CrossRef]
143. Sharma, S.; Parasher, K.; Mukherjee, P.; Sharma, Y.P. Cryopreservation of a Threatened Medicinal Plant, Valeriana Jatamansi Jones, Using Vitrification and Assessment of Biosynthetic Stability of Regenerants. *Cryo Lett.* **2021**, *42*, 300–308.
144. Fahy, G.M.; Macfarlane, D.R.; Angell, C.A.; Meryman, H.T. Vitrification as an approach to cryopreservation. *Cryobiology* **1984**, *21*, 407–426. [CrossRef]
145. Kim, H.-H.; Lee, J.-K.; Yoon, J.-W.; Ji, J.-J.; Nam, S.-S.; Hwang, H.-S.; Cho, E.-G.; Engelmann, F. Cryopreservation of garlic bulbil primordia by the droplet-vitrification procedure. *Cryo Lett.* **2006**, *27*, 143–153.
146. Kartha, K.; Leung, N.; Mroginski, L. In vitro Growth Responses and Plant Regeneration from Cryopreserved Meristems of Cassava (Manihot esculenta Crantz). *Zeitschrift für Pflanzenphysiologie* **1982**, *107*, 133–140. [CrossRef]
147. Ellis, D.; Skogerboe, D.; Andre, C.; Hellier, B.; Volk, G. Implementation of garlic cryopreservation techniques in the national plant germplasm system. *Cryo Lett.* **2006**, *27*, 99–106.
148. Tanaka, D.; Niino, T.; Isuzugawa, K.; Hikage, T.; Uemura, M. Cryopreservation of shoot apices of in-vitro grown gentian plants: Comparison of vitrification and encapsulation-vitrification protocols. *Cryo Lett.* **2004**, *25*, 167–176.
149. Kaczmarczyk, A.; Shvachko, N.; Lupysheva, Y.; Hajirezaei, M.-R.; Keller, E.R.J. Influence of alternating temperature preculture on cryopreservation results for potato shoot tips. *Plant Cell Rep.* **2008**, *27*, 1551–1558. [CrossRef] [PubMed]
150. Schaefer-Menuhr, A.; Schumacher, H.-M.; Mix-Wagner, G. Long-term storage of old potato varieties by cryopreservation of meristems in liquid nitrogen. *Landbauforsch. Voelkenrode* **1994**, *44*, 301–313.
151. Martinez-Montero, M.E.; Martinez, J.; Engelmann, F. Cryopreservation of sugarcane somatic embryos. *Cryo Lett.* **2008**, *29*, 229–242.
152. Senula, A.D.; Keller, E.R.J.; Sanduijav, T.; Yohannes, T. Cryopreservation of cold-acclimated mint (Mentha spp.) shoot tips using a simple vitrification protocol. *Cryo Lett.* **2007**, *28*, 1–12.
153. Bettoni, J.C.; Marković, Z.; Bi, W.; Volk, G.M.; Matsumoto, T.; Wang, Q.-C. Grapevine Shoot Tip Cryopreservation and Cryotherapy: Secure Storage of Disease-Free Plants. *Plants* **2021**, *10*, 2190. [CrossRef]

154. Panis, B.; Nagel, M.; Houwe, I.V.D. Challenges and Prospects for the Conservation of Crop Genetic Resources in Field Genebanks, in In Vitro Collections and/or in Liquid Nitrogen. *Plants* **2020**, *9*, 1634. [CrossRef] [PubMed]
155. Gámez-Pastrana, R.; González-Arnao, M.T.; Martínez-Ocampo, Y.; Engelmann, F. Thermal events in calcium alginate beads during encapsulation dehydration and encapsulation-vitrification protocols. *Acta Hortic.* **2011**, *908*, 47–54. [CrossRef]
156. Funnekotter, B.; Mancera, R.L.; Bunn, E. Advances in understanding the fundamental aspects required for successful cryopreservation of Australian flora. *Vitr. Cell. Dev. Biol. Anim.* **2017**, *53*, 289–298. [CrossRef]
157. Funnekotter, B.; Bunn, E.; Mancera, R.L. Cryo-mesh: A simple alternative cryopreservation protocol. *Cryo Lett.* **2017**, *38*, 155–159.
158. Nadarajan, J.; Pritchard, H.W. Biophysical Characteristics of Successful Oilseed Embryo Cryoprotection and Cryopreservation Using Vacuum Infiltration Vitrification: An Innovation in Plant Cell Preservation. *PLoS ONE* **2014**, *9*, e96169. [CrossRef]
159. Bruňáková, K.; Zámečník, J.; Urbanová, M.; Čellárová, E. Dehydration status of ABA-treated and cold-acclimated *Hypericum perforatum* L. shoot tips subjected to cryopreservation. *Thermochim. Acta* **2011**, *525*, 62–70. [CrossRef]
160. Šesták, J.; Zamecnik, J. Can clustering of liquid water and thermal analysis be of assistance for better understanding of biological germplasm exposed to ultra-low temperatures. *J. Therm. Anal. Calorim.* **2007**, *88*, 411–416. [CrossRef]

Review

Cryopreservation of Woody Crops: The Avocado Case

Chris O'Brien [1,*], Jayeni Hiti-Bandaralage [1], Raquel Folgado [2], Alice Hayward [1], Sean Lahmeyer [2], Jim Folsom [2] and Neena Mitter [1]

[1] Centre for Horticultural Science, Queensland Alliance for Agriculture and Food Innovation, The University of Queensland, St Lucia, QLD 4072, Australia; j.hitibandalarage@uq.edu.au (J.H.-B.); a.hayward@uq.edu.au (A.H.); n.mitter@uq.edu.au (N.M.)
[2] The Huntington Library, Art Museum, and Botanical Gardens, 1151 Oxford Road, San Marino, CA 91108, USA; rfolgado@huntington.org (R.F.); slahmeyer@huntington.org (S.L.); jfolsom@huntington.org (J.F.)
* Correspondence: c.obrien4@uq.edu.au

Abstract: Recent development and implementation of crop cryopreservation protocols has increased the capacity to maintain recalcitrant seeded germplasm collections via cryopreserved in vitro material. To preserve the greatest possible plant genetic resources globally for future food security and breeding programs, it is essential to integrate in situ and ex situ conservation methods into a cohesive conservation plan. In vitro storage using tissue culture and cryopreservation techniques offers promising complementary tools that can be used to promote this approach. These techniques can be employed for crops difficult or impossible to maintain in seed banks for long-term conservation. This includes woody perennial plants, recalcitrant seed crops or crops with no seeds at all and vegetatively or clonally propagated crops where seeds are not true-to-type. Many of the world's most important crops for food, nutrition and livelihoods, are vegetatively propagated or have recalcitrant seeds. This review will look at ex situ conservation, namely field repositories and in vitro storage for some of these economically important crops, focusing on conservation strategies for avocado. To date, cultivar-specific multiplication protocols have been established for maintaining multiple avocado cultivars in tissue culture. Cryopreservation of avocado somatic embryos and somatic embryogenesis have been successful. In addition, a shoot-tip cryopreservation protocol has been developed for cryo-storage and regeneration of true-to-type clonal avocado plants.

Keywords: vitrification; ex situ conservation; long-term conservation; embryogenic; shoot tips; plant biodiversity

1. Introduction

Globally plants are recognized as a vital component of biodiverse ecosystems, the carbon cycle, food production and the bioeconomy. An estimated 7000 species of plants provide food, fiber, fuel, shelter and medicine [1]. Plant genetic diversity is the foundation of crop improvement [2] and a primary target of conservation efforts. The two major approaches to conserve plant genetic resources are ex situ and in situ conservation [3]. In situ conservation involves the designation, management and monitoring of target taxa where they are encountered [4]. It protects an endangered plant species in its natural habitat. In situ techniques are described as protected areas, e.g., genetic reserve, on-farm and home garden conservation. Ex situ conservation involves the sampling, transfer and storage of target taxa from the collecting area [4]. Ex situ techniques include seed, in vitro (tissue culture and cryopreservation), DNA and pollen storage; field gene banks and botanic garden conservation. In vitro storage using tissue culture and cryopreservation techniques can deliver valuable tools to achieve a positive conservation outcome for genetic resources.

The majority of conservation programs focus on seed storage [5]. Many of the world's major food plants produce orthodox seeds which undergo maturation drying and are tolerant to extensive desiccation and can be stored dry at low temperature [6]. Seed storage under dry and cool conditions is the most widely adopted method for long-term ex situ conservation at relatively low costs [7]. About 45% of the accessions stored as seeds are cereals, followed by food legumes [15%], forages [9%] and vegetables [7%] [8]. However, seeds of many woody perennial plants are recalcitrant, e.g., *Juglans* spp. (walnut) [9], *Hevea brasiliensis* (rubber tree) and *Artocarpus heterophyllus* (jackfruit) [10]. Thus, they are difficult to maintain in seed banks. Additionally, seed-based conservation efforts miss clonal lineages that form the foundation of woody perennial agriculture [9]. Crops such as *Persea americana* Mill. (avocado), have recalcitrant seeds that are shed at relatively high moisture content, thus cannot undergo drying to facilitate long-term storage [6,11]. In addition, species that are seedless, e.g., *Musa* spp. (edible banana); or crops vegetatively propagated as their seeds are not true-to-type, e.g., *Manihot esculenta* (cassava), *Malus domestica* Borkh (Apple). and *Citrus* spp. (citrus); are not storable through seeds. Field, in vitro and cryopreserved collections provide an alternative [7].

Field gene banks maintain living collections [12]. They are advantageous as physiological attributes and characteristics of the accessions such as plant habit, yield, tree height and disease resistance can be evaluated periodically [13]; however, there are several limitations posed; high maintenance cost, intensive labor and land requirements, pressure of natural calamities, risk of biotic and abiotic stresses as well as funding sources and economic decisions limiting the level of accession replication to maintain genetic diversity.

Tissue culture maintains plant material collections employing growth retardants [14], reduced light [15] or reduced temperature [16] to achieve slow growth, normally in sterile conditions. Plant germplasm storage via these methods has been increased with more in vitro protocols being developed for a vast number of plant species [17–19]. These approaches are used for large-scale micropropagation, reproduction purposes including embryo rescue, ploidy manipulations, protoplast fusions and somatic embryogenesis and are appropriate tools for short- and mid-term storage of plant genetic resources [7]. These methods allow for physical evaluation of material, rapid multiplication and plant establishment when needed, still, very costly to maintain due to space, consumables and labor inputs [20].

Plant cryopreservation (storage at $-196 \pm 1\ °C$) is a technique whereby plant tissues are preserved at ultra-low temperatures without losing viability [21]. It is the most relevant technology that provides safe long-term conservation of biological material as it maintains ex vivo biological function, does not induce genetic alterations [22] and provides long-term stable storage. Thus, it serves as an ultimate back-up of plant accessions for long-term storage, and material is generally not withdrawn from cryotanks unless it is necessary to use for research such as genetic manipulations [23] or in vitro culture [24]. A wide range of plant tissue can be cryopreserved, e.g., pollen, seeds, shoot tips, dormant buds, cell suspensions, embryonic cultures, somatic and zygotic embryos and callus tissue [25,26]. Recent uses of cryopreservation including cryotherapy to eradicate pathogens, such as phytoplasmas, viruses and bacteria in plants [27,28] is gaining a lot of attention [23]. Samples are normally given a short exposure to LN and surviving cells are regenerated from meristematic tissue which is pathogen free [28]. Cryotherapy has been used successfully in eradicating virus infections in several species with economic importance, such as *Prunus* spp. (plum), *Musa* spp. (banana), *Vitis vinifera* (grape), *Fragaria ananassa* (strawberry), *Solanum tuberosum* (potato), *Rubus idaeus* (raspberry) and *Allium sativum* (garlic) [28]. This review will look at conservation approaches for woody plants, focusing on avocado as a case study.

2. Field Repositories of Woody Crops

Field based germplasm conservation maintains living plants and serves as a source of plant genetic variation. Plants represented in these collections are current and historic cultivars, breeding material, landraces and sometimes wild relatives [9]. All of these are important to maintain for future development of new cultivars with superior growth characteristics or resistance to pest and diseases. Field repositories have the advantage that researchers can physically evaluate and characterize the accessions for parameters such as yield, tree height and disease resistance [29,30]. Table 1 summarizes some examples of woody crops that are held as field repositories. However, the field repositories require an adequate area of land and continuous maintenance as well as on-going funding. They are also vulnerable to loss from natural disasters and damage caused by pests and diseases. This makes it important to potentiate field germplasm conservation with other methods which address some of these concerns.

Table 1. Some examples of field repositories maintaining living collections of economically important crops.

Country	Field Repositories	Genus/Species	Reference
USA	USDA—Geneva NY, Davis CA, Riverside CA	*Malus domestia* Borkh. (apple) *Vitis vinifera* L. (grape) *Actinidia deliciosa* (kiwifruit) *Diospyros* spp. (persimmon) *Ficus carica* L. (fig) *Juglans* spp. (walnut) *Olea europaea* L. (olive) *Pistacia vera* L. (pistachio) *Punica granatum* L. (pomegranate) *Citrus* spp. (citrus) *Prunus* spp. (plum)	[31]
USA	Tropical Botanical Garden	*Artocarpus altilis* (breadfruit)	[32]
Germany	German Fruit Gene bank	*Malus* spp. (apple) *Prunus avium* (cherry) *Prunus domestica* (plum) *Rubus* spp. (raspberry)	[12,33]
United Kingdom	National Fruit Collection	*Malus domestica* Borkh. (apple) *Prunus domestica* (plum) *Pyrus communis* L. (pear) *Prunus avium* (cherry)	[34]

3. In Vitro Conservation

Different in vitro storage methods are employed depending on the storage duration required [17,35], i.e., in vitro culture for short- and medium-term storage and cryopreservation for long-term storage. Many reviews have been carried out to determine success [35–38] and standards established for managing field and in vitro germplasm gene banks [39,40]. These standards ensure effective, safe and efficient conservation of genetic resources. Due to the success of in vitro conservation techniques, many in vitro gene banks have been established nationally and internationally [41,42] (Table 2).

Table 2. Some examples of cryo-storage gene banks maintaining collections of economically important crops.

Country	Gene Bank	Genus/Species	Accessions Held	Reference
France	Institute of Research Development	*Coffea* spp. (coffee)	~500	[12]
Columbia	International Centre for Tropical Agriculture	*Manihot esculenta* (cassava)	5690	[43]
Japan	National Institute of Agrobiological Sciences	*Morus* spp. (mulberry) *Juncus effusus* (rush)	~1000 50	[12]
Japan	Shimane Agriculture Research Centre	*Wasabi japonica* M. (Japanese horseradish)	40	[12]
USA	National Clonal Germplasm Repository	*Malus* spp. (apple) *Pyrus* spp. (pear) *Rubus* spp. (raspberry) *Vitis* spp. (grape)	6073 131 57 1405	[44,45]
Belgium	Bioversity International Transit Centre	*Musa* spp. (banana)	1600	[7]

4. Plant Cryopreservation of Somatic Embryos and Shoot Tips

Cryopreservation of plants covers the entire plant kingdom from herbs and vines to shrubs and trees. The growth may be annual, biennial or perennial and the climate arctic; temperate, sub-tropical or tropical. A range of responses can occur within these groups and they are not always useful groupings for evaluating cryopreservation strategies [46]. The choice of material used, depends on the conservation goal, e.g., seeds and embryos capture species diversity; whereas shoot tips and dormant buds capture specific genotypes [47]. The most commonly used material to cryopreserve is apical meristems. They are at less risk of genetic variations due to their organized structure and are made up of small unvacuolated cells generally having a small vascular system [48]. In species that are recalcitrant and maintained in living field repositories, long-term cryopreservation storage of shoot tips can offer an alternative back-up as compared to seed storage which is only short-term [24].

Cryopreservation has several steps: (1) initial excision of the germplasm; (2) desiccation or pre-culture on osmotic media to reduce water content; (3) cryoprotection through exposure to cryoprotective agents; (4) cryopreservation in LN; (5) re-warming; and (6) unloading of cryoprotective agents and recovery of germplasm after cryopreservation [49]. The most critical step of cryopreservation is avoiding the intracellular and extracellular water that can lead to damage of cells during freezing [21]. Crystal formation, without extreme reduction of cellular water, can only be prevented though 'vitrification' i.e., the physical process of transition of an aqueous solution into an amorphous and glassy state (non-crystalline state) [50].

4.1. Methods to Reduce Water Content

Concentrated intracellular solute is a pre-requisite for successful cryopreservation and can be achieved with the following methods (Table 3) either individually or in combination [50–53].

Table 3. Methods to reduce water content.

Dehydration Method	Uses
Desiccation	(1) Air drying of explants in laminar flow hood or using flow of compressed air. (2) Dehydration of explants in a desiccator with silica gel.
Cryoprotectants	(1) Penetrating cryoprotectants, e.g., dimethyl sulfoxide (DMSO) and glycerol act by replacing intracellular water. (2) Non-penetrating cryoprotectants, e.g., sucrose, polyvinylpyrrolidone (PVP) and polyethylene glycol (PEG), display different osmotic potential inside and outside the cells.
Freeze-induced dehydration	Preferential freezing of extracellular water by slow cooling at a rate of 0.5–2 °C per min creates a hypotonic surrounding for the cell, resulting in outflow of cellular water.
Pre-conditioning of donor plant or explant	Including DMSO abscisic acid, sucrose, polyols or proline in the pre-culture medium or low temperature treatment to induce tolerance to dehydration and freezing.

Cryopreservation protocols using vitrification solutions typically involve a two-step cryoprotection process: (1) loading sometimes called osmoprotection is achieved by incubation in loading solution; and (2) dehydration using vitrification solution [52]. Loading solutions are commonly used to improve permeation of the cryoprotectant through cell membrane, it also induces tolerance to dehydration, which will be imposed by vitrification solutions. A common loading solution used is 2 M glycerol + 0.4 M sucrose [52]. Vitrification solutions contain chemicals that are high in concentration, e.g., ethylene glycol, glycerol and DMSO which have been reported as toxic to plant tissue [54]. It is therefore important to establish minimum exposure time to vitrification solutions in order to dehydrate tissue sufficiently to undergo cryopreservation and avoid damage effects to plant tissue [55,56].

Application of cryoprotectants is the most widely used method in cryopreservation protocols. Cryoprotectants that are penetrating in nature are able to reduce cell water at temperatures sufficiently to minimize the damaging effect of the concentrated solutes on the cells [57]. Whereas non-penetrating cryoprotectants osmotically "squeeze" water from the cells during the initial phases of freezing at temperatures between −10 and −20 °C [57]. Many authors have developed mixtures of cryoprotectants (Table 4) since the discovery of their benefits in protecting cells during the cryogenic process [54,58–61]. The most commonly used cryoprotectants for plant cells are PVS2 [59] and PVS3 [58].

Table 4. Some examples of cryoprotectants used for plant tissue.

Cryoprotectant	Composition
PVS1	30% w/v glycerol, 15% w/v EG, 5% w/v sucrose, 15% w/v DMSO [61]
PVS2	30% w/v glycerol, 15% w/v DMSO, 15% w/v EG and 15% sucrose [59]
PVS3	50% w/v glycerol and 50% w/v sucrose [58]
PVS4	35% w/v glycerol, 20% w/v EG and 20.5% M sucrose [62]
VSL+	20% w/v glycerol, 10% w/v DMSO, 30% w/v EG, 15% sucrose and 10 mM $CaCl_2$ [63]
VSL	20% w/v glycerol, 10% w/v DMSO, 30% w/v EG, 5% sucrose and 10 mM $CaCl_2$ [63]
Steponkus	50% w/v EG, 15% sorbitol, 6.0% bovine serum albumin, 13.7% sucrose [64]
Towill	35% EG, 6.8% w/v DMSO, 10% PEG 8000 and 13.7% sucrose [65]
Fahy	20% DMSO, 20% formamide, 15% propylene glycol [66]

4.2. Cryopreservation Methods

Presently there is no one method of cryopreservation that can be applied to a diverse range of plant species. Many cryopreservation methods (Table 5) have been developed for shoot tips and somatic embryos depending on the plant species used [17]; namely, vitrification, droplet-vitrification, encapsulation-vitrification, encapsulation-dehydration, dehydration, pre-growth, pre-growth-dehydration and D-cryoplate and V-cryoplate, a modification of the encapsulation-vitrification and droplet-vitrification [52,67,68].

Table 5. Some examples of cryopreservation methods, techniques and applications used.

Method	Technique	Application	Survival/Recovery	Reference
Vitrification	Pre-culture of cultures on basal medium supplemented with cryoprotectants, pre-treatment with loading solution, dehydration with PVS, rapid freezing rewarming.	Cocoa secondary somatic embryos	74.5% survival with 5- day pre-culture on 0.5 M sucrose followed by 60 min dehydration in PVS2 treatment for 1 h at 0 °C.	[69]
Droplet-vitrification	Resembles vitrification in all steps with only difference that materials are cryopreserved on foil strips in drops of vitirification solution.	*Hancornia speciosa* Gomes (rubber tree) shoot tips	43% regrowth with pre-culture on basal + proline (0.193 M) for 24 h in the dark at 25 °C and PVS2 15 min at 0 °C.	[70]
Encapsulation-vitrification	Sodium alginate beads are formed and explants are encapsulated in them and dehydrated in PVS before freezing.	*Olea europaea* (olive) somatic embryos	64% regrowth after 4 day pre-culture in sucrose; PVS2 treatment for 3 h treatment and rapid freezing.	[71]
		Parkia speciosa Hassk. (stink bean) shoot tips	Pre-culture on MS + trehalose (5% w/v) for 3 days; PVS2 for 1 h at 0 °C.	[72]

Table 5. Cont.

Method	Technique	Application	Survival/Recovery	Reference
Encapsulation-dehydration	Sodium alginate-encapsulated cultures are dehydrated osmotically with high concentrations of sucrose for 1–7 days and/or desiccated in an air current before slow cooling to –80 °C and then immersed in LN.	*Olea europaea* (olive) somatic embryos	40% regrowth following 4 days of sucrose pre-growth, desiccation and freezing.	[71]
		Prunus armeniaca (apricot) shoots	Recovered after treated with 0.5 M sucrose for 2 days followed by air dehydration for 2 h and frozen in LN.	[73]
Dehydration	Samples are dehydrated by either air current, silica gels, or incubation with cryoprotectant, followed by rapid freezing or two-step freezing.	*Juglans nigra* (walnut) embryo axes	Dried in a laminar flow hood until 5–15% moisture content and 100% recovery after LN.	[74]
Pre-growth and pre-growth-dehydration	Samples are cultured on media containing cryoprotectants such as DMSO, dehydrated and then frozen slowly or rapidly.	*Garcinia mangostana* L. (mangosteen) shoot tips	50% MS + sucrose (0.6 M) + 5% DMSO for 2 days	[75]
V-cryoplate	Modification of encapsulation-vitrification and droplet-vitrification. Dehydration is performed using vitrification solution PVS2.	*Morus alba* (mulberry) shoot tips	87% regrowth, 13 lines pre-cultured at 25 °C for 1 day on MS medium containing 0.3 M sucrose. PVS2 solution for 30 min at 25 °C.	[76]
D-cryoplate	Modification of encapsulation-vitrification and droplet-vitrification. Dehydration is achieved using the air current of the laminar flow cabinet or silica gel.	*Diospyros kaki* (persimmon) shoot tips	Average 87% regrowth, 10 lines 1–3 months cold acclimatization, 3 °C pre-cultured on 0.3 M sucrose, 2 days at 25 °C, laminar flow 30 min at 25 °C.	[77]

4.2.1. Vitrification

Vitrification can include the pre-culture of samples on medium supplemented with sucrose, then treated with a loading solution normally high in sucrose molarity [52] (e.g., a mixture of sucrose and glycerol), dehydration with a vitrification solution such as PVS2 or PVS3, rapid cooling, rewarming, and plant recovery by removing cryoprotectants [78].

4.2.2. Droplet-Vitrification

Droplet-vitrification is a modification of vitrification [79]; treating explants with loading (usually 2 M glycerol and 0.4 M sucrose) and vitrification solutions; cooling them ultra-rapidly in a droplet of vitrification solution either PVS2 or PVS3 placed on an alfoil strip [49] with a droplet of cryoprotectant added before immersion in LN. The alfoil strip helps with the ultra-rapid cooling (about 4000–5000 °C min^{-1}) and re-warming (3000–4500 °C min^{-1}) of samples due to the good conductivity of thermal current of aluminum [80]. The removal of the cryoprotectant is achieved during re-warming stage by using an unloading solution usually with high level of sucrose 1.2 M, then transferred to recovery and regeneration media [25,55]. Droplet vitrification combines the use of highly concentrated vitrification solutions with ultra-fast cooling and re-warming rates [81] shown to be critical for survival [82]. For high success in survival and recovery of shoot tips after LN it is vital that samples are sufficiently dehydrated by the vitrification solution in order to vitrify while rapidly cooling in LN [83].

4.2.3. Encapsulation-Vitrification and Encapsulation-Dehydration

Encapsulation-vitrification and encapsulation-dehydration have been successfully applied to cryopreserve shoot tips of woody species of crops, such as, *Malus* (apple) [84,85], *Pyrus* (pear), *Morus* (mulberry) [84], *Vitis* (grape) [86] and *Poncirus trifoliata* × *Citrus sinensis* (Chinese bitter orange) [87,88]. Dissected shoot tips or somatic embryos are suspended in a solution of sodium alginate. Beads (4–5 mm in size) are then formed using a truncated pipette tip and pipetted into a solution of $CaCl_2$ where they are allowed to set for 30 min [52]. For encapsulation-vitrification, once beads are formed with explant inside, they are then dehydrated in PVS solutions such as PVS2 or PVS3 prior to immersion in LN. Although encapsulation is time-consuming, it eases manipulation due to alginate beads being relatively large in size [52]. For the encapsulation-dehydration technique instead of dehydration with PVS solutions beads are dehydrated in a laminar flow hood or under silica gel before immersion in LN [52].

4.2.4. Dehydration

Of all the methods explained, dehydration is the simplest, as it involves just the dehydration of explants followed by direct immersion in LN. Embryonic axes or zygotic embryos extracted from seeds are mainly used. Desiccation is usually achieved by the air current of a laminar airflow cabinet or over silica gel. Dehydration using a vitrification solution removes intracellular water from cells and permits intracellular solution to undergo phase transition from liquid phase into an amorphous phase upon rapid cooling [52]. Cryoprotectant mixtures are commonly used as vitrification solution, such as PVS2 and PVS3.

4.2.5. Pre-Growth and Pre-Growth-Dehydration

In pre-growth and pre-growth-dehydration, explants are first exposed and grown on media containing cryoprotectants, dehydrated by air under a laminar flow cabinet or with silica gel, and then frozen rapidly. Depending on the plant species optimal conditions can vary greatly.

4.2.6. D-cryoplate and V-cryoplate

D-cryoplate and V-cryoplate use special aluminium cryoplates which have been developed (length 37 mm, width 7 mm and a thickness of 0.5 mm with 10 wells). An alginate solution containing 2% (w/v) sodium alginate in calcium-free MS basal medium with 0.4 M sucrose is poured over the cryo -plate. Samples are placed in wells and more sodium alginate solution is poured over the top to cover them. In V-cryoplate, dehydration is performed using the vitrification solution PVS2, while in D cryo-plate, dehydration is achieved using the air current of the laminar flow cabinet or silica gel [89]. After dehydration cryo-plates are immersed in LN. The main advantages of V-cryoplate and D-cryoplate is that handling of specimens is easy and quick because only the cryo-plates are manipulated [89].

5. The Avocado Case

5.1. Background

Avocado (*Persea americana* Mill.), a high-value fruit found in almost all tropical and sub-tropical regions of the world [90,91] belongs to the plant family Lauraceae [92], genus *Persea* [93]. Mexico is thought to be the center of origin of the species [94]. The genus *Persea* has about 400 to 450 species consisting of the currently often recognized genera *Alseodaphne* Nees, *Apollonias* Nees, *Dehaasia* Blume, *Machilus* Nees, *Nothaphoebe* Blume, *Persea* Mill. and *Phoebe* Nees. There are eight sub-species of *P. americana* including *P. americana* var. nubigena (Williams) Kopp, *P. americana* var. steyermarkii Allen, *P. americana* var. zenymyerii Schieber and Bergh, *P. americana* var. floccosa Mez, *P. americana* var. tolimanensis Zentmyer and Schieber, *P. americana* var. drymifolia Blake, *P. americana* var. guatemalensis Williams, *P. americana* var. americana Mill. [91,95]. Genetic diversity within the genus *Persea*, the sub-genera *Persea* and *Eriodaphne* and the species *P. americana* is large and is threatened by the progressive loss of tropical and sub-tropical forests [95]. This genetic diversity can serve as a resource in crop improvement [96–98] and plays an important role both ecologically and culturally.

The three recognized ecological races of *P. americana* [99]; are the Mexican race, *P. americana* var. drymifolia, adapted to the tropical highlands; the Guatemalan race, *P. americana* var. guatemalensis, adapted to medium elevations in the tropics; and the West Indian race, *P. americana* var. americana, adapted to the lowland humid tropics [100]. The ability of the three main races to withstand cold conditions varies; the West Indian race cannot tolerate temperatures below 15 °C, the Guatemalan race can tolerate cooler temperatures of −3 to −1 °C, and the Mexican race withstands temperatures as low as −7 °C exhibiting the highest cold tolerance [101–103]. They have distinctive characteristics; e.g., plant habit, leaf chemistry, peel texture, fruit color, disease and salinity tolerance [104]. The Guatemalan and Mexican races and their hybrids are very important for conservation and future breeding programs [97]. Cultivars classified as pure Guatemalan and Mexican races and Mexican × Guatemalan hybrids have been shown to have more diversity than those of pure West Indian race and Guatemalan × West Indian hybrid cultivars [97]. In Mexico and Central America, avocado trees grow under highly varied ecological conditions and natural selection over thousands of years has produced vast populations [97]. This serves as an essential source of varied attributes that are not among horticulturally available items [105].

The main avocado sold throughout the world, 'Hass', is a medium sized pear-shaped fruit with dark purplish black leathery skin [106]. Its commercial value is due to its superior taste, size, shelf-life, high growing yield, and in some areas, year-round harvesting [107]. The precise breeding history of 'Hass ' is unknown however, it is reported to be 61% Mexican and 39% Guatemalan [108]. This finding is supported by a study that analyzed the complete genome sequences of a 'Hass' individual and a representative of the highland Mexican landrace, *Persea americana* var. drymifolia; as well as genome sequencing data for other Mexican individuals, Guatemalan and West Indian accessions [108]. Analyses of admixture and introgression highlighted the hybrid origin of 'Hass', pointed to its Mexican and Guatemalan progenitor races and showed 'Hass' contained Guatemalan introgression in approximately one-third of its genome [108]. In Australia, 'Hass', represents 80% of total production [109] with 2019/20 producing 87,546 tonnes of avocados, an increase of 2% more than the previous season's 85,546 tonnes [109]. This increased consumer demand is due to its popularity as a healthy food; often referred to as a superfood due to its beneficial nutrients, vitamins, minerals, fiber and healthy fats [110,111]. Consumer market value of Australian fruit sold domestically was worth ~$845 m in 2019/20 [109].

Due to the vast range of climates and conditions in our eight major avocado growing regions, avocados are produced all year round [109]. Avocado trees propagated by seed, take approximately 4–6 years to bear fruit, in some cases they can take 10 years to come into bearing [111]. Avocado trees are partially able to self-pollinate. Their flowers behave in synchronous dichogamy, flowers are perfect, bearing both male and female parts, however the periods of maleness and femaleness are temporarily distinct to enhance the likelihood of outcrossing [112,113]. The resultant progeny is highly heterozygous in the desirable parent tree characteristics [114]. New cultivars are normally derived from chance seedlings or mutations due to the difficult nature of breeding programs, which are costly, time-consuming and under threat of abiotic and biotic stresses. Nevertheless, the avocado industry's goal is to preserve superior cultivars for commercial production. Thus, to meet this goal, avocado is propagated clonally through grafting with breeding programs based on both scion and rootstock cultivars. The threat of Ambrosia beetle species and its symbiont fungus Laurel Wilt disease to the avocado field gene banks and commercial industry in Florida, California, and Israel is a glaring example of a biotic stress that could destroy the industry [115]. For scion cultivars the focus is on high yield [116], extending harvest season, regular bearing tendencies and disease resistance e.g., Anthracnose [117], Cercospora spot [118] and Verticillium wilt [117]. Rootstocks are often selected for dwarf size [119], salinity tolerance, adaptation to alkaline soil [119,120] and pest and disease resistance [120] such as *Phytophthora cinnamomi* Rands and *Rosellina necatrix* [121]. Clonal rootstocks are thought to be the only rootstocks for the future for achieving sustainable productivity gains [122–124]. These influence the total productivity of the plant in terms of yield and health. Rootstocks from Mexico, 'Orizaba 3', 'Antigua' and 'Galvan', show a universal adaptation to multiple soil stress problems. The last two, also, have tolerance to *P. cinnamomi* [96]. Many breeding programs have concentrated on the development of new rootstocks such as 'Dusa', 'Bounty' and 'Velvick' [125] to help the industry overcome these threats [126]. 'Dusa's popularity has increased significantly since the mid-2000s. It is a common standard against which other *P. cinnamomi* tolerant rootstocks are compared in international breeding programs. It has been reported to bear fruit even under heavy *P. cinnamomi* disease pressure and has higher yields than many other rootstocks [127]. 'Bounty' is often selected for its *P. cinnamomi* tolerance and ability to survive in wet soils [127].

5.2. Avocado Conservation

5.2.1. Global Germplasm Repositories

Field living germplasm collections (Table 6) and (Figure 1), are currently the most used conservation method, but funding and threats from natural calamities; pest and diseases are a problem.

Table 6. Avocado germplasm maintained as field repositories throughout the world.

Country	Germplasm Repositories	No. of Accessions	References
USA	The Huntington San Marino CA	56 *Persea americana* accessions 4 wild *Persea* spp (6 accessions)	[128]
USA	Riverside University CA	~230 avocado scion accessions ~15 wild *Persea* spp. ~246 avocado rootstock accessions	[129] [129,130]
USA	National Genetic Resources Program, Miami, Florida	*P. americana* (167 accessions) and *P. schiedeana* (1 accession)	[44,131]
USA	The Sub-Tropical Horticulture Research Station, Miami, Florida	~400 avocado accessions	[132]
Mexico	National Research Institute of Forestry and Livestock in Guanajuato	500 accessions belonging to *P. americana*: Mexican and Guatemalan races. Related species: *P. schiedeana, P. cinerascens, P. floccosa, P nubigena*	[133]
Mexico	State of Mexico of the Fundación Salvador Sanchez Colin-CICTAMEX, S.C.	800 accessions of avocado and related species. Mexican, Guatemalan, West Indian races, *P. americana* var. costaricensis race materials.	[133]
Mexico	Coatepec Harinas and Temascaltepec; State of Mexico	Wild relatives: *Beilschmiedia anay, B. miersii, P. schiedeana, P. longipes, P. cinerascens, P. hintonni, P. floccosa, P. tolimanensis, P. steyermarkii, P. nubigena, P. lingue, P. donnell-smithii, P. parvifolia, P. chamissonis, Persea* spp.	[133]
Ghana	University of Ghana Forest and Horticultural Crops Research Centre	110 local land races and 5 varieties from South Africa ('Hass', 'Fuerte', 'Ryan', 'Ettinger' and 'Nabal'	[134]
Israel	Volcanic Centre in Bet Dagan	194 trees, propagated from 148 accessions	[96]
Spain	The Experimental Station 'La Mayora' in Malaga	75 avocado accessions	[132,135]
Cuba	N/A	210 genotypes	[132]
Chile	N/A	4 botanical breeds of *P. americana*: var. drymifolia, var. guatemalensis, var. jacket and var. costaricencis	[132]
Australia	Maroochydore Research Station	46 avocado accessions	[136]
Nigeria		8 avocado accessions	[137]
Brazil	Brasilia, in the Federal District, depending on the Embrapa Research Institute	30 avocado accessions	[138]
Brazil	Conceicao do Almeida and Juazeiro collections, both in the Bahia State	22 avocado accessions	[138]
Brazil	Piracicaba, in the Sao Paulo State	33 avocado accessions	[138]
Brazil	Jaboticabal, in the Sao Paulo State	7 avocado accessions	[138]

Figure 1. One of the 56 avocado accessions being maintained in The Huntington Botanical Gardens [in San Marino, California USA] living germplasm collection.

5.2.2. Cryopreservation of Avocado Somatic Embryos

To preserve global avocado diversity; development of improved technologies for avocado conservation, breeding/improvement and propagation is essential. In vitro somatic embryogenesis has direct importance to these objectives [139,140]. Somatic embryogenesis is the process by which somatic cells give rise to totipotent embryogenic cells capable of becoming complete plants [141]. Somatic embryogenesis can be a robust tool to regenerate genetically clonal plants from single cells chosen from selected plant material, or genetically engineered cells [142]. Somatic embryogenic cultures are generally highly heterogeneous since they consist of embryos at different developmental stages [143]. Though heterozygous in nature when regenerated using zygotic embryos as explants, cryopreservation of avocado somatic embryos offers an attractive pathway to conserve avocado germplasm. Recovery of plantlets from somatic embryos and clonal multiplication in vitro is an essential step for commercial application of this technology to crop improvement [144].

Somatic embryogenesis in avocado was first achieved using immature zygotic embryos of cv 'Hass' [145]. Studies have reported that the embryogenic capacity of avocado was highly genotype dependent [146]. To improve somatic embryogenesis previous studies have shown that several factors are vital for success, (1) composition of media, (2) hormone type and concentration, (3) type and concentration of gelling agent and (4) light intensity [147]. Morphogenic competence of somatic embryos has been reported to be lost 3–4 months after induction depending on the genotype [145,148]. In addition, the main factor limiting conversion of somatic embryos into plantlets is incomplete maturation [149]. Studies have found that there are two types of regeneration that occur after maturation; unipolar (only shoot apex or root) and bipolar (both shoot apex and root).

Shoots regenerated from unipolar embryos can either be rooted or rescued using in vitro micrografting [150]. Studies have shown that the percentage of high-quality bipolar embryos from avocado somatic embryos was extremely low at 2–3% and was genotype dependent [145,150,151]. This low rate of somatic embryo conversion is currently the main bottleneck in avocado regeneration via somatic embryogenesis [144]. A study described an in vitro induction and multiplication system for somatic embryos of avocado, across four cultivars, which remained healthy and viable for 11 months, on a medium used for mango somatic embryogenesis [139]. Furthermore for one of the cultivars, cultivar 'Reed', a two-step regeneration system was developed that resulted in 43.3% bipolar regeneration [139].

Cryopreservation of avocado somatic embryos has been successful for various cultivars (Table 7). The effect of cryogenic storage on five avocado cultivars ('Booth 7', 'Hass', 'Suardia', 'Fuerte' and 'T362') using two cryopreservation protocols (controlled-rate freezing and vitrification) was investigated [152]. In terms of controlled-rate freezing, three out of five embryogenic cultivars were successfully cryopreserved with a recovery of 53 to 80%. Using vitrification, cultivar 'Suardia' showed 62% recovery whereas 'Fuerte' had only a 5% recovery. When the droplet-vitrification technique was used, two 'Duke-7' embryogenic cell lines showed viability ranging from 78 to 100% [153]. Protocols employed in both studies cannot be applied in general to multiple cultivars and optimization of loading sucrose concentrations and plant vitrification solution 2 (PVS2), temperature and times need more intensive research.

Table 7. Summary of successfully applied cryopreservation techniques to avocado somatic embryos. * Recovery is defined as any somatic embryo clump which was proliferating into new callus clumps.

Cryopreservation Technique	Cultivars	* Recovery Percentages
Vitrification	'Suardia'	62%
	'Fuerte'	5% [152]
	'A10'	91%
	'Reed'	73%
	'Velvick'	86%
	'Duke 7'	80% [144]
Slow freezing	'Suardia'	60–80%
	'T362'	4–53%
	'Fuerte'	73–75% [152]
Droplet vitrification	'A10'	100%
	'Reed'	85%
	'Velvick'	93% [144]
	Two lines of 'Duke 7'	78–100% [153]

5.2.3. Shoot-Tip Cryopreservation of Avocado

Cryopreservation is a secure and cost-effective method for long-term storage of avocado. It provides a high degree of genetic stability in maintaining avocado collections for the long-term compared to other conservation methods. Shoot-tip cryopreservation conserves 'true-to-type' avocado plant tissue. It is ideal for preserving a core selection of avocado genotypes, for example, with superior characteristics, disease and pest resistance, rarity, drought and salinity tolerance. In one study, it was shown that axillary buds of Mexican and Guatemalan races were viable through fluorescein diacetate staining after dehydration with sterile air and being treated with cryopreservation solutions; however, shoot regeneration was not achieved with the cryopreserved material [154]. Another study, showed that dehydration at 60 min with sterile air and 30 min in PVS4 at 0 °C produced normal plant development and 100% survival was obtained after 30, 45 and 60 days [155].

5.2.4. Critical Factors Identified for Successful Cryopreservation of Avocado Shoot-Tips

Although still cultivar-dependent, in vitro protocols have been established for multiple cultivars of avocado [111] advancing cryopreservation of avocado. Droplet vitrification can be considered as a "generic" cryopreservation protocol for hydrated tissues, such as in vitro cultures [49,156]. Vitrification-based procedures offer practical advantages in comparison to classical freezing techniques and are more appropriate for complex organs e.g., avocado shoot tips, which contain a variety of cell types, each with unique requirements under conditions of freeze-induced dehydration [157]. A problem associated with cryopreservation is formation of lethal ice crystals. To overcome this vitrification makes use of the physical phase called 'vitrification', i.e., solidification of a liquid forming an amorphous 'or glassy' structure [7] to avoid ice crystal formation of a watery solution. Glass is viscous and stops all chemical reactions that require molecular diffusion, which leads to dormancy and stability over time [158]. Samples can be vitrified and rapidly supercooled at low temperatures and form in a solid metastable glass with crystallization [66]. For procedures that involve vitrification, cell dehydration occurs using a concentrated cryoprotective media and/or air desiccation and is performed first before rapid freezing in LN [157]. It is important that cells are not damaged or injured during the vitrification process and are vitrified enough to sustain immersion in LN [24]. As a result, all factors that affect intracellular ice formation are avoided [157].

Oxidative stress is a common and often severe problem in plant tissue [159,160] of most woody plant species, such as avocado. Therefore, it is important to optimize regrowth conditions of extracted avocado shoot tips to prevent browning when developing an in vitro cryopreservation protocol. Browning of cell tissue takes place as the cytoplasm and vacuoles are mixed and phenolic compounds readily become oxidized by air, peroxidase or polyphenol oxidase. Oxidization of phenolic compounds inhibit enzyme activity and result in darkening of the culture medium and subsequent lethal browning of explants [161]. The antioxidant ascorbic acid (ASA) or vitamin C (ASA) occurs naturally in plants, in plant tissue and meristems [162]. It has many roles in a plant's physiological processes but mainly in its defense against oxidative damage resulting from aerobic metabolism, photosynthesis, pollutants and other stresses caused by the environment [163]. Wounding of avocado tissue can lead to an increase in reactive oxygen species (ROS) within the shoot therefore affecting the viability. ROS are highly reactive molecules and have been shown to cause damage in cells. Many molecules are considered as ROS, some of which include oxygen-free radical species and reactive oxygen non-radical derivatives [48]. The most common ROS species found in plants are superoxide (O_2^-), hydroperoxyl (OOH), hydroxyl radical (OH) and singlet oxygen (O_2) [48]. ASA has an important role in the detoxification of ROS species both enzymatically or non-enzymatically [164]. It can do this by scavenging a singlet oxygen, hydrogen peroxide, superoxide and hydroxyl radical [163].

It has been reported by several authors that the addition of antioxidants can help increase the viability of plants by suppressing browning which leads to shoot tip death [83,165–169]. By maintaining a higher antioxidant level protection improved post cryopreservation [166]. It has been reported that in *Actinidia* spp. (kiwifruit) the addition of ASA in regrowth media improved the survival after cryopreservation by reducing lipid peroxidation [83]. The addition of ASA to pre-culture media, loading solution, unloading solution and regrowth media significantly increased regrowth of shoot tips of *Rubus* spp. (raspberry) [168]. A recent study found treating *Persea americana* cv 'Reed' (avocado), with varying concentrations of different antioxidants (ASA, polyvinylpyrrolidone [PVP], citric acid and melatonin) reduced browning caused when extracting shoot tips. The type of antioxidant and concentration had an effect on viability, vigor and health of the shoots [170].

Avocado is highly susceptible to osmotic stresses imposed by cryoprotectants which are high in osmolarity. Cold sensitive species such as avocado are likely to be positively responsive to vitrification treatments during cryopreservation if optimizations are done carefully [171]. In order to improve on tolerance to cryoprotectants and increase permeation of the cryoprotectant through the cell membrane and induce tolerance to dehydration caused by vitrification solutions, a pre-step called 'loading' is used [52]. Loading is achieved by incubating tissues for 10–20 min in solutions composed of glycerol and sucrose [48]. This loading step is particularly useful for plant species, that are sensitive to direct exposure to cryoprotectants due to dehydration intolerance and osmotic stresses [48]. However, use of loading solution alone for avocado shoot tips is not adequate to induce tolerance to cryoprotectants, and other pre-treatments/pre-culture such as osmotic conditioning with sugars and cold acclimatization are necessary [172].

Pre-culturing shoot tips with a high sugar enriched media has been reported previously by several authors [173–175] to increase the viability post-cryopreservation by better pre-conditioning the shoot. Also, time of incubation in pre-culture solutions was critical to ensuring survival and high regrowth rates [55,176]. There have been attempts to use alternative sources of sugar in pre-culture media, such as, sorbitol or mannitol [177–180], glucose and fructose; all have shown no negative effects on post-cryopreservation survival [181]. However, most researchers prefer to use sucrose as the sugar source when adding to pre-culture media [181]. Sucrose has been found to be more beneficial in pre-culture as compared to sorbitol and mannitol as these two sugars were unable to support regrowth of olive somatic embryos [182]. However, when 0.2 M sorbitol was combined with 5% DMSO it was an effective cryoprotectant for embryogenic tissue of *Pinus roxburghii* Sarg. (chir pine) [183]. Sucrose is an excellent glass former and is able to stabilize membranes and proteins [184]. Sucrose stimulates the production of other elements such as proline, glycine betaine, glycerol and polyamines, which have colligative as well as non-colligative effects [185,186]. Of the above-mentioned sugars [187], glycerol [188], proline [189] and glycine betaine [190] have proved their cryoprotectant ability, whereas polyamines are known for their antioxidant properties. Therefore, these compounds play a vital role in protecting the cells during cryopreservation. It has also been shown that pre-culturing in high sucrose media enhances the acclimatization process to low temperature and stimulates osmotic dehydration [47].

Water availability and temperature are influenced by environmental variables and are major determinants of plant growth and development [191]. Most tropical and sub-tropical species have little to no freezing tolerance, however, temperate plant species have evolved some form of cold tolerance [191,192]. It has been shown in temperate plants that they have the genetic ability to increase cold tolerance significantly when exposed to environmental cues that signal the arrival of winter [193]. Many plants can increase their tolerance to the cold by exposure to lower temperatures, generally with temperatures below 10 °C [193]. This process is referred to as cold hardening or cold acclimatization (CA) and requires days to weeks for full development [50,193,194]. Several biochemical, physiological and metabolic functions are altered in plants by low temperature as well as gene expression [195]. Expression of cold induced genes include those that control the function of cell membranes to stabilize and protect themselves against freezing injury [196]. Freezing tolerance can be increased by 2–8 °C in spring annuals, 10–30 °C in winter annuals and 20–200 °C in tree species [193]. Cold acclimatization can help improve the regrowth rates of in vitro plants, improve regeneration rates [197]. Cold acclimatization has been used as an in vitro pre-treatment on donor plants before shoot tip extraction [198] in developing cryopreservation protocols in plants such as *Malus domestica* Borkh (apple), *Malus sieversii* (Ledeb.) (wild apple) and *Phoenix dactylifera* (date palm) [199,200]. Cold acclimatization with or without ABA significantly improved the survival of *Rubus* spp. [201]. Abscisic acid (ABA) pre-treatment alone could not increase the survival of plants grown under warm conditions after cryopreservation, but the survival tripled when cold acclimatization was combined with ABA pre-treatment [201]. High sucrose (0.3 M) or low temperature (10 °C) incubation treatments primed in vitro plants of cvs 'Reed' and 'Velvick' shoot tips to tolerate cryoprotectant (PVS2) treatments but was cultivar-specific [202].

6. Conclusions

Field living germplasm collections are currently the only conservation method for avocado, but funding and threats from natural calamities; pest and diseases are a problem. Cryopreservation is an invaluable tool that could be utilized in conjunction with field repositories to securely preserve this important horticultural crop. There have been significant improvements within the cryopreservation platform to preserve *Persea* spp. germplasm [202–204]. Studies have shown that cryopreservation of somatic embryos offers usefulness in conserving *Persea* germplasm biodiversity [144,152,153]. An important factor for somatic embryos is that regeneration can be achieved after exposure to LN to ensure that protocols can be effectively applied for conservation programs [176]. Cryopreservation of somatic embryos is valuable as it is readily retrievable for further biotechnology manipulations as well as storage of biotechnology products such as genetically transformed lines [23,205].

To date, although cultivar-dependent, in vitro multiplication protocols have been established for maintaining multiple avocado cultivars in tissue culture from mature glasshouse cuttings [111]. This can be used to supply new plants to avocado farmers, meeting a critical issue that is preventing the expansion of industry, the shortage of available avocado trees. Twenty thousand in vitro plants can be maintained in a 10 sqm tissue culture room saving on land, fertilizer, pesticides promoting an environmentally sustainable and efficient method of multiplication of avocado plants.

Development of the in vitro shoot-tip cryopreservation protocol was highly dependent on the availability of this reliable in vitro multiplication and regeneration protocol. For the first time studies [202–204] have shown that in vitro cryopreservation using droplet-vitrification for mature material of two avocado cultivars have been successful. Correctly treating avocado shoot tips with the ideal pre-treatment before LN is vital for a successful outcome [202]. It was identified that the use of 100 and 250 mg L^{-1} of ASA can effectively reduce browning of freshly extracted avocado shoot tips [170,202]. High sucrose and cold pre-treatments are effective in increasing survivability following cryoprotectant incubation of avocado shoot tips. While pre-treatments are effective for avocado, the type of pre-treatment needed and the degree of effectiveness was cultivar-specific [202]. This can be directly linked to the genetics of the two cultivars which display varying tolerance to cold and salinity in their natural growing environments; namely, cv 'Velvick' from West Indian race (no cold tolerance) and cv 'Reed' from Guatemalan race (moderate cold tolerance) [204]. The type of cryoprotectant and exposure time to the cryoprotectant was also essential in obtaining morphologically normal and vigorous plants [204]. Avocado shoots that survived LN grew into full plants ready for rooting after 24 weeks [204]. Cultivar 'Reed' shoots were successfully rooted [206] and after 8 weeks, plantlets were ready to be acclimatized in a University of Queensland glasshouse (Figure 2). These plants will be screened for growth parameters and yield in a field trial at Duranbah, Queensland. Shoot tips from cv 'Velvick' are currently in the rooting stage.

In vitro multiplication and in vitro cryopreservation protocols provide another set of tools that can be used to preserve global avocado diversity to improve conservation germplasm collections, breeding and propagation. Somatic embryogenesis, cryopreservation of somatic embryos and shoot tips, have the ability to be adapted to lead to the establishment of a global Cryo-Bank conserving avocado biodiversity and offering a source of disease-free genetic material. They provide useful tools for further optimization of the species and other woody plant species facing similar challenges in conservation. Shoot tip cryopreservation is ideal for preserving a core selection of avocado genotypes, for example, with superior characteristics, disease and pest resistance, rarity, drought and salinity tolerance. Shoot tip cryopreservation of avocado is a major breakthrough and this work can pave the way for storing a core collection of *Persea* spp. for true-to-type avocado shoot tip preservation.

Figure 2. Shoot tips of cv 'Reed' treated with VSL and revived from LN growing in a glasshouse.

Author Contributions: C.O.: Writing and review of literature — original draft, Drafting and production. J.H.-B.: Drafting and production. R.F.: Drafting and production. S.L.: Drafting and production. A.H.: Drafting and production. J.F.: Drafting and production. N.M.: Drafting and production. All authors have read and agreed to the published version of the manuscript.

Funding: Chris O'Brien is supported by an Australian Commonwealth Government Research Training Program (RTP) Scholarship and funding from The Huntington Library, Art Museum, and Botanical Gardens as well as funding from Advance Queensland Innovation Partnerships Project Avocado Tissue-Culture: From Lab-to-Orchard (AQIP06316-17RD2).

Institutional Review Board Statement: Not applicable.

Informed Consent Statement: Not applicable.

Data Availability Statement: Not applicable.

Acknowledgments: The Queensland Alliance for Agriculture and Food Innovation (QAAFI) is a research institute of The University of Queensland (UQ), supported by the Queensland Government Department of Agriculture and Fisheries. Chris O'Brien is supported by an Australian Commonwealth Government Research Training Program (RTP) Scholarship and funding from The Huntington Library, Art Museum, and Botanical Gardens as well as funding from Advance Queensland Innovation Partnerships Project Avocado Tissue-Culture: From Lab-to-Orchard (AQIP06316-17RD2). The authors would also like to acknowledge the following people for providing information on avocado germplasm collections Mary Lu Arpaia, Eric Focht, Patricia Manosalva, Alejandro Barrientos-Priego, Tatiana Cantuarias, Elizabeth Dann, Ricardo Goenaga. We also acknowledge Christina Walters, Kim E. Hummer and Gan-Yuan Zhong for information on clonal germplasm repositories.

Conflicts of Interest: The authors declare no conflict of interest.

References

1. Wilson, E. *The Diversity of Life*; Penguin: London, UK, 1992.
2. Harlan, J.R.; de Wet, J.M. Toward a rational classification of cultivated plants. *Taxon* **1971**, *20*, 509–517. [CrossRef]
3. Engels, J.; Ebert, A.; Thormann, I.; De Vicente, M. Centres of crop diversity and/or origin, genetically modified crops and implications for plant genetic resources conservation. *Genet. Resour. Crop Evol.* **2006**, *53*, 1675–1688. [CrossRef]

4. Maxted, N.; Guarino, L.; Myer, L.; Chiwona, E. Towards a methodology for on-farm conservation of plant genetic resources. *Genet. Resour. Crop Evol.* **2002**, *49*, 31–46. [CrossRef]
5. Haidet, M.; Olwell, P. Seeds of success: A national seed banking program working to achieve long-term conservation goals. *Nat. Areas J.* **2015**, *35*, 165–173. [CrossRef]
6. Engelmann, F.; Engels, J. Technologies and strategies for *ex situ* conservation. In *Managing Plant Genetic Diversity*; Engels, J.M., Ramanatha, R.U., Brown, A.H.D., Jackson, M.T., Eds.; CAB International: Wallingford, UK, 2002; pp. 89–103. ISBN 9780851995229. [CrossRef]
7. Panis, B.; Nagel, M. Challenges and Prospects for the Conservation of Crop Genetic Resources in Field Genebanks, in In Vitro Collections and/or in Liquid Nitrogen. *Plants* **2020**, *9*, 1634. [CrossRef]
8. Tsioumani, E. The State of the World's Biodiversity for Food and Agriculture: A Call to Action? *Environ. Policy Law* **2019**, *49*, 110–112. [CrossRef]
9. Migicovsky, Z.; Warschefsky, E.; Klein, L.L.; Miller, A.J. Using living germplasm collections to characterize, improve, and conserve woody perennials. *Crop Sci.* **2019**, *59*, 2365–2380. [CrossRef]
10. Normah, M.N.; Sulong, N.; Reed, B.M. Cryopreservation of shoot tips of recalcitrant and tropical species: Advances and Strategies. *Cryobiology* **2019**, *87*, 1–14. [CrossRef]
11. Berjak, P.; Pammenter, N. Seed recalcitrance-current perspectives. *S. Afr. J. Bot.* **2001**, *67*, 79–89. [CrossRef]
12. Niino, T.; Arizaga, M.V. Cryopreservation for preservation of potato genetic resources. *Breed. Sci.* **2015**, *65*, 41–52. [CrossRef] [PubMed]
13. Engelmann, F. Use of biotechnologies for the conservation of plant biodiversity. *Vitr. Cell. Dev. Biol. Plant* **2011**, *47*, 5–16. [CrossRef]
14. Acedo, V.; Arradoza, C. *In vitro* conservation of yam germplasm. *Philipp. J. Crop Sci.* **2005**. [CrossRef]
15. Reed, B.; Bell, R. In Vitro Tissue Culture of Pear: Advances in Techniques for Micropropagation and Germplasm Preservation. In Proceedings of the VIII International Symposium on Pear 596, Ferrara—Bologna, Italy, 31 December 2002; pp. 412–418, ISBN 9066058668. [CrossRef]
16. Bertrand-Desbrunais, A.; Noirot, M.; Charrier, A. Slow growth *in vitro* conservation of coffee (*Coffea* spp.). *Plant Celltissue Organ Cult.* **1992**, *31*, 105–110. [CrossRef]
17. Engelmann, F. In vitro conservation methods. In *Biotechnology and Plant Genetic Resources: Conservation and Use*; Callow, J.A., Ford-Lloyd, B.V., Newburry, J.H., Eds.; CABI: Wallingford, UK, 1997; pp. 119–162. ISBN 0851991424.
18. George, E.F.; Sherrington, P.D. Plant propagation by tissue culture: Handbook and directory of commercial laboratories. In *Journal of Basic Microbiology*; Schmauder, H.P., Ed.; Exegetics Ltd.: Eversley, UK, 1984; ISBN 0950932507. [CrossRef]
19. Reed, B.M.; Gupta, S.; Uchendu, E.E. In vitro genebanks for preserving tropical biodiversity. In *Conservation of Tropical Plant Species*, 1st ed.; Normah, M.N., Chin, H.F., Reed, B.M., Eds.; Springer New York: New York, NY, USA, 2013; pp. 77–106. ISBN 978-1-4614-3776-5. [CrossRef]
20. Bhojwani, S.S.; Razdan, M.K. *Plant Tissue Culture: Theory and Practice*; Elsevier: Amsterdam, The Netherlands, 1986; p. 0444596151.
21. Benson, E.E. Cryopreservation Theory. In *Plant Cryopreservation: A Practical Guide*; Reed, B.M., Ed.; Springer: New York, NY, USA, 2008; pp. 15–32. ISBN 978-0-387-72275-7. [CrossRef]
22. Harding, K. Genetic integrity of cryopreserved plant cells: A review. *CryoLetters* **2004**, *25*, 3–22.
23. Engelmann, F. Cryopreservation of embryos: An overview. In *Plant Embryo Culture, Methods in Molecular Biology (Methods and Protocols.)*; Thorpe, T.A., Yeung, E.C., Eds.; Humana Press: Totowa, NJ, USA, 2011; Volume 710, pp. 155–184. ISBN 978-1-61737-987-1. [CrossRef]
24. Bi, W.L.; Pan, C.; Hao, X.Y.; Cui, Z.H.; Kher, M.; Marković, Z.; Wang, Q.C.; Teixeira da Silva, J. Cryopreservation of grapevine (*Vitis* spp.)—A review. *Vitr. Cell. Dev. Biol. Plant* **2017**, *53*, 449–460. [CrossRef]
25. Benelli, C.; de Carlo, A.; Engelmann, F. Recent Advances in the cryopreservation of shoot-derived germplasm of economically important fruit trees of Actinidia, Diospyros, Malus, Olea, Prunus, Pyrus and Vitis. *Biotechnol. Adv.* **2013**, *31*, 175–185. [CrossRef]
26. Engelmann, F. Germplasm collection, storage, and conservation. In *Plant Biotechnology and Agriculture. Prospects for 21st Century*; Altman, A., Hagegawa, A., Eds.; Academic Press: San Diego, CA, USA, 2012; pp. 255–267. [CrossRef]
27. Bettoni, J.C.; Costa, M.D.; Gardin, J.P.P.; Kretzschmar, A.A.; Pathirana, R. Cryotherapy: A new technique to obtain grapevine plants free of viruses. *Rev. Bras. De Frutic.* **2016**, *38*. [CrossRef]
28. Wang, Q.; Valkonen, J.P. Cryotherapy of shoot tips: Novel pathogen eradication method. *Trends Plant Sci.* **2009**, *14*, 119–122. [CrossRef]
29. Barrientos-Priego, A.F.; Borys, M.W.; Escamilla-Prado, E.; Ben-Ya'acov, A.; De La Cruz-Torres, E.; Lopez-Lopez, L. A study of the avocado germplasm resources, 1988–1990. IV. Findings in the Mexican Gulf Region. In Proceedings of the Second World Avocado Congress, Orange, CA, USA, 21–26 April 1992; pp. 551–558.
30. Benz, B. The conservation of cultivated plants. *Nat. Educ. Knowl.* **2012**, *3*, 4. Available online: https://www.nature.com/scitable/knowledge/library/the-conservation-of-cultivated-plants-80059198/ (accessed on 18 January 2021).
31. GRIN-Global, U.S. National Plant Germplasm System. Available online: https://npgsweb.ars-grin.gov/gringlobal (accessed on 4 January 2020).
32. NTBG Breadfruit Institute. To Promote the Conservation, Study, and Use of Breadfruit for Food and Reforestation. Available online: https://ntbg.org/breadfruit/ (accessed on 5 January 2021).

33. Hanke, M.V.; Höfer, M.; Flachowsky, H.; Peil, A. *Fruit Genetic Resources Management: Collection, Conservation, Evaluation and Utilization in Germany*; International Society for Horticultural Science (ISHS): Leuven, Belgium, 2014; pp. 231–234, 2406–6168. [CrossRef]
34. National-Fruit-Collection National Fruit Collection held at Brogdale Farm. Available online: http://www.nationalfruitcollection.org.uk/ (accessed on 4 January 2020).
35. Withers, L.A.; Engelmann, F. In vitro conservation of plant genetic resources. In *Agricultural Biotechnology*; Altman, A., Ed.; Marcel Dekker: New York, NY, USA, 1998; pp. 57–88. ISBN 0-8247-9439-7.
36. Ashmore, S.E. *Status Report on the Development and Application of In Vitro Techniques for the Conservation and Use of Plant Genetic Resources*; International Plant Genetic Resources Institute: Rome, Italy, 1997; ISBN 9290433396.
37. Mandal, B.; Tyagi, R.; Pandey, R.; Sharma, N.; Agrawal, A. In Vitro Conservation Of Germplasm of Agri-Horticultural Crops at NBPGR: An Overview. In *Conservation of Plant Genetic Resources In Vitro*; Razdan, M.K., Cocking, E.C., Eds.; Science Publishers Inc.: Hauppauge, NY, USA, 2000; Volume 2, pp. 297–307.
38. Reed, B.M.; Denoma, J.; Wada, S.; Postman, J. Micropropagation of pear (*Pyrus* spp.). *Methods Mol. Biol.* **2013**, *11013*, 3–18. [PubMed]
39. FAO. *Genebank Standards: For Plant Genetic Resources for Food and Agriculture*; Food and Agriculture Organization of the United Nations: Rome, Italy, 2014; ISBN 978-92-5-108262-1.
40. Reed, B.M.; Engelmann, F.; Dulloo, M.E.; Engels, J.M.M. Technical Guidelines for the Management of Field and in vitro Germplasm Collections. In *IPGRI Handbooks for Genebanks No. 7*; International Plant Genetic Resources Institute: Rome, Italy, 2004; Volume 472, ISBN 9290436409.
41. Benson, E.; Harding, K.; Debouck, D.G.; Dumet, D.; Escobar, R.; Mafla, G.; Panis, B.; Panta, A.; Tay, D.; Houwe, I. *Refinement and Standardization of Storage Procedures for Clonal Crops. Global Public Goods Phase 2: Part 1. Project Landscape and General Status of Clonal Crop in Vitro Conservation Technologies*; System-Wide Genetic Resources Programme (SGRP): Rome, Italy, 2011; ISBN 929043905X.
42. Benson, E.; Harding, K.; Debouck, D.G.; Dumet, D.; Escobar, R.; Mafla, G.; Panis, B.; Panta, A.; Tay, D.; Houwe, I. *Refinement and Standardization of Storage Procedures for Clonal Crops. Global Public Goods Phase 2. Part 2: Status of in Vitro Conservation Technologies for: Andean Root and Tuber Crops, Cassava, Musa, Potato, Sweetpotato and Yam*; System-Wide Genetic Resources Programme (SGRP): Rome, Italy, 2011; ISBN 9290439068.
43. CGAIR Cassava Diversity. Available online: https://ciat.cgiar.org/what-we-do/crop-conservation-and-use/cassava-diversity/ (accessed on 5 January 2021).
44. Gutierrez, B. The Agricultural Research Service United States Department of Agriculture. *Personal Communication*, USDA-ARS Plant Genetic Resources Unit: Geneva, NY, USA, 2020.
45. Walters, C. (The United States Department of Agriculture). *Personal Communication*, USDA Agricultural Genetic Resources Preservation Research: Fort Collins, CO, USA, 2020.
46. Reed, B. Choosing and applying cryopreservation protocols to new plant species or tissues. In Proceedings of the I International Symposium on Cryopreservation in Horticultural Species 908, Leuven, Belgium, 5–8 April 2009; pp. 363–372.
47. Reed, B.M. Plant Cryopreservation: A Practical Guide. Reed, B.M., Ed.; Springer New York: New York City, NY, USA, 2008; ISBN 978-0-387-72275-7.
48. Kaczmarczyk, A.; Funnekotter, B.; Menon, A.; Phang, P.Y.; Al-Hanbali, A.; Bunn, E.; Mancera, R. Current issues in plant cryopreservation. In *Current Frontiers in Cryobiology*; Katkov, I.I., Ed.; In Tech: Rijeka, Croatia, 2012; pp. 417–438. ISBN 978-9535101918.
49. Streczynski, R.; Clark, H.; Whelehan, L.M.; Ang, S.-T.; Hardstaff, L.K.; Funnekotter, B.; Bunn, E.; Offord, C.A.; Sommerville, K.D.; Mancera, R.L. Current issues in plant cryopreservation and importance for *ex situ* conservation of threatened Australian native species. *Aust. J. Bot.* **2019**, *67*, 1–15. [CrossRef]
50. Panis, B.; Lambardi, M. Status of cryopreservation technologies in plants (crops and forest trees). In *The Role of Biotechnology in Exploring and Protecting Agricultural Genetic Resources*; Ruane, J., Sonnino, A., Eds.; Food & Agriculture Organization: Rome, Italy, 2006; pp. 61–78. ISBN 92-5-105480-0.
51. Pérez, R.M. Cryostorage of Citrus embryogenic cultures. In *Somatic Embryogenesis in Woody Plants*; Jain, S.M., Gupta, P.K., Newton, R.J., Eds.; Springer: Dordrecht, The Netherlands, 2000; Volume 67, pp. 687–705. ISBN 978-90-481-5508-8. [CrossRef]
52. Sakai, A.; Engelmann, F. Vitrification, encapsulation-vitrification and droplet-vitrification: A review. *CryoLetters* **2007**, *28*, 151–172. [PubMed]
53. Tanino, K.K.; Chen, T.H.H.; Fughigami, L.H.; Weiser, C.J. Metabolic alterations associated with abscisic acid-induced frost hardiness in bromegrass suspension culture cells. *Plant Cell Physiol.* **1990**, *31*, 505–511. [CrossRef]
54. Kim, H.-H.; No, N.-Y.; Shin, D.-J.; Ko, H.-C.; Kang, J.-H.; Cho, E.-G.; Engelmann, F. Development of alternative plant vitrification solutions to be used in droplet-vitrification procedures. *Acta Hortic.* **2011**, *908*, 181–186. [CrossRef]
55. Azimi, M.; O'Brien, C.; Ashmore, S.; Drew, R. Cryopreservation of papaya germplasm. *Acta Hortic.* **2005**, *692*, 43–50. [CrossRef]
56. Yamada, T.; Sakai, A.; Matsumura, T.; Higuchi, S. Cryopreservation of apical meristems of white clover (*Trifolium repens* L.) by vitrification. *Plant Sci.* **1991**, *78*, 81–87. [CrossRef]
57. McGann, L.E. Differing actions of penetrating and nonpenetrating cryoprotective agents. *Cryobiology* **1978**, *15*, 382–390. [CrossRef]
58. Nishizawa, S.; Sakai, A.; Amano, Y.; Matsuzawa, T. Cryopreservation of asparagus (*Asparagus officinalis* L.) embryogenic suspension cells and subsequent plant regeneration by vitrification. *Plant Sci.* **1993**, *91*, 67–73. [CrossRef]

59. Sakai, A.; Kobayashi, S.; Oiyama, I. Cryopreservation of nucellar cells of navel orange (*Citrus sinensis* Osb. var. brasiliensis Tanaka) by vitrification. *Plant Cell Rep.* **1990**, *9*, 30–33. [CrossRef]
60. Uragami, A.; Sakai, A.; Nagai, M. Cryopreservation of asparagus (*Asparagus-officinalis* L.) cultured *in vitro*. *Jpn. Agric. Res. Q.* **1993**, *27*, 112–115.
61. Uragami, A.; Sakai, A.; Nagai, M.; Takahashi, T. Survival of cultured cells and somatic embryos of *Asparagus officinalis* cryopreserved by vitrification. *Plant Cell Rep.* **1989**, *8*, 418–421. [CrossRef] [PubMed]
62. Matsumoto, T. Studies on Cryopreservation of In Vitro Grown Apical Meristems of Wasabi (Wasabia japonica M.). Doctorial Thesis, Bulletin of Shimane Agricultural Experiment Station, Shimane, Japan, 1999.
63. Suzuki, M.; Tandon, P.; Ishikawa, M.; Toyomasu, T. Development of a new vitrification solution, VSL, and its application to the cryopreservation of gentian axillary buds. *Plant Biotechnol. Rep.* **2008**, *2*, 123. [CrossRef]
64. Langis, R.; Schnabel, B.; Earle, E.; Steponkus, P. Cryopreservation of *Brassica campestris* L. cell suspensions by vitrification. *Cryoletters* **1989**. Available online: https://iifiir.org/en/fridoc/cryopreservation-of-brassica-campestris-l-cell-suspensions-by-88804 (accessed on 21 February 2021).
65. Towill, L.E. Cryopreservation of isolated mint shoot tips by vitrification. *Plant Cell Rep.* **1990**, *9*, 178–180. [CrossRef]
66. Fahy, G.M.; MacFarlane, D.; Angell, C.; Meryman, H. Vitrification as an approach to cryopreservation. *Cryobiology* **1984**, *21*, 407–426. [CrossRef]
67. Engelmann, F. Importance of cryopreservation for the conservation of plant genetic resources. Cryopreservation of tropical plant germplasm: Current research progress and application. In *Proceedings of the an International Workshop, Tsukuba, Japan, 31 October – 1 November 1998*; International Plant Genetic Resources Institute (IPGRI): Maccarese, Italy, 2000; pp. 8–20. ISBN 9290434-287.
68. Yamamoto, S.-i.; Rafique, T.; Priyantha, W.S.; Fukui, K.; Matsumoto, T.; Niino, T. Development of a Cryopreservation Procedure Using Aluminium Cryo-plates. *Cryoletters* **2011**, *32*, 256–265. [PubMed]
69. Adu-Gyamfi, R.; Wetten, A. Cryopreservation of cocoa (*Theobroma cacao* L.) somatic embryos by vitrification. *CryoLetters* **2012**, *33*, 494–505.
70. de Oliveira Prudente, D.; Paiva, R.; Nery, F.C.; de Oliveira Paiva, P.D.; Alves, J.D.; Máximo, W.P.F.; Silva, L.C. Compatible solutes improve regrowth, ameliorate enzymatic antioxidant systems, and reduce lipid peroxidation of cryopreserved Hancornia speciosa Gomes lateral buds. *Vitr. Cell. Dev. Biol. Plant* **2017**, *53*, 352–362. [CrossRef]
71. Shibli, R.; Al-Juboory, K. Cryopreservation of 'Nabali' olive (*Olea europea* l.) somatic embryos by encapsulation-dehydration and encapsulation-vitrification. *CryoLetters* **2000**, *21*, 357–366. [PubMed]
72. Nadarajan, J.; Mansor, M.; Krishnapillay, B.; Staines, H.J.; Benson, E.E.; Harding, K. Applications of differential scanning calorimetry in developing cryopreservation strategies for Parkia speciosa, a tropical tree producing recalcitrant seeds. *CryoLetters* **2008**, *29*, 95–110.
73. Soliman, H.I. Cryopreservation of *in vitro*-grown shoot tips of apricot (*Prunus armeniaca* L.) using encapsulation-dehydration. *Afr. J. Biotechnol.* **2013**, *12*, 1419–1430.
74. Ballesteros, D.; Pence, V.C. Survival and growth of embryo axes of temperate trees after two decades of cryo-storage. *Cryobiology* **2019**, *88*, 110–113. [CrossRef] [PubMed]
75. Ibrahim, S.; Normah, M. The survival of in vitro shoot tips of Garcinia mangostana L. after cryopreservation by vitrification. *Plant Growth Regul.* **2013**, *70*, 237–246. [CrossRef]
76. Yamamoto, S.; Rafique, T.; Sekizawa, K.; Koyama, A.; Ichihashi, T.; Niino, T. Development of an effective cryopreservation protocol using aluminum cryo-plates for in vitro-grown shoot tips of mulberries (Morus spp.) originated from the tropics and subtropics. *Sanshi Konchu Biotec (J. Insect Biotechnol. Sericology)* **2012**, *81*, 57–62.
77. Matsumoto, T.; Yamamoto, S.-i.; Fukui, K.; Rafique, T.; Engelmann, F.; Niino, T. Cryopreservation of persimmon shoot tips from dormant buds using the D cryo-plate technique. *Hortic. J.* **2015**. [CrossRef]
78. Kim, H.H.; Kim, J.B.; Baek, H.J.; Cho, E.G.; Chae, Y.A.; Engelmann, F. Evolution of DMSO concentration in garlic shoot tips during a vitrification procedure. *CryoLetters* **2004**, *25*, 90–100.
79. Kartha, K.; Leung, N.; Mroginski, L. *In vitro* growth responses and plant regeneration from cryopreserved meristems of cassava (*Manihot esculenta* Crantz). *Z. Für Pflanzenphysiol.* **1982**, *107*, 133–140. [CrossRef]
80. Bettoni, J.C.; Bonnart, R.; Volk, G.M. Challenges in implementing plant shoot tip cryopreservation technologies. *Plant Cell Tissue Organ Cult.* **2020**. [CrossRef]
81. Panis, B.; Piette, B.; Swennen, R. Droplet vitrification of apical meristems: A cryopreservation protocol applicable to all Musaceae. *Plant Sci.* **2005**, *168*, 45–55. [CrossRef]
82. Benson, E.; Harding, K.; Debouck, D.G.; Dumet, D.; Escobar, R.; Mafla, G.; Panis, B.; Panta, A.; Tay, D.; Houwe, I. *Refinement and Standardization of Storage Procedures for Clonal Crops. Global Public Goods Phase 2. Part 3: Multi-Crop Guidelines for Developing in Vitro Conservation Best Practices for Clonal Crops*; System-Wide Genetic Resources Programme (SGRP): Rome, Italy, 2011; ISBN 9290438339.
83. Mathew, L.; McLachlan, A.; Jibran, R.; Burritt, D.J.; Pathirana, R. Cold, antioxidant and osmotic pre-treatments maintain the structural integrity of meristematic cells and improve plant regeneration in cryopreserved kiwifruit shoot tips. *Protoplasma* **2018**, *255*, 1065–1077. [CrossRef]
84. Niino, T.; Sakai, A. Cryopreservation of alginate-coated in vitro-grown shoot tips of apple, pear and mulberry. *Plant Sci.* **1992**, *87*, 199–206. [CrossRef]

85. Paul, H.; Daigny, G.; Sangwan-Norreel, B. Cryopreservation of apple (Malus× domestica Borkh.) shoot tips following encapsulation-dehydration or encapsulation-vitrification. *Plant Cell Rep.* **2000**, *19*, 768–774. [CrossRef]
86. Wang, Q.; Tanne, E.; Arav, A.; Gafny, R. Cryopreservation of *in vitro*-grown shoot tips of grapevine by encapsulation-dehydration. *Plant Celltissue Organ Cult.* **2000**, *63*, 41–46. [CrossRef]
87. Wang, Q.; Batuman, Ö.; Li, P.; Bar-Joseph, M.; Gafny, R. Cryopreservation of in vitro-grown shoot tips of 'Troyer' citrange [Poncirus trifoliata (L.) Raf.× Citrus sinensis (L.) Osbeck.] by encapsulation-dehydration. *Plant Cell Rep.* **2002**, *20*, 901–906. [CrossRef]
88. Wang, Q.; Batuman, Ö.; Li, P.; Bar-Joseph, M.; Gafny, R. A simple and efficient cryopreservation of in vitro-grown shoot tips of Troyer' citrange [Poncirus trifoliata (L.) Raf.× Citrus sinensis (L.) Osbeck.] by encapsulation-vitrification. *Euphytica* **2002**, *128*, 135–142. [CrossRef]
89. Niino, T.; Yamamoto, S.; Matsumoto, T.; Engelmann, F.; Valle Arizaga, M.; Tanaka, D. *Development of V and D Cryo-Plate Methods as Effective Protocols for Cryobanking*; International Society for Horticultural Science (ISHS): Leuven, Belgium, 2019; pp. 249–262, 2406–6168. [CrossRef]
90. Kopp, L. A taxonomic revision of the genus Persea in the Western Hemisphere (Persea: Lauraceae). Revisión taxonómica del género Persea en el hemisferio occidental (Persea: Lauraceae). *Garden* **1966**, *14*, 1–120.
91. Scora, R.W.; Bergh, B.O. Origin of and Taxonomic Relationships within the Genus Persea. In Proceedings of the Second World Avocado Congress, Orange, CA, USA, 21–26 April 1992; Volume 2, pp. 505–574.
92. Bergh, B.; Ellstrand, N. Taxonomy of the avocado. *Calif. Avocado Soc. Yearb.* **1986**, *70*, 135–145.
93. Rohwer, J.G.; Li, J.; Rudolph, B.; Schmidt, S.A.; van der Werff, H.; Li, H.-w. Is Persea (Lauraceae) monophyletic? Evidence from nuclear ribosomal ITS sequences. *Taxon* **2009**, *58*, 1153–1167. [CrossRef]
94. Storey, W.; Bergh, B.; Zentmyer, G. The origin, indigenous range and dissemination of the avocado. *Calif. Avocado Soc. Yearb.* **1986**, *70*, 127–133.
95. Pliego-Alfaro, F.; Palomo-Ríos, E.; Mercado, J.; Pliego, C.; Barceló-Muñoz, A.; López-Gómez, R.; Hormaza, J.; Litz, R. Persea americana avocado. *Biotechnol. Fruit Nut Crop.* **2020**, 258–281. [CrossRef]
96. Ben-Ya'acov, A.; Bufler, G.; Barrientos-Priego, A.; De La Cruz-Torres, E.; López-López, L. A Study of Avocado Germplasm Resources, 1988–1990. I. General Description of the International Project and its Findings. In Proceedings of the Second World Avocado Congress, Orange, CA, USA, 21–26 April 1992; Volume 2, pp. 535–541.
97. Ge, Y.; Zhang, T.; Wu, B.; Tan, L.; Ma, F.; Zou, M.; Chen, H.; Pei, Y.; Liu, Y.; Chen, Z.; et al. Genome-wide assessment of avocado germplasm determined from specific length amplified fragment sequencing and transcriptomes: Population structure, genetic diversity, identification, and application of race-specific markers. *Genes* **2019**, *10*, 215. [CrossRef] [PubMed]
98. O'Brien, C.; Hiti-Bandaralage, J.C.H.; Hayward, A.; Mitter, N. Avocado (*Perse, americana* Mill.). In *Step Wise Protocols for Somatic Embryogenesis of Important Woody Plants*; Jain, S.M., Gupta, P.K., Eds.; Springer Cham: cham, Switzerland, 2018; Volume 85, pp. 305–328. ISBN 978-3-319-79086-2. [CrossRef]
99. Furnier, G.; Cummings, M.; Clegg, M. Evolution of the avocados as revealed by DNA restriction fragment variation. *J. Hered.* **1990**, *81*, 183–188. [CrossRef]
100. Popenoe, W. The avocado—a horticultural problem. *Trop Agric.* **1941**, *18*, 3–7.
101. Crane, J.H.; Balerdi, C.F.; Maguire, I. Avocado growing in the Florida home landscape. *Hort. Sci. Dept. Fla. Coop. Ext. Serv. Inst. Food Agric. Sci. Univ. Florida. Circ.* **2007**, *1034*, 1–12.
102. Krezdorn, A. Influence of rootstock on cold hardiness of avocados. *Proc. Fla. State Hort. Soc.* **1973**, *86*, 346–348.
103. Mickelbart, M.V.; Arpaia, M.L. Rootstock influences changes in ion concentrations, growth, and photosynthesis of 'Hass' avocado trees in response to salinity. *J. Am. Soc. Hortic. Sci.* **2002**, *127*, 649–655. [CrossRef]
104. Chen, H.; Morrell, P.L.; Ashworth, V.E.; de La Cruz, M.; Clegg, M.T. Tracing the geographic origins of major avocado cultivars. *J. Hered.* **2009**, *100*, 56–65. [CrossRef]
105. Barrientos-Priego, A.F.; López-López, L. Historia y genética del aguacate. *Télizd. Y Moraa.(Comps.). El Aguacate Y Su Manejo Integrado. 2ª (Ed.) Ed. Mundi-Prensa. Df México* **2000**, 19–31. Available online: http://www.avocadosource.com/Journals/CICTAMEX/CICTAMEX_1998/cictamex_1998_33-51.pdf (accessed on 12 February 2021).
106. Ayala-Silva, T.; Ledesma, N. Avocado history, biodiversity and production. In *Sustainable Horticultural Systems*; Nandwani, D., Ed.; Springer: Cham, Switzerland, 2014; pp. 157–205. ISBN 978-3-319-06903-6. [CrossRef]
107. Köhne, S. Selection of Avocado Scions and breeding of rootstocks in South Africa. In Proceedings of the New Zealand and Australia Avocado Grower's Conference, Tauranga, New Zealand, 20–22 September 2005.
108. Rendón-Anaya, M.; Ibarra-Laclette, E.; Bravo, A.M.; Lan, T.; Zheng, C.; Carretero-Paulet, L.; Perez-Torres, C.A.; Chacón-López, A.; Hernandez-Guzmán, G.; Chang, T.-H. The avocado genome informs deep angiosperm phylogeny, highlights introgressive hybridization, and reveals pathogen influenced gene space adaptation. *Proc. Natl. Acad. Sci. USA* **2019**, *116*, 17081–17089. [CrossRef]
109. Avocados Australia. Facts at a Glance 2019/20 for the Australian Avocado Industry. Available online: https://avocado.org.au/news-publications/statistics/ (accessed on 11 February 2021).
110. Dreher, M.L.; Davenport, A.J. Hass avocado composition and potential health effects. *Crit. Rev. Food Sci. Nutr.* **2013**, *53*, 738–750. [CrossRef] [PubMed]
111. Hiti-Bandaralage, J.C.; Hayward, A.; Mitter, N. Micropropagation of avocado (*Persea americana* Mill.). *Am. J. Plant Sci.* **2017**, *8*, 2898–2921. [CrossRef]

112. Alcaraz, M.; Hormaza, J. Selection of potential pollinizers for 'Hass' avocado based on flowering time and male–female overlapping. *Sci. Hortic.* **2009**, *121*, 267–271. [CrossRef]
113. Davenport, T. Avocado flowering. *Hortic. Rev.* **1986**, *8*, 89.
114. Lavi, U.; Lahav, E.; Degani, C.; Gazit, S.; Hillel, J. Genetic variance components and heritabilities of several avocado traits. *J. Am. Soc. Hortic. Sci.* **1993**, *118*, 400–404. [CrossRef]
115. Arpaia, M.L. *Avocado Brainstorming 2018, Towards a Sustainable Future*; Fairview Hotel: Tzaneen, South Africa, 2018.
116. Bergh, B. Avocado research in Israel. *Calif. Avocado Soc. Yearb.* **1975**, *58*, 103–126.
117. Parkinson, L.; Geering, A. Biosecurity capacity building for the Australian avocado industry. Available online: https://avocado.org.au/public-articles/ta30v2-biosec/ (accessed on 11 February 2021).
118. Darvas, J.; Kotze, J. Avocado fruit diseases and their control in South Africa. *S. Afr. Avocado Grow. Assoc. Yearb.* **1987**, *10*, 117–119.
119. Vega, D.E.S. Propagation in vitro of Rootstocks of Avocado. *Calif. Avocado Soc. Yearb.* **1989**, *73*, 196–199.
120. Pliego-Alfaro, F.; Murashige, T. Possible rejuvenation of adult avocado by graftage onto juvenile rootstocks *in vitro*. *HortScience* **1987**, *22*, 1321–1324.
121. Lorea Hernández, F.G. La familia Lauraceae en el sur de México: Diversidad, distribución y estado de conservación. *Boletín De La Soc. Botánica De México* **2002**, *71*. [CrossRef]
122. Ben-Ya'acov, A. Avocado rootstock-scion relationships. *S. Afr. Avocado Grow. Assoc. Yearb.* **1987**, *10*, 30–32.
123. Wolstenholme, B. Theoretical and applied aspects of avocado yield as affected by energy budgets and carbon partitioning. *S. Afr. Avocado Grow. Assoc. Yearb.* **1987**, *10*, 58–61.
124. Wolstenholme, B. Avocado rootstocks: What do we know; are we doing enough research. *S. Afr. Avocado Grow. Assoc. Yearb.* **2003**, *2003*, 106–112.
125. Smith, L.; Dann, E.; Pegg, K.; Whiley, A.; Giblin, F.; Doogan, V.; Kopittke, R. Field assessment of avocado rootstock selections for resistance to *Phytophthora* root rot. *Australas. Plant Pathol.* **2011**, *40*, 39–47. [CrossRef]
126. Mickelbart, M.V.; Bender, G.S.; Witney, G.W.; Adams, C.; Arpaia, M.L. Effects of clonal rootstocks on 'Hass' avocado yield components, alternate bearing, and nutrition. *J. Hortic. Sci. Biotechnol.* **2007**, *82*, 460–466. [CrossRef]
127. New Zealand avocado Varieties and rootstocks. Available online: https://industry.nzavocado.co.nz/grow/avocado-varieties/ (accessed on 6 February 2021).
128. Folgado, R. (The Huntington Library, Art Museum, and Botanical Gardens, San Marino CA, USA). Personal Communication, 2020.
129. Arpaia, M.L.; Focht, E. (Department of Botany and Plant Sciences, University of California, Riverside, CA, USA). Personal Communication, 2020.
130. Manosalva, P. (Department of Microbiology and Plant Pathology, University of California, Riverside, CA, USA). Personal Communication, 2020.
131. Goenaga, R. (Agricultural Research Service, United States Department of Agriculture, Miami, FL, USA). Personal Communication, 2020.
132. Álvarez, S.P.; Quezada, G.Á.; Arbelo, O.C. Avocado (*Persea americana* Mill.). *Cultiv. Trop.* **2015**, *36*, 111–123. [CrossRef]
133. Barrientos-Priego, A. (Departamento de Fitotecnia, Texcoco de Mora, Mexico). Personal Communication, 2020.
134. Nkansah, G.; Ofosu-Budu, K.; Ayarna, A. Avocado germplasm conservation and improvement in Ghana. In Proceedings of the VII World Avocado Congress 2011, Cairns, Australia, 5–9 September 2011.
135. Alcaraz, M.; Hormaza, J. Molecular characterization and genetic diversity in an avocado collection of cultivars and local Spanish genotypes using SSRs. *Hereditas* **2007**, *144*, 244–253. [CrossRef]
136. Dann, E. (Centre for Horticultural Science Queensland Alliance for Agriculture and Food Innovation, The University of Queensland, St Lucia, Queensland, Australia). Personal Communication, 2020.
137. Borokini, T.I. Conservation Science. *Int. J. Conserv. Sci.* **2013**, *4*, 197–212.
138. Cantuarias-Avilés, T. (Department of Plant Production, University of Sao Paulo, Piracicaba SP, Brazil). Personal Communication, 2020.
139. Encina, C.L.; Parisi, A.; O'Brien, C.; Mitter, N. Enhancing somatic embryogenesis in avocado (Persea americana Mill.) using a two-step culture system and including glutamine in the culture medium. *Sci. Hortic.* **2014**, *165*, 44–50. [CrossRef]
140. Guan, Y.; Li, S.-G.; Fan, X.-F.; Su, Z.-H. Application of somatic embryogenesis in woody plants. *Front. Plant Sci.* **2016**, *7*, 938. [CrossRef]
141. Kulkarni, V.; Suprasanna, P.; Bapat, V. Plant regeneration through multiple shoot formation and somatic embryogenesis in a commercially important and endangered Indian banana cv. Rajeli. *Curr. Sci.* **2006**, 842–846. Available online: http://www.jstor.org/stable/24089199 (accessed on 4 April 2021).
142. Márquez-Martín, B.; Barceló-Muñoz, A.; Pliego-Alfaro, F.; Sánchez-Romero, C. Somatic embryogenesis and plant regeneration in avocado (Persea americana Mill.): Influence of embryogenic culture type. *J. Plant Biochem. Biotechnol.* **2012**, *21*, 180–188. [CrossRef]
143. Jain, S.M.; Ishii, K. *Micropropagation of Woody Plants and Fruits*; Kluwer Academic Publishers: Amsterdam, The Netherlands, 2003; Volume 75, ISBN 1-4020-1135-0.
144. O'Brien, C.; Constantin, M.; Walia, A.; Yiing, J.L.Y.; Mitter, N. Cryopreservation of somatic embryos for avocado germplasm conservation. *Sci. Hortic.* **2016**, *211*, 328–335. [CrossRef]
145. Pliego-Alfaro, F.; Murashige, T. Somatic embryogenesis in avocado (Persea americana Mill.) in vitro. *Plant Celltissue Organ Cult.* **1988**, *12*, 61–66. [CrossRef]

146. Litz, R.; Litz, W. Somatic embryogenesis of avocado (Persea americana) and its application for plant improvement. *Int. Symp. Trop. Subtrop. Fruits* **2000**, *575*. [CrossRef]
147. Mujib, A.; Šamaj, J. *Somatic Embryogenesis*; Springer Science & Business Media: Berlin/Heidelberg, Germany, 2006; Volume 2, p. 3540287175.
148. Mooney, P.; Staden, J.V. Induction of embryogenesis in callus from immature embryos of Persea americana. *Can. J. Bot.* **1987**, *65*, 622–626. [CrossRef]
149. Ammirato, P.V. Organizational events during somatic embryogenesis. *Plant Biol.* **1986**. Available online: https://www.scirp.org/(S(i43dyn45teexjx455qlt3d2q))/reference/ReferencesPapers.aspx?ReferenceID=1114828 (accessed on 7 January 2021).
150. Raharjo, S.; Litz, R.E. Rescue of Genetically Transformed Avocado by Micrografting. In Proceedings of the V World Avocado Congress (Actas V Congreso Mundial del Aguacate), Granada-Málaga, Spain, 19–24 October 2003; pp. 119–122.
151. Witjaksono, Y.; Litz, R. Maturation and germination of avocado (Persea americana Mill.) somatic embryos. *Plant Cell Tissue Organ Cult* **1999**, *58*, 141–148. [CrossRef]
152. Efendi, D.; Litz, R.E. Cryopreservation of Avocado. In Proceedings of the V Congreso Mundial del Aguacate, Actas; Consejería de Agricultura y Pesca, Junta de Andalucía, Sevilla, Spain; 2003; Volume 1, pp. 111–114.
153. Guzmán-García, E.; Bradaï, F.; Sánchez-Romero, C. Cryopreservation of avocado embryogenic cultures using the droplet-vitrification method. *Acta Physiol. Plant.* **2013**, 1–11. [CrossRef]
154. Vargas, V.M. Efecto Fisiológico de Brasinoesteroides y Crioprotectores Sobre Yemas Axilares de Aguacate Criollo Producidas in Vitro. 2008. Available online: http://www.avocadosource.com/WAC7/WAC7_TOC.htm (accessed on 24 February 2021).
155. Vidales-Fernandez, I.; Larios-Guzman, A.; Tapia-Vargas, L.M.; Guillen-Andrade, H.; Villasenor-Ramirez, F. Criopreservación de germoplasma de aguacate. In Proceedings of the VII World Avocado Congress, Cairns, Australia, 5–9 September 2011.
156. Panis, B.; Piette, B.; André, E.; Van den houwe, I.; Swennen, R. *Droplet Vitrification: The First Generic Cryopreservation Protocol for Organized Plant Tissues?* International Society for Horticultural Science (ISHS): Leuven, Belgium, 2011; pp. 157–162, 2406–6168. [CrossRef]
157. Engelmann, F. Plant cryopreservation: Progress and prospects. *Vitr. Cell. Dev. Biol. Plant* **2004**, *40*, 427–433. [CrossRef]
158. Burke, M.J. The glassy state and survival of anhydrous biological systems. In *Membranes, Metabolism and Dry Organisms*; Cornell University Press: Ithaca, NY, USA, 1986; pp. 358–363. ISBN 978-0801419799.
159. Krishna, H.; Sairam, R.; Singh, S.; Patel, V.; Sharma, R.; Grover, M.; Nain, L.; Sachdev, A. Mango explant browning: Effect of ontogenic age, mycorrhization and pre-treatments. *Sci. Hortic.* **2008**, *118*, 132–138. [CrossRef]
160. Uchendu, E.E.; Paliyath, G.; Brown, D.C.; Saxena, P.K. In vitro propagation of North American ginseng (Panax quinquefolius L.). *Vitr. Cell. Dev. Biol. Plant* **2011**, *47*, 710–718. [CrossRef]
161. Preece, J.; Compton, M. Problems with Explant Exudation in Micropropagation. In *High-Tech and Micropropagation I. Biotechnology in Agriculture and Forestry*; Bajaj, Y.P.S., Ed.; Springer: Berlin, Germany, 1991; Volume 17, pp. 168–189. ISBN 978-3-642-76417-2. [CrossRef]
162. Akram, N.A.; Shafiq, F.; Ashraf, M. Ascorbic Acid-A Potential Oxidant Scavenger and Its Role in Plant Development and Abiotic Stress Tolerance. *Front. Plant Sci.* **2017**, *8*, 613. [CrossRef]
163. Shao, H.-B.; Chu, L.-Y.; Lu, Z.-H.; Kang, C.-M. Primary antioxidant free radical scavenging and redox signaling pathways in higher plant cells. *Int. J. Biol. Sci.* **2007**, *4*, 8. [CrossRef] [PubMed]
164. Smirnoff, N.; Wheeler, G.L. Ascorbic acid in plants: Biosynthesis and function. *Crit. Rev. Plant Sci.* **2000**, *19*, 267–290. [CrossRef]
165. González-Benito, M.E.; Kremer, C.; Ibáñez, M.A.; Martín, C. Effect of antioxidants on the genetic stability of cryopreserved mint shoot tips by encapsulation–dehydration. *Plant Celltissue Organ Cult.* **2016**, *127*, 359–368. [CrossRef]
166. Johnston, J.W.; Harding, K.; Benson, E.E. Antioxidant status and genotypic tolerance of Ribes *in vitro* cultures to cryopreservation. *Plant Sci.* **2007**, *172*, 524–534. [CrossRef]
167. Reed, B.M. 4. Are antioxidants a magic bullet for reducing oxidative stress during cryopreservation? *Cryobiology* **2012**, *65*, 340. [CrossRef]
168. Uchendu, E.E.; Leonard, S.W.; Traber, M.G.; Reed, B.M. Vitamins C and E improve regrowth and reduce lipid peroxidation of blackberry shoot tips following cryopreservation. *Plant Cell Rep.* **2010**, *29*, 25. [CrossRef] [PubMed]
169. Wang, Q.; Laamanen, J.; Uosukainen, M.; Valkonen, J. Cryopreservation of *in vitro*-grown shoot tips of raspberry (Rubus idaeus L.) by encapsulation–vitrification and encapsulation–dehydration. *Plant Cell Rep.* **2005**, *24*, 280–288. [CrossRef]
170. O'Brien, C.; Hiti-Bandaralage, J.C.A.; Folgado, R.; Lahmeyer, S.; Hayward, A.; Mitter, N. *Developing a Cryopreservation Protocol for Avocado (Persea americana Mill.) Apical Shoot Tips Using Different Antioxidants*; International Society for Horticultural Science (ISHS): Leuven, Belgium, 2020; pp. 15–22, 2406–6168. [CrossRef]
171. Sharma, S.D. Cryopreservation of Somatic Embryos—An Overview. 2005. Available online: http://nopr.niscair.res.in/bitstream/123456789/5629/1/IJBT%204%281%29%2047-55.pdf (accessed on 12 January 2021).
172. Chang, Y.; Reed, B.M. Pre-culture conditions influence cold hardiness and regrowth of *Pyrus cordata* shoot tips after cryopreservation. *HortScience* **2001**, *36*, 1329–1333. [CrossRef]
173. Feng, C.-H.; Cui, Z.-H.; Li, B.-Q.; Chen, L.; Ma, Y.-L.; Zhao, Y.-H.; Wang, Q.-C. Duration of sucrose pre-culture is critical for shoot regrowth of *in vitro*-grown apple shoot-tips cryopreserved by encapsulation-dehydration. *Plant Celltissue Organ Cult.* **2013**, *112*, 369–378. [CrossRef]

174. Kaczmarczyk, A.; Shvachko, N.; Lupysheva, Y.; Hajirezaei, M.-R.; Keller, E.R.J. Influence of alternating temperature pre-culture on cryopreservation results for potato shoot tips. *Plant Cell Rep.* **2008**, *27*, 1551–1558. [CrossRef] [PubMed]
175. Park, S.U.; Kim, H.H. Cryopreservation of sweet potato shoot tips using a droplet-vitrification procedure. *CryoLetters* **2015**, *36*, 344–352. [PubMed]
176. Ashmore, S.E.; Azimi, M.; Drew, R.A. Cryopreservation trials in *Carica papaya*. *Acta Hortic.* **2001**, *560*, 117–120. [CrossRef]
177. López-López, L.; Barrientos-Priego, A.; Ben-Ya'acov, A. Variabilidad genética de los bancos de germoplasma de aguacate preservados en el Estado de México. *Rev. Chapingo Ser. Hortic.* **1999**, *5*, 19–23.
178. Mikuła, A. Comparison of three techniques for cryopreservation and reestablishment of long-term *Gentiana tibetica* suspension culture. *CryoLetters* **2006**, *27*, 269–282.
179. Mikuła, A.; Tykarska, T.; Kuraś, M. Ultrastructure of *Gentiana tibetica* proembryogenic cells before and after cooling treatments. *CryoLetters* **2005**, *26*, 367–378.
180. Pritchard, H.; Grout, B.; Short, K. Osmotic stress as a pregrowth procedure for cryopreservation: 1. Growth and ultrastructure of sycamore and soybean cell suspensions. *Ann. Bot.* **1986**, *57*, 41–48. [CrossRef]
181. Panis, B.; Strosse, H.; van den Hende, S.; Swennen, R. Sucrose pre-culture to simplify cryopreservation of banana meristem cultures. *CryoLetters* **2002**, *23*, 375–384.
182. Lynch, P.T.; Siddika, A.; Johnston, J.W.; Trigwell, S.M.; Mehra, A.; Benelli, C.; Lambardi, M.; Benson, E.E. Effects of osmotic pre-treatments on oxidative stress, antioxidant profiles and cryopreservation of olive somatic embryos. *Plant Sci.* **2011**, *181*, 47–56. [CrossRef] [PubMed]
183. Malabadi, R.B.; Nataraja, K. Cryopreservation and plant regeneration via somatic embryogenesis using shoot apical domes of mature *Pinus roxburghii* sarg. trees. *Vitr. Cell. Dev. Biol. Plant* **2006**, *42*, 152. [CrossRef]
184. Crowe, L.M. Lessons from nature: The role of sugars in anhydrobiosis. *Comp. Biochem. Physiol. Part A Mol. Integr. Physiol.* **2002**, *131*, 505–513. [CrossRef]
185. Antony, J.J.J.; Keng, C.L.; Mahmood, M.; Subramaniam, S. Effects of ascorbic acid on PVS2 cryopreservation of Dendrobium Bobby Messina's PLBs supported with SEM analysis. *Appl. Biochem. Biotechnol.* **2013**, *171*, 315–329. [CrossRef]
186. Hirsh, A.G. Vitrification in plants as a natural form of cryoprotection. *Cryobiology* **1987**, *24*, 214–228. [CrossRef]
187. Herbert, R.; Vilhar, B.; Evett, C.; Orchard, C.; Rogers, H.; Davies, M.; Francis, D. Ethylene induces cell death at particular phases of the cell cycle in the tobacco TBY-2 cell line. *J. Exp. Bot.* **2001**, *52*, 1615–1623. [CrossRef]
188. Williams, W.P.; Quinn, P.J.; Tsonev, L.I.; Koynova, R.D. The effects of glycerol on the phase behaviour of hydrated distearoylphosphatidylethanolamine and its possible relation to the mode of action of cryoprotectants. *Biochim. Biophys. Acta Bba Biomembr.* **1991**, *1062*, 123–132. [CrossRef]
189. Burritt, D.J. Proline and the cryopreservation of plant tissues: Functions and practical applications. In *Current Frontiers in Cryopreservation*; Katkov, I.I., Ed.; InTech: Rijeka, Croatia, 2012.
190. Cleland, D.; Krader, P.; McCree, C.; Tang, J.; Emerson, D. Glycine betaine as a cryoprotectant for prokaryotes. *J. Microbiol. Methods* **2004**, *58*, 31–38. [CrossRef]
191. Janská, A.; Maršík, P.; Zelenková, S.; Ovesná, J. Cold stress and acclimation—What is important for metabolic adjustment? *Plant Biol.* **2010**, *12*, 395–405. [CrossRef] [PubMed]
192. Arora, R. *Freezing Tolerance and Cold Acclimation in Plants*; Deparment of Horticulture, Iowa State University: Ames, IA, USA, 2010. [CrossRef]
193. Gusta, L.; Trischuk, R.; Weiser, C.J. Plant cold acclimation: The role of abscisic acid. *J. Plant Growth Regul.* **2005**, *24*, 308–318. [CrossRef]
194. Reed, B. Pre-treatment strategies for cryopreservation of plant tissues. In *In Vitro Conservation of Plant Genetic Resources*; Normah, M.N., Narimah, M.K., Clyde, M.M., Eds.; Universiti Kebangsaan: Bangi Selangor, Malaysia, 1996; pp. 73–87.
195. Janmohammadi, M.; Zolla, L.; Rinalducci, S. Low temperature tolerance in plants: Changes at the protein level. *Phytochemistry* **2015**, *117*, 76–89. [CrossRef] [PubMed]
196. Thomashow, M.F. Plant cold acclimation: Freezing tolerance genes and regulatory mechanisms. *Annu. Rev. Plant Biol.* **1999**, *50*, 571–599. [CrossRef] [PubMed]
197. Dumet, D.; Chang, Y.; Reed, B.M.; Benson[1], E.E. Replacement of cold acclimatization with high sucrose pretreatment in black currant cryopreservation. *Satoshi Katomasaya Ishikawamiwako Ito Tatsuo Matsumoto 338* **2000**, *17*, 393.
198. Coelho, N.; González-Benito, M.E.; Martín, C.; Romano, A. Cryopreservation of *Thymus lotocephalus* shoot tips and assessment of genetic stability. *CryoLetters* **2014**, *35*, 119–128.
199. Fki, L.; Bouaziz, N.; Chkir, O.; Benjemaa-Masmoudi, R.; Rival, A.; Swennen, R.; Drira, N.; Panis, B. Cold hardening and sucrose treatment improve cryopreservation of date palm meristems. *Biol. Plant.* **2013**, *57*, 375–379. [CrossRef]
200. Kushnarenko, S.V.; Romadanova, N.V.; Reed, B.M. Cold acclimation improves regrowth of cryopreserved apple shoot tips. *CryoLetters* **2009**, *30*, 47–54.
201. Reed, B.M. Responses to ABA and cold acclimation are genotype dependent for cryopreserved blackberry and raspberry meristems. *Cryobiology* **1993**, *30*, 179–184. [CrossRef]
202. O'Brien, C.; Hiti-Bandaralage, J.; Folgado, R.; Lahmeyer, S.; Hayward, A.; Folsom, J.; Mitter, N. A method to increase regrowth of vitrified shoot tips of avocado (*Persea americana* Mill.): First critical step in developing a cryopreservation protocol. *Sci. Hortic.* **2020**, *266*, 109305. [CrossRef]

203. O'Brien, C.; Hiti-Bandaralage, J.C.; Folgado, R.; Lahmeyer, S.; Hayward, A.; Mitter, N. Developing a cryopreservation protocol for avocado (*Persea americana* Mill.) shoot tips. *Cryobiology* **2018**, *85*, 171. [CrossRef]
204. O'Brien, C.; Hiti-Bandaralage, J.C.A.; Folgado, R.; Lahmeyer, S.; Hayward, A.; Folsom, J.; Mitter, N. First report on cryopreservation of mature shoot tips of two avocado (*Persea americana* Mill.) rootstocks. *Plant Celltissue Organ Cult.* **2020**. [CrossRef]
205. Sánchez-Romero, C.; Márquez-Martín, B.; Pliego-Alfaro, F. Somatic and zygotic embryogenesis in avocado. In *Somatic Embryogenesis*; Springer: Berlin/Heidelberg, Germany, 2006; pp. 271–284.
206. Hiti-Bandaralage, J.; Hayward, A.; O'Brien, C.; Gleeson, M.; Nak, W.; Mitter, N. Advances in Avocado Propagation for the Sustainable Supply of Planting Materials. In *Achieving Sustainable Cultivation of Tropical Fruits*; Burleigh dodds Science Publishing: Cambridge, UK, 2019; pp. 215–238. ISBN 9781786762849.

MDPI
St. Alban-Anlage 66
4052 Basel
Switzerland
Tel. +41 61 683 77 34
Fax +41 61 302 89 18
www.mdpi.com

Plants Editorial Office
E-mail: plants@mdpi.com
www.mdpi.com/journal/plants

www.ingramcontent.com/pod-product-compliance
Lightning Source LLC
LaVergne TN
LVHW070641100526
838202LV00013B/852